DNA Barcoding and Molecular Phylogeny

Subrata Trivedi • Hasibur Rehman •
Shalini Saggu • Chellasamy Panneerselvam •
Sankar K. Ghosh
Editors

DNA Barcoding and Molecular Phylogeny

Second Edition

Editors
Subrata Trivedi
Department of Biology, Faculty of Science
University of Tabuk
Tabuk, Saudi Arabia

Hasibur Rehman
Department of Urology
The University of Alabama at Birmingham (UAB)
Birmingham, AL, USA

Shalini Saggu
Department of Cell, Developmental and Integrative Biology (CDIB)
The University of Alabama at Birmingham (UAB)
Birmingham, AL, USA

Chellasamy Panneerselvam
Department of Biology, Faculty of Science
University of Tabuk
Tabuk, Saudi Arabia

Sankar K. Ghosh
University of Kalyani
Kalyani, Nadia, West Bengal, India

ISBN 978-3-030-50074-0 ISBN 978-3-030-50075-7 (eBook)
https://doi.org/10.1007/978-3-030-50075-7

This Springer imprint is published by the registered company Springer Nature Switzerland AG.
The registered company address is: Gewerbestrasse 11, 6330 Cham, Switzerland

Contents

Part I
DNA Barcoding: Advantages and Significance

Implications and Utility of DNA Barcoding

J. Suriya, M. Krishnan, S. Bharathiraja, V. Sekar, and V. Sachithanandam

Abstract Classical way of practicing taxonomy is in endangered race in the era of genomics. In recent years, taxonomy became fashionable owing to the revolutionary approaches in taxonomy called DNA barcoding. It is a novel approach that has generated optimism in enhancing biodiversity assessments. In DNA barcoding, complete data can be retrieved from a single specimen regardless of life stage or morphological characters. The core idea behind DNA barcoding is the fact of the minor variation of highly conserved region of DNA during the evolution within the species. Sequences that have successfully been utilized for DNA barcoding include cytoplasmic mitochondrial DNA (cox1) and chloroplast DNA (trnL-F, matK, ndhF, atpB, and rbcL) and nuclear DNA (ITS and housekeeping genes). Now it has been used for diverse applications such as biodiversity assessment, life history and ecological studies, forensic analysis, and many more. In this chapter, we discuss the significance and utility of DNA barcoding in various fields.

Keywords DNA Barcoding · Molecular taxonomy · Biodiversity · Mitochondrial genes · Gene sequencing

J. Suriya (✉) · M. Krishnan
Department of Environmental Biotechnology, School of Environmental Sciences, Bharathidasan University, Tiruchirappalli, Tamil Nadu, India

S. Bharathiraja
Department of Biomedical Engineering, Nano-Biomedicine Lab, Pukyong National University, Busan, South Korea

V. Sekar · V. Sachithanandam
National Centre for Sustainable Coastal Management, Ministry of Environment, Forests & Climate Change, Anna University Campus, Chennai, Tamil Nadu, India

S. Trivedi et al. (eds.), *DNA Barcoding and Molecular Phylogeny*,
https://doi.org/10.1007/978-3-030-50075-7_1

1 Introduction

Earth comprises an innumerable diversity of organisms which are mostly uncharacterized (Wilson 2003). The classical techniques used for species identification pave the way for many challenges and problems; henceforth, the taxonomist became an endangered race in the era of genomics. Molecular techniques are promising approaches for species identification because it overcomes the difficulties in the morphology-based identification methods. Now taxonomy became fashionable owing to the innovative approaches in taxonomy called DNA barcoding. DNA barcodes usually comprise short sequences of DNA that can be amplified with the aid of polymerase chain reaction (PCR) and then sequenced for analyzing the species of interest. It has been successfully utilized for the identification of subspecies, eco- and morpho types, cultivars, mutants, clones, and species complex. Molecular operational taxonomic units (MOTU) are used for the identification of dissimilar species, whereas similar species are identified by comparing the barcode of a species of interest with sequences retrieved from other taxa (Floyd et al. 2002). It is used for barcoding of various organisms including freshwater meiobenthos (Markmann and Tautz 2005), soil meiofauna (Blaxter et al. 2004), extinct birds (Lambert et al. 2005), marine organisms (Shander and Willassen 2005), fishes (Ward et al. 2005), and insects (Smith et al. 2005). In particular, DNA barcodes are more potentially applied in biosecurity such as law enforcement and primatology (Lorenz et al. 2005) as well as for surveillance of disease vectors (Besansky et al. 2003) and invasive insects (Armstrong and Ball 2005). Among other species identification approaches, DNA barcoding is a promising technique for species identification due to its expert-authenticated verification system and accuracy (Ajmal Ali et al. 2014).

The two main aims of DNA barcoding are (i) discovery of new species and its identification and (ii) assigning of unknown specimens to species (Hebert et al. 2003). Figure 1 illustrates the different steps involved in the DNA barcoding approach.

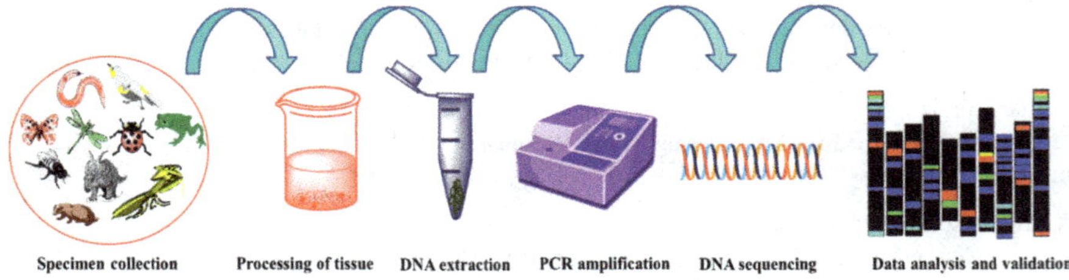

Fig. 1 Steps involved in DNA barcoding process—from specimen to barcode data analysis

2 Significance of DNA Barcoding

Specimen identification using barcoding is usually attained by obtaining short DNA sequence—the "barcode"—from the genome of the species. Then the barcode sequence of the unknown species is compared with the reference library of barcode sequences of known species. If the sequence of the unknown species closely matches with the sequences in the reference library, a species is identified. Novel barcode sequence implies the findings of new species. The DNA barcode gives great support to innumerable scientific domains such as epidemiology, ecology, biomedicine, evolutionary biology, biogeography, and conservation biology and also in bio-industry. For handling large sampling groups, DNA barcoding is the most preferred technique due to its time and cost-effectiveness. In addition to this, DNA barcoding is also employed in the isolation and identification of unknown species which have significance in the medical, ecological, and agronomical fields (Armstrong and Ball 2005; Ball and Armstrong 2006). Moreover, it is used for the detection of patented organisms in agro-biotechnology, either to protect intellectual property rights for bioresources (Taberlet et al. 2007) or to certify the source organism (Rastogi et al. 2007).

The main advantage of DNA barcoding is the acquiring of a large volume of molecular data. It is particularly useful in determining the identity of damaged organisms. DNA barcoding is potentially utilized in the diet analyses, food industry, and forensic sciences in preventing poaching of endangered species and illegal trade. Shark fin trade is the best example for illegal trade in several coastal African countries which is quite a threat to biodiversity conservation. Moreover, DNA barcoding also plays a crucial role in environmental law enforcement.

3 Gene Markers in DNA Barcoding

Different gene regions such as mitochondrial DNA (*COI*), nuclear DNA (*ITS*), and chloroplast DNA (*rbcL, trnL-F, matK, psbA, trnH*, and *psbK*) have been used for DNA barcoding approach; however, cytochrome c oxidase is a universal DNA barcode for animals (Ajmal Ali et al. 2014). More than 95% of species have unique COI barcode sequences. The COI gene has been successfully utilized for the identification of abundance of species such as birds (Kerr et al. 2007), butterflies (Hajibabaei et al. 2006), gastropods (Remigio and Hebert 2003), ants (Smith et al. 2005), springtails (Hogg and Hebert 2004), Protista (Evans et al. 2007), spiders (Greenstone et al. 2005), Crustacea (Costa et al. 2007), and fish (Ward et al. 2005). In plants, mitochondrial DNA revealed gradual substitution rates than animals and also showed intramolecular recombination. Henceforth, *rpoB, rpoC1 272*, and *rbcL* or a section of *matK* markers are used for plant identification owing to its higher rate of evolution.

4 Applications of DNA Barcoding

4.1 *DNA Barcoding in Biodiversity Assessments and Conservations*

Barcoding helps to identify the species quickly and cheaply. According to Hebert et al.'s (2010) estimation, 0.1% of all described animal species are sequenced within 90 min at a cost of $2–5 per specimen, whereas morphological analyses required several months of field work at a cost of $100 per species (Hebert and Gregory 2005; Stoeckle and Hebert 2008). Genetic tools are more useful in the identification of cryptic and invasive species (Stoeckle 2008), management of coral reefs (Neigel et al. 2007), and fisheries (Swartz et al. 2008). Discovery of cryptic diversity plays an important role in new species identification by DNA barcoding. DNA barcoding is also used for the identification of new species (Pauls et al. 2010). It identifies the innumerable species remain unexplored (Wilson 1994). For adequate description of the species, DNA barcoding along with traditional taxonomic methods is used (Prendini 2005).

DNA barcoding conserves the species by enhancing local biodiversity assessments to give priority for conservation areas, by giving data about phylogenetic diversity and evolutionary histories. The main advantage of DNA barcoding in biodiversity conservation is assessing the biodiversity in a time and cost-effective manner where financial resources are limited. This is particularly crucial because the majority of described biodiversity is in developing countries, where resources for biodiversity assessments are lacking. Francis et al. (2010) and Ward et al. (2008) have conserved several species with the aid of barcoding in Southeast Asia and Australia, respectively, and also Neigel et al. (2007) identified many juvenile and larval species for which morphological data are absent.

4.1.1 DNA Barcoding on Bacteria

Most of the bacterial species are cryptic in nature; henceforth, DNA barcoding holds promise for identifying bacteria at species level (Begerow et al. 2010). Phytoplasmas are bacterial phytopathogens which cause losses in agricultural production. Makarova et al. (2012) used elongation factor Tu (tuf) gene to develop a universal DNA barcode for phytoplasma identification. Portion of the DnaA replication initiation factor (RIF) has been used to discriminate plant-associated bacterial pathogens at the species level in *Xanthomonas, Pseudomonas*, and *Xylella* (Schneider et al. 2011). Natalie (2013) used DNA barcoding to evaluate the consequences of fire on microbial communities in chaparral soils and also to compare the microbes in the burnt and unburned soil samples. Thao et al. (2013) used fluorescent proteins to differentiate selected enterobacterial species rapidly in DNA barcoding. Multilocus sequence typing (MLST) approaches are widely applied for prokaryote identification like *Wolbachia* strains (Baldo et al. 2006). DNA barcode approaches produce phylogenetic hypotheses for entire bacterial communities with the aid of universal primers (Wang et al. 2012), which allows the ecologists to give answer to the fundamental questions regarding their distribution and assembly (Mouquet et al. 2012). Links et al. (2012) evaluated the efficiency of 16S rRNA and cpn60 genes as bacterial barcodes. They suggested that the cpn60 is the suitable barcode for bacteria than 16S rRNA due to its large barcode gap. However, 16S rRNA gene is mostly used as the barcode for the identification of bacterial and archaeal communities. Enormous studies have used this molecular marker as a barcode to quantify bacterial community from environmental samples (Hugenholtz et al. 1998). Those results implied that we have only studied small part of the bacterial community that exists in the environments.

4.1.2 DNA Barcoding on Fungi

Identification of fungi by morphological characters needs well-trained experts due their very complex nature (filamentous fungi). Fungi occupy one of the largest kingdom; however, large numbers of them are not described yet (Mora et al. 2011). It is very tricky to identify numerous fungi due their unculturable nature (Begerow et al. 2010). In the clinical laboratory, proper identification of fungal pathogens continues to be a difficult task. It is crucial to identify disease-causing agents for detecting novel therapeutic agents. In such cases, molecular identification holds promise in the identification of fungal species. Various genetic loci mostly used for fungi identification include 28S rRNA large subunit (LSU) (Scorzetti et al. 2002), β-tubulin (Balajee et al. 2007), internal transcribed spacer (ITS) of the ribosomal DNA (rDNA) (Rainer and de Hoog 2006), and multilocus sequencing (Ngamskulrungroj et al. 2009). Various essential fungal groups were identified using different methods on the basis of their sequence variation (Cerqueira et al. 2014). Several studies indicated that COI is suitable only for few genera such as *Penicillium* (Seifert et al. 2007). Multiple copies of COI gene are present in the *Fusarium, Aspergillus niger*, and Basidiomycota groups; henceforth, COI is unfit for the identification of fungi at the species level (Gilmore et al. 2009; Vialle et al. 2009). In addition, PCR amplification and sequencing are very difficult (Dentinger et al. 2011) and needed nested primers for the amplification of the entire region of COI (Gilmore et al. 2009). These results pave the way for the use of multiple genetic loci for fungal species identification (Roe et al. 2010). Multi-gene approaches are commonly used for phylogenetic studies. Geiser et al. (2007) used diverse sequences such as ITS, cox1, calmodulin, and β-tubulin for black aspergilli species identification, and they reported that either calmodulin or β-tubulin could be a suitable barcode for species identification.

4.1.3 DNA Barcoding on Terrestrial Species

Among terrestrial ecosystems, soil represents one of the most diversified habitats with the majority of species that form terrestrial biodiversity. About 25% of the organisms on earth are represented by soil organisms (Decaëns et al. 2013). One gram of soil may host several thousands of diverse bacterial and fungal species, and a square meter of soil may contain several hundreds of different arthropods species (Decaëns et al. 2013). Floyd et al. (2002) employed DNA barcoding approach for the identification of soil nematodes. Hebert et al. (2004) identified 40% north American avian species with the aid of barcoding. Following this study, several scientists employed DNA barcoding tool to assess avian species of the Neotropics (Kerr et al. 2009), the Palearctic (Johnsen et al. 2010), the Indomalayan region (Lohman et al. 2010), and Australasia (Patel et al. 2010). Canadian insects were described by Hebert et al. (2016) through the barcoding approach. Francis et al. (2010) evaluated the potential of DNA barcodes to understand and to conserve mammals' diversity in Southeast Asia, and they found that within the region 50% of mammal species richness is underestimated. Moreover, this approach has been shown to be effective in the identification of earthworms (Chang and James 2011). All of them described a highly structured variation of the barcode in amphibians (Smith et al. 2008), reptiles (Nagy et al. 2012), small mammals (Lu et al. 2012), bats (De Pasquale and Galimberti 2014), squirrels (Ermakov et al. 2015), primates (Ruiz-García et al. 2014), and rodents (Galan et al. 2012).

4.1.4 DNA Barcoding on Freshwater Species

Freshwater occupies only a relatively small proportion of the earth; however, freshwater habitats are home to innumerable species (Dudgeon et al. 2006) and extensively used by human beings. Henceforth, it is very crucial to recurrently evaluate their ecological health and develop potential monitoring approaches. Rossi and Mantelatto (2013), Udayasuriyan et al. (2015), and Jose et al. (2016) used MT-COI gene for the identification of freshwater prawns. Finn et al. (2014) discovered the loss of a baetid mayfly due to the loss of alpine glaciers. Hurwood et al. (2014) applied mtDNA sequences to explore the freshwater prawn diversity from western India to western Java. Lim et al. (2016) and Kermarrec et al. (2014) utilized NGS to assess metazoan and diatom communities, respectively. A large number of scientists such as Hubert et al. (2008), Nwani et al. (2011), Bhattacharjee and Ghosh (2014), Chakraborty and Ghosh (2014), Lakra et al. (2016), and Keskin and Atar (2013) applied DNA barcoding for the identification of freshwater fishes, whereas Corse et al. (2010), Carreon-Martinez et al. (2011), and Jo et al. (2014) employed DNA barcoding to find out freshwater carnivorous fish diets.

4.1.5 DNA Barcoding of Marine Species

Marine has rich biodiversity compared to terrestrial and freshwater ecosystem, because it covers more than 70% of our planet. The marine ecosystem is the suitable habitat for many micro- and macroorganisms, both flora and fauna. Thirty-four animal phyla are found in the marine, among the total of 35 animal phyla (Gray 1997). The occurrence of cryptic species is prevalent in marine ecosystems. The major difficulty in the marine ecosystem is connecting larval stages with the adult forms. DNA barcoding can effectively connect the larval stages to adult forms. DNA barcoding has been utilized to find out invasive species threat to marine biodiversity (Molnar et al. 2008). Indicator species barcoding can be successful in the assessment and reduction of marine pollution including coastal pollution. The main aim of barcoding is the discovery of new species.

Seagrasses possess several valuable secondary compounds like phenolic, rosmarinic, and zosteric acids which are used in traditional medicines as antioxidants and potential antifouling agent. Various molecular markers are used for seagrass identification such as trnK introns and rbcL for Zostera (Les et al. 2002), nuclear ITS for *Halophila* (Waycott et al. 2009), and ITS1, 5.8S rDNA, and ITS2 for *Halophila* (Uchimura et al. 2008). Mangroves have enormous economic and ecological significance. This unique and dynamic habitat is increasingly depleted and threatened by natural as well as manmade disaster. So, we urge to conserve this ecosystem. DNA barcoding serves to produce phylogenetic information which is helpful in developing unified mangrove management plan worldwide (Daru et al. 2013). UNESCO world heritage listed the Sunderbans as the only largest halophytic mangrove forest (http://whc.unesco.org/en/list) in the world. A large number of estuarine and marine species come to this ecosystem to breed and the juveniles stay back to utilize its natural resources (Trivedi et al. 2013).

Morphological techniques are ineffective in the identification of different species of red marine macro algae. Usually mitochondrial COI gene and 23S rRNA's UPA (universal plastid amplicon) domain V gene are used for the detection of diverse species of Kallymeniaceae family red alga. COI is a more sensitive marker and pave the way for the discovery of *Euthora timburtonii* (Clarkston and Saunders 2010). Moreover, it is very tricky to identify Gracilaria morphologically, and DNA barcoding can be very effective to identify Gracilaria at the species level (Kim et al. 2010). Genes of 16S rRNA and 23S rRNA were used as the barcode for the isolation and identification of a novel microalga from the Indian Ocean (Ahmad et al. 2013).

The pteropods are ecologically important marine species due to their susceptibility to ocean acidification. *Diacavolinia* pteropod barcoding showed that the Atlantic specimens contain only single monophyletic species and also revealed species-level difference between the Pacific and Atlantic populations (Maas et al. 2013). Two hundred and twenty-seven Canadian marine mollusk species were identified with the aid of DNA barcoding (Layton et al. 2014). Chen et al. (2011) potentially utilized barcoding approach to identify 60 venerid species from the coast of China. Marine oyster has significant economic value. Oysters are mostly differentiated on the basis of their shell morphology; however, the habitat alters the shell morphology greatly (Tack et al. 1992). In such cases, DNA barcoding holds promise to identify oyster at the species level. Sponge Barcoding Project, http://www.spongebarcoding.org, is much helpful for developing workflow to scrutinize a large number of sponge.

FISH-BOL (http://www.fishbol.org) and SHARKBOL (http://www.sharkbol.org) are the two main fish global barcoding initiatives. DNA barcoding is not only efficient for the identification of whole fish, but it also very helpful in the identification of eggs, larvae, fins, and fillets which are very difficult to identify based on their morphology. Three hundred and ninety-one ornamental fish species isolated from 8 coral reef locations were identified with the help of DNA barcoding (Steinke et al. 2009). Many scientists have potentially utilized DNA barcoding approach for the identification of marine fish throughout the world (Trivedi et al. 2014; Ardura et al. 2013; Weigt et al. 2012). Mammalia Barcode of Life (http://www.mammaliabol.org) is launched for barcoding of mammals including the marine mammals. Alfonsi et al. (2013) carried out the study along the French Atlantic coast and found that DNA barcoding combined with a stranding network can be very useful in monitoring marine mammal diversity.

4.2 Metabarcoding

In recent years, soil metabarcoding has drawn the attention of the scientists, resulting in a large number of publications. This is because of easy handling and reduction of cost and time in NGS platforms and also many bioinformatics pipelines are developed for analysis of metadata (Yang et al. 2013). Researchers have successfully utilized this approach to shed light on innumerable hitherto unexplored biodiversity from unexplored areas. The two main objectives of the soil biodiversity are (1) to assess the soil biodiversity's structure and functions, for example, ecological roles, and (2) to obtain the vacillations of soil biodiversity in different environmental conditions for ascertaining protective measurements. These two objectives can be easily fulfilled by using metabarcoding.

For example, metabarcoding is used for assessing the spatial distribution of soil biodiversity. The soil DNA can be stored for prolonged periods of time (Lauber et al. 2010) which offers to obtain soil biodiversity over time in relation to climatic change (Dumbrell et al. 2011).

4.3 Detection of Mislabeling and Food Piracy

DNA barcoding is potentially applied for the identification, authentication, and safety assessment of food, particularly for cooked, processed, or smoked products. This molecular approach allows us to detect the origin of certain food products

(Galimberti et al. 2013). Lowenstein et al. (2009) collected Japanese delicacy tuna sushi from various US restaurants and found out the presence of endangered species, fraud, and also a health hazard. Two hundred and fifty-four Canadian seafood samples were analyzed for mislabeling and the results divulged that 41% of the samples were mislabeled (Hanner et al. 2011). Holmes et al. (2009) identified shark fins that were illegally marketed by fishers in Australia. A study was conducted on 62 sea food samples in Malaysia to ascertain mislabeling. Among these samples, 16% were mislabeled (Chin et al. 2016). Wong and Hanner (2008) detected the mislabeling of 25% of the North American sea food with the aid of DNA barcoding.

DNA barcodes have been used to identify the mislabeling of toxic pufferfish in a Chicago market as monkfish (Cohen et al. 2009). It has been used by several researchers as a forensic tool for the traceability of edible fish (Barbuto et al. 2010; Smith et al. 2008). Newmaster et al. (2013) investigated herbal product substitution and contamination to protect the consumers from health risks associated with food adulteration. Among 44 samples, 30 samples were substituted and only two companies had herbal products without any substitution among 12 companies. Herbal tea contamination was detected by Stoeckle et al. (2011) using DNA barcoding. Parvathy et al. (2014, 2015) have potentially used DNA barcoding for detecting chili and plant-based adulteration in black pepper powder and in turmeric powder. DNA barcoding-based species identification is applied to the verification/ certification of mushroom-containing dietary supplements (Raja et al. 2017). Vassou et al. (2015) collect 13 species of Sida from market samples for the identification of *Sida cordifolia*, used for treating neurological disorders. They found that none of the market samples belonged to the *S. cordifolia.* These types of substitutions not only fail to give the expected effect but may also give undesirable health effects. COX1 is employed for mammalian meat traceability (Luo et al. 2011) and avian meat product identification (Hebert et al. 2004).

4.4 Controlling Agricultural Pest and Predators

DNA barcoding is much helpful for agricultural scientists in the exact and rapid identification of agriculturally important insect pests and their predators. DNA barcoding was used by Li et al. (2012) to identify Noctuidae pests. Smith et al. (2013) reported that more than 20,000 sequences of microgastrine wasps were produced by scientists. Nagoshi et al. (2011) identified armyworms in Florida with the aid of DNA barcoding. This approach is also applied for the identification of scale insects (Kondo et al. 2008), Ectoedemia (Nieukerken et al. 2012), Nearctic Muscidae (Renaud et al. 2012), storage pests (Yang et al. 2013), and aphid cryptic species differentiation (Rebijith et al. 2013). Mostly spiders are used to regulate the population of insect pests. Barrett and Hebert (2005) used DNA barcoding for spider identification. Marie-Stephane et al. (2012) utilized COI, Cytb, 12SrRNA, and ITSS markers for *Typhlodromus pyri* population identification. European beetle in particular German fauna was assigned to known species by using CO1 gene (Hendrich et al. 2015). Woodcock et al. (2013) assessed the beetle diversity in Canada, Manitoba, and Churchill to develop barcode library for subarctic region beetles by using DNA molecular markers for management of pest control.

4.5 Identifying Disease Vectors and Parasites

Accurate identification of parasites and their vectors is crucial in vector-borne disease surveillance programs. It is very difficult to identify many parasites morphologically; in addition, morphological expertise in parasite and its vector identification is scarce (Besansky et al. 2003). DNA barcoding overwhelms this problem which allows non-taxonomists to identify disease vectors and to understand and control disease-carrying pathogens and pests (http://www.barcoding.si.edu/PDF/CBOL). Disease-causing vectors such as mosquitoes are potentially identified with the help of DNA barcoding (Wang et al. 2012; Kumar et al. 2013; Ashfaq et al. 2014). Kumar et al. (2007) confirmed 62 mosquito species in DNA barcoding among 63 morphologically different species from India. Bourke et al. (2013) reported that DNA barcoding is a potential tool in the identification of diverse species of Brazilian mosquitoes.

Dhananjeyan et al. (2010) and Naddaf et al. (2012) also successfully employed DNA-based method molecular markers for the identification of mosquitoes in India and Iran, respectively. CO1 gene was used for the identification of damaged specimen of German mosquitoes with reference to three Ades species reported in Europe (Kruger et al. 2014).

4.6 Identification of Medical Plants

Medicinal plants were used as preventive/protective agent for various diseases; people use medicinal plants throughout the world. However, adulteration is the major problem for users and it is important to authenticate the plant species. DNA barcoding is a promising tool in the authentication of medicinal plants to discriminate the good one from adulterants. Vassou et al. (2015) applied DNA barcoding for the identification of *Sida cordifolia. Peucedanum praeruptorum*, a traditional medicinal plant, was identified by Zhou et al. (2014) using DNA barcoding. *Ginkgo biloba* is exploited for treating dementia, Alzheimer's disease, and Parkinson's disease which was authenticated by using DNA barcoding (Little 2014). An antimutagenic and antioxidant fruit *Phoenix dactylifera was* distinguished by Enan and Ahamed (2014). Hou et al. (2013) applied DNA barcoding for the identification of traditional Chinese medicinal plant *Lonicera japonica*. Wong et al. (2013) evaluated the efficiency of seven DNA barcodes for discriminating closely related medicinal *Gentiana* species and their adulterants. MATK gene was used as a molecular marker for the characterization of *Croton bonplandianus* (Chandramohan et al. 2013), whereas Rai et al. (2012) used *ITS2* gene for detecting substitution in the medicinal plant *Asparagus racemosus* and for the identification of *Cinnamomum osmophloeum*, respectively.

4.7 Monitoring Water Quality

Diatoms are commonly used for the assessments of water quality; however, species-level identification is very difficult because it needs in-depth knowledge of the organisms under investigation and it is also a time-consuming approach. To overcome this bottleneck, Zimmermann et al. (2015) applied metabarcoding of diatoms via NGS sequencing for the evaluation of water quality. DNA barcodes of stream macroinvertebrates were used for monitoring water quality by Sweeney et al. (2011). Macroinvertebrate diversity is used as indicator for assessing ecosystem health in AZTI's Marine Biotic Index (AMBI). Genetic-based AMBI was used for faster and cheaper marine monitoring and health assessment of marine ecology (Aylagas et al. 2014).

4.8 DNA Barcoding on Forensic Sciences and Conservation of Endangered Species

Application of DNA barcoding in the investigations of animal cruelty and poaching and illegal collection and trade of flora and fauna has increased recently. Naro-Maciel et al. (2010) employed DNA barcoding for globally threatened marine turtles. DNA barcoding is a potential tool for detecting poaching of Indian Peafowl, Chinese Sika deer subspecies, Roe deer, Guanaco, Crocodile, Reedbuck, Lowland tapir, and Cypriot mouflon (Gupta et al. 2005; Wu et al. 2005; An et al. 2007; Marín et al. 2009; Eaton et al. 2010; Dalton and Kotze 2011; Sanches et al. 2011; Barbanera et al. 2012). Indo-Chinese spitting cobra, scarlet macaw, sturgeons and paddlefish, elephant, and Southeast Asian monitor lizards were detected by several researchers with the aid of DNA barcoding (Shivji et al. 2005; Baker et al. 2010; Filonzi et al. 2010; Gupta et al. 2011) also employed DNA barcoding for detecting illegal killing of tiger.

DNA barcoding can be used to identify endangered sea turtles by assessing turtle meat, carcasses, or eggs that are illegally traded (Vargas et al. 2009). Laramie et al. (2015) used DNA barcoding for evaluating the distribution of an endangered salmonid in Methow and Okanogan Subbasins of the Upper Columbia River, which span the border between Washington, USA, and British Columbia, Canada. Eva et al. (2016) detected endangered Mekong giant catfish *Pangasianodon gigas* by applying DNA barcode technique. Panprommin and Panprommin (2017) assessed molecular markers for the identification of a critically endangered species like

Trigonostigma somphongsi in Thailand. Bhattacharyya et al. (2015) employed ISSR and DAMD molecular markers for assessing *Dendrobium nobile*, an endangered medicinal orchid. COI was utilized by Dubey et al. (2011) for the identification of some endangered Indian snake species.

5 Conclusion

DNA barcoding is mainly focused on animals and it is high time for giving importance to barcoding of plants and protists. The major drawback of plant barcoding is the absence of universal barcode gene in plants. In spite of some bottlenecks, DNA barcoding approach has successfully been applied for assessing and conserving biodiversity in the massive and diverse ecosystem, detecting mislabeling and food piracy, agricultural pest controlling, disease vector identification, medical plants identification, conservation of endangered species, monitoring water quality, and also employed in forensic palynology. In conclusion, DNA barcoding is a crucial approach which expands our knowledge by assessing innumerable species in an inexpensive and time effective manner. It increases the communication between diverse scientific communities such as phylogeneticists, taxonomists, and population geneticists.

Acknowledgments The authors are grateful to the University Grants Commission (UGC), New Delhi—Dr. D.S. Kothari Post-Doctoral Fellowship—for their financial support.

References

Ahmad I, Fatma Z, Yazdani SS (2013) DNA barcode and lipid analysis of new marine algae potential for biofuel. Algal Res 2:10–15

Ajmal Ali M, Gyulai G, Hidvégi N, Kerti B, Al Hemaid FM, Pandey AK, Lee J (2014) The changing epitome of species identification – DNA barcoding. Saudi J Biol Sci 21:204–231

Alfonsi E, Meheust E, Fuchs S (2013) The use of DNA barcoding to monitor the marine mammal biodiversity along the French Atlantic coast. In: Nagy ZT, Backeljau T, De Meyer M, Jordaens K (eds) DNA barcoding: a practical tool for fundamental and applied biodiversity research, vol 365. Pensoft ZooKeys, Moscow, pp 5–24

An J, Lee M-Y, Min M-S, Lee M-H, Lee H (2007) A molecular genetic approach for species identification of mammals and sex determination of birds in a forensic case of poaching from South Korea. Foren Sci Int 167:59–61

Ardura A, Planes S, Garcia-Vazquez E (2013) Applications of DNA barcoding to fish landings: authentication and diversity assessment. In: Nagy ZT, Backeljau T, De Meyer M, Jordaens K (eds) DNA barcoding: a practical tool for fundamental and applied biodiversity research, vol 365. Pensoft ZooKeys, Moscow, pp 49–65

Armstrong KF, Ball SL (2005) DNA barcodes for biosecurity: invasive species identification. Philos Trans R Soc Lond Ser B Biol Sci 360(1462):1813–1823

Ashfaq M, Hebert PDN, Mirza JH, Khan AM, Zafar Y, Mirza S (2014) Analyzing mosquito (Diptera: Culicidae) diversity in Pakistan by DNA barcoding. PLoS One 9:e97268

Aylagas E, Borja A, Rodríguez-Ezpeleta N (2014) Environmental status assessment using DNA metabarcoding: towards a genetics based marine biotic index (gAMBI). PLoS One 9(3):e90529

Baker CS, Steel D, Choi Y, Lee H, Kim KS, Choi SK et al (2010) Genetic evidence of illegal trade in protected whales links Japan with the US and South Korea. Biol Lett 6(5):647–650. https://doi.org/10.1098/rsbl.2010.0239

Balajee SA, Houbraken J, Verweij PE, Hong SB, Yaghuchi T, Varga J, Samson RA (2007) *Aspergillus* species identification in the clinical setting. Stud Mycol 59:39–46

Baldo L, Hotopp JCD, Jolley KA et al (2006) Multilocus sequence typing system for the endosymbiont *Wolbachia pipientis*. Appl Environ Microbiol 72(11):7098–7110

Ball SL, Armstrong KF (2006) DNA barcodes for insect pest identification: a test case with tussock moths (Lepidoptera: Lymantriidae). Can J For Res 36:337–350

Barbanera F, Guerrini M, Beccani C, Forcina G, Anayiotos P, Panayides P (2012) Case report. Conservation of endemic and threatened wildlife: molecular forensic DNA against poaching of the Cypriot mouflon (Ovis orientalis ophion, Bovidae). Foren Sci Int Genet 6:671–675

Barbuto M, Galimberti A, Ferri E, Labra M, Malandra R, Galli P et al (2010) DNA barcoding reveals fraudulent substitutions in shark seafood products: the Italian case of "palombo" (Mustelus spp.). Food Res Int 43:376–381

Barrett RDH, Hebert PDN (2005) Identifying spiders through DNA barcodes. Can J Zool 83:481–491

Begerow D, Nilsson H, Unterseher M, Maier W (2010) Current state and perspectives of fungal DNA barcoding and rapid identification procedures. Appl Microbiol Biotechnol 87(1):99–108

Besansky NJ, Severson DW, Ferdig MT (2003) DNA barcoding of parasites and invertebrate disease vectors: what you don't know can hurt you. Trends Parasitol 19:545–546

Bhattacharjee MJ, Ghosh SK (2014) Design of mini-barcode for catfishes for assessment of archival biodiversity. Mol Ecol Res 14:469–477

Bhattacharyya P, Kumaria S, Tandon P (2015) Applicability of ISSR and DAMD markers for phyto-molecular characterization and association with some important biochemical traits of Dendrobium nobile, an endangered medicinal orchid. Phytochemistry 117:306–316

Blaxter M, Elsworth B, Daub J (2004) DNA taxonomy of a neglected animal phylum: an unexpected diversity of tardigrades. Proc R Soc B 271:189–192

Bourke BP, Oliveira TP, Suesdek L, Bergo ES, Sallum MA (2013) A multi-locus approach to barcoding in the anopheles strodei subgroup (Diptera: Culicidae). Parasit Vectors 6:111

Carreon-Martinez L, Johnson TB, Ludsin SA, Heath DD (2011) Utilization of stomach content DNA to determine diet diversity in piscivorous fishes. J Fish Biol 78:1170–1182

Cerqueira GC, Arnaud MB, Inglis DO, Skrzypek MS, Binkley G, Simison M, Miyasato SR, Binkley J, Orvis J, Shah P, Wymore F, Sherlock G, Wortman JR (2014) The *Aspergillus* genome database: multispecies curation and incorporation of RNA-Seq data to improve structural gene annotations. Nucl Acids Res 42:D705–D710

Chakraborty M, Ghosh SK (2014) An assessment of the DNA barcodes of Indian freshwater fishes. Gene 537:20–28

Chandramohan A, Divya SR, Dhanarajan MS (2013) Matk gene based molecular characterization of medicinal plant—Croton bonplandianum Baill. Int J Biosci Res 2:1–7

Chang C-H, James S (2011) A critique of earthworm molecular phylogenetics. Pedobiologia 54S: S3–S9

Chen J, Li Q, Kong L (2011) How DNA barcodes complement taxonomy and explore species diversity: the case study of a poorly understood marine fauna. PLoS One 6(6):e21326

Chin TC, Bakar AA, Abidin DHZ, Nor SAM (2016) Detection of mislabelled seafood products in Malaysia by DNA barcoding: improving transparency in food market. Food Control 64:247–256

Clarkston BE, Saunders GW (2010) A comparison of two DNA barcode markers for species discrimination in the red algal family Kallymeniaceae (Gigartinales, Florideophyceae), with a description of Euthora timburtonii sp. Botany 88:119–131

Cohen NJ, Deeds JR, Wong ES, Hanner RH, Yancy HF, White KD et al (2009) Public health response to puffer fish (Tetrodotoxin) poisoning from mislabeled product. J Food Prot 72:810–817

Corse E, Costedoat C, Chappaz R, Pech N, Martin J, Gilles A (2010) A PCR-based method for diet analysis in freshwater organisms using 18S rDNA barcoding on faeces. Mol Ecol Resour 10:96–108

Costa FO, de Waard JR, Boutillier J, Ratnasingham S, Dooh RT, Hajibabaei M, Hebert PDN (2007) Biological identifications through DNA barcodes: the case of the Crustacea. Can J Fish Aquat Sci 64:272–295

Dalton DL, Kotze A (2011) DNA barcoding as a tool for species identification in three forensic wildlife cases in South Africa. Foren Sci Int 207:e51–e54

Daru BH, Yessoufou K, Mankga LT, Davies TJ (2013) A global trend towards the loss of evolutionarily unique species in mangrove ecosystems. PLoS One 8(6):e66686

De Pasquale PP, Galimberti A (2014) New records of the Alcathoe bat, *Myotis alcathoe* (Vespertilionidae) for Italy. Barbastella 7:1

Decaëns T, Porco D, Rougerie R, Brown GG, James SW (2013) Potential of DNA barcoding for earthworm research in taxonomy and ecology. Appl Soil Ecol 65:35–42

Dentinger BTM, Didukh MY, Moncalvo J-M (2011) Comparing COI and ITS as DNA barcode markers for mushrooms and allies (*Agaricomycotina*). PLoS One 6:e25081

Dhananjeyan KJ, Paramasivan R, Tewari SC, Rajendran R, Thenmozhi V, Jerald V, Leo S et al (2010) Molecular identification of mosquito vectors using genomic DNA isolated from egg-shells, larval and pupal exuvium. Trop Biomed 27:47–53

Dubey B, Meganathan PR, Haque I (2011) DNA mini-barcoding: an approach for forensic identification of some endangered Indian snake species. Foren Sci Int: Genet 5:181–184

Dudgeon D, Arthigton AH, Gessner MO et al (2006) Freshwater biodiversity: importance, threats, status and conservation challenges. Biol Rev Camb Philos Soc 81:163–182

Dumbrell AJ, Ashton PD, Aziz N, Feng G, Nelson M, Dytham C, Fitter AH, Helgason T (2011) Distinct seasonal assemblages of arbuscular mycorrhizal fungi revealed by massively parallel pyrosequencing. New Phytol 190:794–804

Eaton MJ, Meyers GL, Kolokotronis S-O, Leslie MS, Martin AP, Amato G (2010) Barcoding bushmeat: molecular identification of central African and south American harvested vertebrates. Conserv Genet 11:1389–1404

Enan MR, Ahamed A (2014) DNA barcoding based on plastid matK and RNA polymerase for assessing the genetic identity of date (Phoenix dactylifera L.) cultivars. Genet Mol Res 13:3527–3536

Ermakov OA, Simonov E, Surin VL, Titov SV, Brandler OV, Ivanova NV, Borisenko AV (2015) Implications of hybridization, NUMTs, and overlooked diversity for DNA barcoding of Eurasian ground squirrels. PLoS One 10(1):e0117201

Eva B, Harmony P, Thomas G, Francois G, Alice V, Claude M, Tonya D (2016) Trails of river monsters: detecting critically endangered Mekong giant catfish Pangasianodon gigas using environmental DNA. Glob Ecol Conser 7:148–156

Evans KM, Wortley AH, Mann DG (2007) An assessment of potential diatom "barcode" genes (cox1, rbcL, 18S and ITS rDNA) and their effectiveness in determining relationships in Sellaphora (Bacillariophyta). Protist 158:349–364

Filonzi L, Chiesa S, Vaghi M, Marzano FN (2010) Molecular barcoding reveals mislabeling of commercial fish products in Italy. Food Res Int 43:1383–1388

Finn DS, Zamora-Muñoz C, Múrria C, Sáinz-Bariáin M, Alba-Tercedor J (2014) Evidence from recently deglaciated mountain ranges that *Baetis alpinus* (Ephemeroptera) could lose significant genetic diversity as alpine glaciers disappear. Freshw Sci 33(1):207–216

Floyd R, Abebe E, Papert A, Blaxter M (2002) Molecular barcodes for soil nematode identification. Mol Ecol 11:839–850

Francis CM, Borisenk AV, Ivanova NV, Eger JL, Lim BK, Guillen-Servent A, Kruskop SV, Mackie I, Hebert PDN (2010) The role of DNA barcode in understanding and conservation of mammal diversity in Southeast Asia. PLoS One 5(9):e12575

Galan M, Pagès M, Cosson JF (2012) Next-generation sequencing for rodent barcoding: species identification from fresh, degraded and environmental samples. PLoS One 7(11):e48374

Galimberti A, De Mattia F, Losa A, Bruni I, Federici S, Casiraghi M, Martellos S, Labra M (2013) DNA barcoding as a new tool for food traceability. Food Res Int 50:55–63

Geiser DM, Klich MA, Frisvad JC, Peterson SW, Varga J, Samson RA (2007) The current status of species recognition and identification in Aspergillus. Stud Mycol 59:1–10

Gilmore SR, Gräfenhan T, Louis-Seize G, Seifert KA (2009) Multiple copies of cytochrome oxidase 1 in species of the fungal genus *Fusarium*. Mol Ecol Resour 9(Suppl S1):90–98

Gray JS (1997) Marine biodiversity: patterns, threats and conservation needs. Biodivers Conserv 6:153–175

Greenstone MH, Rowley DL, Heimbach U, Lundgren JG, Pfannenstiel RS, Rehner SA (2005) Barcoding generalist predators by polymerase chain reaction: carabids and spiders. Mol Ecol 14:3247–3266

Gupta SK, Bhagavatula J, Thangaraj K, Singh L (2011) Case report. Establishing the identity of the massacred tigress in a case of wildlife crime. Foren Sci Int: Genet 5:74–75

Gupta SK, Verma SK, Singh L (2005) Molecular insight into a wildlife crime: the case of a peafowl slaughter. Foren Sci Int 154:214–217

Hajibabaei M, Janzen DH, Burns JM, Hallwachs W, Hebert PDN (2006) DNA barcodes distinguish species of tropical Lepidoptera. Proc Natl Acad Sci U S A 103:968–971

Hanner R, Becker S, Ivanova NV, Steinke D (2011) FISH-BOL and seafood identification: geographically dispersed case studies reveal systemic market substitution across Canada. Mitochondrial DNA 22(S1):106–122

Hebert PDN, Cywinska A, Ball SL, de Waard JR (2003) Biological identifications through DNA barcodes. Proc Biol Sci 270(1512):313–321

Hebert PDN, de Waard JR, Landry J-F (2010) DNA barcodes for 1/1,000 of the animal kingdom. Biol Lett 6:359–362

Hebert PDN, Gregory TR (2005) The promise of DNA barcoding for taxonomy. Syst Biol 54:852–859

Hebert PDN, Ratnasingham S, Zakharov EV, Telfer AC, Levesque-Beaudin V, Milton MA, Pedersen S, Jannetta P, de Waard JR (2016) Counting animal species with DNA barcodes: canadian insects. Phil Trans R Soc B 371:20150333

Hebert PDN, Stoeckle MY, Zemlak TS, Francis CM (2004) Identification of birds through DNA barcodes. PLoS Biol 2:1657–1663

Hendrich L, Moriniere J, Haszprunar G, Hebert PDN, Hausmann A, Kohler F, Balke M (2015) A comprehensive DNA barcode database for central European beetles with a focus on Germany: adding more than 3500 identified species to BOLD. Mol Ecol Resour 15:795–818

Hogg ID, Hebert PDN (2004) Biological identification of springtails (Hexapoda: Collembola) from the Canadian Arctic, using mitochondrial DNA barcodes. Can J Zool 82:749–754

Holmes BH, Steinke D, Ward RD (2009) Identification of shark and ray fins using DNA barcoding. Fish Res 95:280–288

Hou DY, Song JY, Shi LC, Ma XC, Xin TY, Han JP (2013) Stability and accuracy assessment of identification of traditional Chinese materia medica using DNA barcoding: a case study on Flos Lonicerae japonicae. Biomed Res Int 2013:549037

Hubert N, Hanner RH, Holm E, Mandrak NE, Taylor EB, Burridge M et al (2008) Identifying Canadian freshwater fishes through DNA barcodes. PLoS One 3:e2490

Hugenholtz P, Goebel BM, Pace NR (1998) Impact of culture independent studies on the emerging phylogenetic view of bacterial diversity. J Bacteriol 180(18):4765–4774

Hurwood DA, Dammannagoda S, Krosch MN, Jung H, Salin KR, Youssef MA-BH, de Bruyn M, Mather PB (2014) Impacts of climatic factors on evolution of molecular diversity and the natural distribution of wild stocks of the giant freshwater prawn (*Macrobrachium rosenbergii*). Freshw Sci 33(1):217–231

Jo H, Gim J-A, Jeong K-S, Kim H-S, Joo G-J (2014) Application of DNA barcoding for identification of freshwater carnivorous fish diets: is number of prey items dependent on size class for Micropterus salmoides? Ecol Evol 4(2):219–229

Johnsen A, Rindal E, Ericson PGP et al (2010) DNA barcoding of Scandinavian birds reveals divergent lineages in trans-Atlantic species. J Ornithol 151:565–578

Jose D, Nidhin B, Anil Kumar KP, Pradeep PJ, Harikrishnan M (2016) A molecular approach towards the taxonomy of fresh water prawns Macrobrachium striatum and M. equidens (Decapoda, Palaemonidae) using mitochondrial markers. Mitochondrial DNA 27(4):2585–2593

Kermarrec L, Franc A, Rimet F, Chaumeil P, Frigerio J-M, Humbert J-F, Bouchez A (2014) A next-generation sequencing approach to river biomonitoring using benthic diatoms. Freshw Sci 33 (1):349–363

Kerr KCR, Lijtmaer DA, Barreira AS et al (2009) Probing evolutionary patterns in Neotropical birds through DNA barcodes. PLoS One 4:e4379

Kerr KCR, Stoeckle MY, Dove CJ, Weigt LA, Francis CM, Hebert PDN (2007) Comprehensive DNA barcode coverage of north American birds. Mol Ecol Notes 7:535–543

Keskin E, Atar HH (2013) DNA barcoding commercially important fish species of Turkey. Mol Ecol Resour 13(5):788–797

Kim MS, Yang MY, Cho GY (2010) Applying DNA barcoding to Korean Gracilariaceae (Rhodophyta) Cryptogamie. Algologie 31(4):387–401

Kondo T, Gullan PJ, Williams DJ (2008) Coccidolog. The study of scale insects (Hemiptera: Sternorrhyncha: Coccoidea). Revista Corpoica – Ciencia Y Tecnologia Agropecuaria 9:55–61

Kruger A, Obermayr U, Czajka C, Bueno-Mari R, Jost A, Rose A (2014) COI sequencing for invasive mosquito surveillance in Germany reveals genetically divergent specimens near Aedes geniculatus (Diptera: Culicidae). J Eur Mosq Control Assoc 32:22–26

Kumar NP, Krishnamoorthy N, Sahu SS, Rajavel AR, Sabesan S, Jambulingam P (2013) DNA barcodes indicate members of the Anopheles fluviatilis (Diptera: Culicidae) species complex to be conspecific in India. Mol Ecol Resour 13:354–361

Kumar NP, Rajavel AR, Natarajan R, Jambulingam P (2007) DNA barcodes can distinguish species of Indian mosquitoes (Diptera: Culicidae). J Med Entomol 44:1–7

Lakra WS, Singh M, Goswami M, Gopalakrishnan A, Lal KK, Mohindra V, Sarkar UK, Punia PP, Singh KV, Bhatt JP, Ayyappan S (2016) DNA barcoding Indian freshwater fishes. Mitochondrial DNA 27(6):4510–4517

Lambert DM, Baker A, Huynen L, Haddrath O, Hebert PDN, Millar CD (2005) Is a large-scale DNA-based inventory of ancient life possible? J Hered 96:279–284

Laramie MB, Pilliod DS, Goldberg CS (2015) Characterizing the distribution of an endangered salmonid using environmental DNA analysis. Biol Conserv 183:29–37

Lauber CL, Zhou N, Gordon JI, Knight R, Fierer N (2010) Effect of storage conditions on the assessment of bacterial community structure in soil and human-associated samples. FEMS Microbiol Lett 307:80–86

Layton KKS, Martel AL, Hebert PDN (2014) Patterns of DNA barcode variation in Canadian marine molluscs. PLoS One 9(4):e95003

Les DH, Moody ML, Jacobs SWL, Bayer RJ (2002) Systematics of seagrasses (Zosteraceae) in Australia and New Zealand. Syst Bot 27:468–484

Li J, Han H, Gao Q, Jin Q, Chi M, Wu C, Zhang A (2012) Species identification of Noctuidae (Insecta: Lepidoptera) with DNA barcoding of support vector machine and neighbor-joining method. J Biosafety 21:308–314

Lim NKM, Tay YC, Srivathsan A, Tan JWT, Kwik JTB, Baloğlu B, Meier R, Yeo DCJ (2016) Next-generation freshwater bioassessment: eDNA metabarcoding with a conserved metazoan primer reveals species-rich and reservoir-specific communities. R Soc Open Sci 3:160635

Links MG, Dumonceaux TJ, Hemmingsen SM, Hill JE (2012) The chaperonin-60 universal target is a barcode for bacteria that enables de novo assembly of metagenomic sequence data. PLoS One 7(11):e49755

Little DP (2014) Authentication of Ginkgo biloba herbal dietary supplements using DNA barcoding. Genome 57:513–516

Lohman DJ, Ingram KK, Prawiradilaga DM et al (2010) Cryptic genetic diversity in "widespread" southeast Asian bird species suggests that Philippine avian endemism is gravely underestimated. Biol Conserv 143:1885–1890

Lorenz JG, Jackson WE, Beck JC, Hanner R (2005) The problems and promise of DNA barcodes for species diagnosis of primate biomaterials. Philos Trans R Soc B 360:1869–1877

Lowenstein JH, Amato G, Kolokotronis S-O (2009) The real maccoyii: identifying tuna sushi with DNA barcodes – contrasting characteristic attributes and genetic distances. PLoS One 4(11): e7866

Lu L, Chesters D, Zhang W, Li G, Ma Y, Ma H, Song X, Wu H, Meng F, Zhu C, Liu Q (2012) Small mammal investigation in spotted fever focus with DNA-barcoding and taxonomic implications on rodents species from Hainan of China. PLoS One 7(8):e43479

Luo AR, Zhang AB, Ho SYW, Xu WJ, Zhang YZ, Shi WF et al (2011) Potential efficacy of mitochondrial genes for animal DNA barcoding: a case study using eutherian mammals. BMC Genomics 12:84

Maas AE, Blanco-Bercial L, Lawson GL (2013) Reexamination of the species assignment of Diacavolinia pteropods using DNA barcoding. PLoS One 8(1):e53889

Makarova O, Contaldo N, Paltrinieri S, Kawube G, Bertaccini A, Nicolaisen M (2012) DNA barcoding for identification of "*Candidatus phytoplasmas*" using a fragment of the elongation factor Tu gene. PLoS One 7(12):e52092

Marie-Stephane T, Mireille O, Serge K (2012) An integrative morphological and molecular diagnostics for Typhlodromus pyri (Acari: Phytoseiidae). Zool Scr 41:68–78

Marín JC, Saucedo CE, Corti P, González BA (2009) Application of DNA forensic techniques for identifying poached guanacos (*Lama guanicoe*) in Chilean Patagonia. J Foren Sci 54 (5):1073–1076

Markmann M, Tautz D (2005) Reverse taxonomy: an approach towards determining the diversity of meiobenthic organisms based on ribosomal RNA signature sequences. Philos Trans R Soc B 360:1917–1924

Molnar JL, Gamboa RL, Revenga C (2008) Assessing the global threat of invasive species to marine biodiversity. Front Ecol Environ 6:485–492

Mora C, Tittensor DP, Adl S, Simpson AGB, Worm B (2011) How many species are there on earth and in the ocean? PLoS Biol 9:e1001127

Mouquet N, Devictor V, Meynard CN et al (2012) Ecophylogenetics: advances and perspectives. Biol Rev 87(4):769–785

Naddaf SR, Oshaghi MA, Vatandoost H (2012) Confirmation of two sibling species among Anopheles fluviatilis mosquitoes in south and southeastern Iran by analysis of cytochrome oxidase I gene. J Arthropod-Borne Dis 6:144–150

Nagoshi RN, Meagher RL, Brambila J (2011) Use of DNA barcodes to identify invasive armyworm Spodoptera species in Florida. J Insect Sci 11:154

Nagy ZT, Sonet G, Glaw F (2012) First large-scale DNA barcoding assessment of reptiles in the biodiversity hotspot of Madagascar, based on newly designed COI primers. PLoS One 7(3): e34506

Naro-Maciel E, Reid B, Fitzsimmons NN, Le M, Desalle R, Amato G (2010) DNA barcodes for globally threatened marine turtles: a registry approach to documenting biodiversity. Mol Ecol Resour 10:252–263

Natalie JWW (2013) Determining the microbial diversity in chaparral soils before and after wildfires through DNA barcoding. Project Number-S1119, California State Science Fair, Project Summary

Neigel J, Domingo A, Stake J (2007) DNA barcoding as a tool for coral reef conservation. Coral Reefs 26:487–499

Newmaster SG, Grguric M, Shanmughanandhan D, Ramalingam S, Ragupathy S (2013) DNA barcoding detects contamination and substitution in north American herbal products. BMC Med 11:222

Ngamskulrungroj P, Gilgado F, Faganello J, Litvintseva AP, Leal AL, Tsui KM, Mitchell TG, Vainstein MH, Meyer W (2009) Genetic diversity of the *Cryptococcus* species complex suggests that *Cryptococcus gattii* deserves to have varieties. PLoS One 4:e5862

Nieukerken EJV, Doorenweerd C, Stokvis FR, Dick SJ, Groenenberg DSJ (2012) DNA barcoding of the leaf-mining moth subgenus Ectoedemias. Str. (Lepidoptera: Nepticulidae) with COI and EF1-a: two are better than one in recognizing cryptic species. Contr Zool 81:1–24

Nwani CD, Becker S, Braid HE, Ude EF, Okogwu OI, Hanner R (2011) DNA barcoding discriminates freshwater fishes from southeastern Nigeria and provides river system-level phylogeographic resolution within some species. Mitochondrial DNA 22(Suppl 1):43–51

Panprommin D, Panprommin N (2017) Assessment of the DNA barcoding for identification of Trigonostigma somphongsi, a critically endangered species in Thailand. Biochem Systemat Ecol 70:200–204

Parvathy VA, Swetha VP, Sheeja TE, Leela NK, Chempakam B, Sasikumar B (2014) DNA barcoding to detect chilli adulteration in traded black pepper powder. Food Biotechnol 28:25–40

Parvathy VA, Swetha VP, Sheeja TE, Sasikumar B (2015) Detection of plant-based adulterants in turmeric powder using DNA barcoding. Pharm Biol 53:1774–1779

Patel S, Waugh J, Millar CD, Lambert DM (2010) Conserved primers for DNA barcoding historical and modern samples from New Zealand and Antarctic birds. Mol Ecol Res 10:431–438

Pauls SU, Blahnik RJ, Zhou X, Wardwell CT, Holzenthal RW (2010) DNA barcode data confirm new species and reveal cryptic diversity in Chilean Smicridea (Smicridea) (trichoptera: hydropsychidae). J N Am Benthol Soc 29:1058–1074

Prendini L (2005) Comment on "identifying spiders through DNA barcodes". Can J Zool 83:498–504

Rai PS, Bellampalli R, Dobriyal RM, Agarwal A, Satyamoorthy K, Narayana DBA (2012) DNA barcoding of authentic and substitute samples of herb of the family Asparagaceae and Asclepiadaceae based on the ITS2 region. J Ayurveda Integr Med 3:136–140

Rainer J, de Hoog GS (2006) Molecular taxonomy and ecology of *Pseudallescheria*, *Petriella* and *Scedosporium prolificans* (*Microascaceae*) containing opportunistic agents on humans. Mycol Res 110:151–160

Raja HA, Baker TR, Little JG, Oberlies NH (2017) DNA barcoding for identification of consumer-relevant mushrooms: a partial solution for product certification? Food Chem 214:383–392

Rastogi G, Dharne MS, Walujkar S, Kumar A, Patole MS, Shouche YS (2007) Species identification and authentication of tissues of animal origin using mitochondrial and nuclear markers. Meat Sci 76:666–674

Rebijith KB, Asokan R, Kumar NKK, Krishna V, Chaitanya BN, Ramamurthy VV (2013) DNA barcoding and elucidation of cryptic aphid species (Hemiptera: Aphididae) in India. Bull Entomol Res 3:601–610

Remigio EA, Hebert PDN (2003) Testing the utility of partial COI sequences for phylogenetic estimates of gastropod relationships. Mol Phylogenet Evol 29:641–647

Renaud AK, Savage J, Adamowicz SJ (2012) DNA barcoding of northern Nearctic Muscidae (Diptera) reveals high correspondence between morphological and molecular species limits. BMC Ecol 12:24

Roe AD, Rice AV, Bromilow SE, Cooke JE, Sperling FA (2010) Multilocus species identification and fungal DNA barcoding: insights from blue stain fungal symbionts of the mountain pine beetle. Mol Ecol Resour 10:946–959

Rossi N, Mantelatto FL (2013) Molecular analysis of the freshwater prawn Macrobrachium olfersii (Decapoda, Palaemonidae) supports the existence of a single species throughout its distribution. PLoS One 8(1):e54698

Ruiz-García M, Pinedo-Castro M, Shostell JM (2014) How many genera and species of woolly monkeys (Atelidae, Platyrrhine, primates) are there? The first molecular analysis of *Lagothrix flavicauda*, an endemic Peruvian primate species. Mol Phylogenet Evol 79:179–198

Sanches A, Perez WAM, Figueiredo MG, Rossini BC, Cervini M, Galetti PM Jr et al (2011) Wildlife forensic DNA and lowland tapir (Tapirus terrestris) poaching. Conserv Genet Resour 3:189–193

Schneider KL, Marrero G, Alvarez AM, Presting GG (2011) Classification of plant associated bacteria using RIF, a computationally derived DNA marker. PLoS One 6(4):e18496

Scorzetti G, Fell JW, Fonseca A, Statzell-Tallman A (2002) Systematics of basidiomycetous yeasts: a comparison of large subunit D1/D2 and internal transcribed spacer rDNA regions. FEMS Yeast Res 2:495–517

Seifert KA, Samson RA, Dewaard JR, Houbraken J, Lévesque CA, Moncalvo JM, Louis-Seize G, Hebert PD (2007) Prospects for fungus identification using *CO1* DNA barcodes, with *Penicillium* as a test case. Proc Natl Acad Sci U S A 104:3901–3906

Shander C, Willassen E (2005) What can biological barcoding do for marine biology. Mar Biol Resour 1:79–83

Shivji MS, Chapman DD, Pikitch EK, Raymond PW (2005) Genetic profiling reveals illegal international trade in fins of the great white shark Carcharodon carcharias. Conser Genet 6 (6):1035–1039

Smith MA, Fernandez-Triana JL, Eveleigh E, Gomez J, Guclu C, Hallwachs W, Hebert PDN et al (2013) DNA barcoding and the taxonomy of Microgastrinae wasps (hymenoptera, Braconidae): impacts after 8 years and nearly 20000 sequences. Mol Ecol Res 13:168–176

Smith MA, Fisher BL, Hebert PDN (2005) DNA barcoding for effective biodiversity assessment of a hyperdiverse arthropod group: the ants of Madagascar. Philos Trans R Soc B 360:1825–1834

Smith PJ, McVeagh SM, Steinke D (2008) DNA barcoding for the identification of smoked fish products. J Fish Biol 72:464–471

Smith MA, Poyarkov NA, Hebert PDN (2008) CO1 DNA barcoding amphibians: take the chance, meet the challenge. Mol Ecol Resour 8:235–246

Steinke D, Zemlak TS, Hebert PDN (2009) Barcoding Nemo: DNA-based identifications for the ornamental fish trade. PLoS One 4(7):e6300

Stoeckle MY (2008) Blog: DNA identifies invasive parasitic wasp' in the barcode of life blog. http://phe.rockefeller.edu/barcode/blog/2008/07/07/dna-identifies-invasive-parasitic-wasps/. Accessed 26 Feb 2011

Stoeckle MY, Gamble CC, Kirpekar R, Yung G, Ahmed S, Little DP (2011) Commercial teas highlight plant DNA barcode identification successes and obstacles. Nat Sci Rep 1:42

Stoeckle MY, Hebert PDN (2008) Barcode of life: DNA tags help classify animals. Sci Am 298 (10):39–43

Swartz ER, Mwale M, Hanner R (2008) A role for barcoding in the study of African fish diversity and conservation. South Afr J Sci 104(4):293–298

Sweeney BW, Battle JM, Jackson JK (2011) Can DNA barcodes of stream macroinvertebrates improve descriptions of community structure and water quality? J North Am Benthol Soc 30 (1):195–216

Taberlet P, Coissac E, Pompanon F, Gielly L, Miquel C, Valentini A, Vermat T, Corthier G, Brochmann C, Willerslev E (2007) Power and limitations of the chloroplast trnL (UAA) intron for plant DNA barcoding. Nucleic Acids Res 35(3):e14

Tack JF, Berghe E, Polk PH (1992) Ecomorphology of Crassostrea cucullata (born, 1778) (Ostreidae) in a mangrove creek (Gazi, Kenya). Hydrobiologia 247:109–117

Thao L, James T, Alexandra B et al (2013) Bar-coded Enterobacteria: an undergraduate microbial ecology laboratory module. Am J Educ Res 1(1):26–30

Trivedi S, Affan R, Alessa AHA et al (2014) DNA barcoding of Red Sea fishes from Saudi Arabia – the first approach. DNA Barcodes 2:17–20

Trivedi S, Ghosh SK, Choudhury A (2013) DNA sequence of cytochrome c oxidase subunit 1 (COI) region of an oyster Saccostrea cucullata collected from Sunderbans. J Environ Sociobiol 10(1):77–81

Uchimura M, Faye EJ, Shimada S, Inoue T, Nakamura Y (2008) A reassessment of Halophila species (Hydrocharitaceae) diversity with special reference to Japanese representatives. Bot Mar 51:258–268

Udayasuriyan R, Saravana Bhavan P, Vadivalagan C, Rajkumar G (2015) Efficiency of different COI markers in DNA barcoding of freshwater prawn species. J Entomol Zool Stud 3(3):98–110

Vargas SM, Araujo FCF, Santos FR (2009) DNA barcoding of Brazilian sea turtles (Testudines). Genet Mol Biol 32(3):608–612

Vassou SL, Kusuma G, Parani M (2015) DNA barcoding for species identification from dried and powdered plant parts: a case study with authentication of the raw drug market samples of Sida cordifolia. Gene 559:86–93

Vialle A, Feau N, Allaire M, Didukh M, Martin F, Moncalvo JM, Hamelin RC (2009) Evaluation of mitochondrial genes as DNA barcode for *Basidiomycota*. Mol Ecol Resour 9(Suppl S1):99–113

Wang G, Li C, Guo X, Xing D, Dong Y (2012) Identifying the main mosquito species in China based on DNA barcoding. PLoS One 7:e47051

Wang J, Soininen J, He J, Shen J (2012) Phylogenetic clustering increases with elevation for microbes. Environ Microbiol Rep 4(2):217–226

Ward RD, Holmes BH, White WT, Last PR (2008) DNA barcoding Australasian chondrichtyans: results and potential uses in conservation. Mar Freshw Res 59(1):57–71

Ward RD, Zemlak TS, Innes BH, Last PR, Hebert PDN (2005) DNA barcoding Australia's fish species. Philos Trans R Soc Lond B 360:1847–1857

Waycott M, Duarte CM, Carruthers TJB et al (2009) Accelerating loss of seagrasses across the globe threatens coastal ecosystems. Proc Natl Acad Sci U S A 106:12377–12381

Weigt LA, Baldwin CC, Driskell A (2012) Using DNA barcoding to assess Caribbean reef fish biodiversity: expanding taxonomic and geographic coverage. PLoS One 7(7):e41059

Wilson EO (ed) (1994) Biodiversity. National Academy Press, Washington, DC

Wilson EO (2003) The encyclopaedia of life. Trends Ecol Evol 18:77–80

Wong KL, But PPH, Shaw PC (2013) Evaluation of seven DNA barcodes for differentiating closely related medicinal Gentiana species and their adulterants. Chin Med 8:16

Wong EHK, Hanner RH (2008) DNA barcoding detects market substitution in north American seafood. Food Res Int 41:828–837

Woodcock TS, Boyle EE, Roughley RE, Kevan PG, Labbee RN, Smith ABT, Goulet H et al (2013) The diversity and biogeography of the Coleoptera of Churchill: insights from DNA barcoding. BMC Ecol 13:40

Wu H, Wan Q-H, Fang S-G, Zhang S-Y (2005) Application of mitochondrial DNA sequence analysis in the forensic identification of Chinese sika deer subspecies. Foren Sci Int 148:101–105

Yang Q, Zhao S, Kucerova Z, Stejskal V, Opit G, Qin M, Cao Y, Li F et al (2013) Validation of the 16S rDNA and COI DNA barcoding technique for rapid molecular identification of stored product Psocids (Insecta: Psocodea: Liposcelididae). J Econ Entomol 106:419–425

Zhou J, Wang W, Liu M, Liu Z (2014) Molecular authentication of the traditional medicinal plant Peucedanum praeruptorum and its substitutes and adulterants by DNA – barcoding technique. Pharmacogn Mag 10:385–390

Zimmermann J, Glöckner G, Jahn R, Enke N, Gemeinholzer B (2015) Metabarcoding vs. morphological identification to assess diatom diversity in environmental studies. Mol Ecol Resour 15(3):526–542

Significance of DNA Barcoding in Avian Species: Tracing the History and Building the Future

Farhina Pasha

Abstract Avian species or subspecies have almost identical barcode of life; therefore, when a new unidentified sample is encountered, all it takes for its recognition is a plumage. Although a feather is dead keratin, at the end of this feather some dead skin cells are attached which are the source for DNA extraction and further COI gene sequence amplification. Not only the faecal sample from a fleeting bird has lots of intestinal epithelial cells, but it also forms a great sample for DNA extraction. The DNA barcode so obtained can be then compared to the databases available publicly, such as ABBI/BOLD. Despite all the milestones that DNA barcoding has achieved, there are certain issues or limitations, which must be clearly recognized and resolved. Furthermore, there is a need to conduct extensive studies in tropical taxa and those with limited dispersal. These studies must be undertaken with broader taxonomic lineage and wider geographic boundaries to discover all sister taxa as well.

Keywords DAN Barcoding · BOLD · COI · Avian species · Moa · Bird strike

1 Introduction

Fossils are not only interesting discoveries but they also allow to peep into the past biodiversity and evolutionary changes occurring in that species. There are many techniques to identify the species and the changes that occur in the species. 'DNA barcoding', the much-talked about technique, which has marked its existence in the world of taxonomy and identification of species, using single mtDNA gene cytochrome *c* oxidase I (COI), has partially been appreciated and partially criticized for the same reason that identification is based on a single gene. Main characteristics of DNA barcoding include the following: (1) The technique uses fragments of DNA for

F. Pasha (✉)
Department of Biology, Faculty of Sciences, University of Tabuk, Tabuk, Saudi Arabia
e-mail: fbasha@ut.edu.sa

S. Trivedi et al. (eds.), *DNA Barcoding and Molecular Phylogeny*,
https://doi.org/10.1007/978-3-030-50075-7_2

identification or characterization of any species (2) It can be used for all stages of life and life forms, i.e. the technique can be performed on the cells directly collected from the organism or can be done on the fossils (3) it unveils the look-alike species (4) The process is fast and helps in recognition of known species and detection of new once and can recognize the relation between two species (5) it forms standardized DNA barcode libraries, which makes it feasible to identify species: native or invasive, rare or abundant, endangered or nascent. It can be well phrased that 'DNA barcoding has proliferated new leaves of hopes to the tree of life'.

2 Significance of DNA Barcoding in Aves

DNA barcoding is not only useful to identify or categorize the organisms, but it can also be conducted on the fossils having traces of DNA. DNA of the fossil or extinct organisms can be collected from the museums all around the world. The museums are full of specimens, which have preserved birds' skin and feather and that of other animals too. These can be of significant importance as DNA extraction samples. By using DNA barcoding, the correlation can be established between the birds from the past and the birds from the modern era. This hopefully will solve the cryptic species identification and of course trace original species taxonomic position.

Earlier, the sequence diversity in a 648-bp region of the mtDNA, cytochrome c oxidase I (COI), was used for the DNA barcode to identify the animal species. This study tested the effectiveness of a COI barcode in establishing the bird species as well. They tested 260 avian species from North America and found that all the 260 avian species sampled had a unique COI sequence and 130 species had two or more than two specimens present as observed during testing, which was confirmed, as their COIs were similar or identical to other species. The COI variation amongst the species was 0.43% whereas between the species was reported to be 7.93%. Therefore, their results established a positive wave for DNA-barcoding technique as a milestone in the identification of new species as well as establishing more precise and accurate taxonomy. They also anticipated a 'standard sequence threshold' called as 'Barcoding gap', which was 10 times the mean intraspecific variation for the group under study (Hebert et al. 2004).

3 Indexing the Ancient Biota

DNA barcoding can be used for the identification of different species of plants as well as animals of the present world. It also helps to give insight into the past species and their correlation with the present. DNA barcoding can be conducted on the DNA isolated from the fossils. In an effort to decipher the ancient life via DNA barcoding, Lambert et al. (2005) sequenced 26 subfossil Moa bones by COI gene analysis. The result showed a unique barcode and intraspecific COI sequence with a variance range of 0–1.24%. For deciphering novel species, a standard sequence threshold of 2.7% COI sequence was set. These results were in confirmation with those of Hebert's (2004). Employing this value, six different moa groups were identified. With the standard sequence threshold of 1.24%, 10 moa species were identified amongst the known groups with one deviation. This probably was previously unidentified species. It is also inferred that due to slow rate of growth and reproduction, the interspecies variation was also very slow.

In another study, Ewan Grant-Mackie (2006), a mathematician, determined the DNA barcodes of pūkeko, takahē, moho, Samoan swamp hens, Solomon Island swamp hens, Tongan swamp hens, and Australian swamp hens from North Island and South Island. The samples of moho and Polynesian (Samoan, Solomon Island, and Tongan) swamp were collected in 1920. He conducted the DNA barcoding of the mitochondrial cytochrome c oxidase subunit 1 (CO1) gene. The DNA-barcoding results indicated that the north island pukeko were of Australian origin and those birds came to North Island by flying and the south island pukeko were of Polynesian origin.

Avian species or subspecies have almost identical barcode of life; therefore, when a new unidentified sample is encountered, all it takes for its recognition is a plumage. Although a feather is dead keratin, at the end of this feather some dead skin cells are attached which are the source for DNA traces and COI gene sequence amplification. Not only feathers but also the faecal sample from a fleeting bird having lots of intestinal epithelial cells forms a great sample for DNA extraction. The DNA barcode so obtained can be then compared to the databases available publicly, such as ABBI/BOLD. Similarly, if the fossils of some ancient birds are available, then the DNA extracted from the bone samples can be used for DNA barcoding. Methodology for the analysis of Avian samples is exhibited in Fig. 1.

4 The Present Scenario

The success of this technique is evidenced as there are over 198 publications from 2004 to 2017. In a study, a desert Wheatear spotted on the Scottish Coast in winter 2013 had its origin in Gobi Desert; a Yellow Wagtail found around the sewage farm in northeast Russia and a Black Redstart which was identified by its faecal matter left on the beachfront during migration all came from China (Martin 2013).

In another study, ornithologist Craig Symes closely observed the white-winged flufftail (*Sarothrura ayresi*). White-winged flufftail is so rare that it has been spotted in 15 sites only in South Africa in the past 136 years. Ornithologists believe that there are only 250 left, so the bird comes under critically endangered species. Very little information is available about this bird and it is still not clear whether they are two different species in South Africa and Ethiopia or they have migrated from one place to other as this small water bird with big feet was also spotted in Ethiopia, the only country after South Africa (Dalton et al. 2016). After careful genetic and

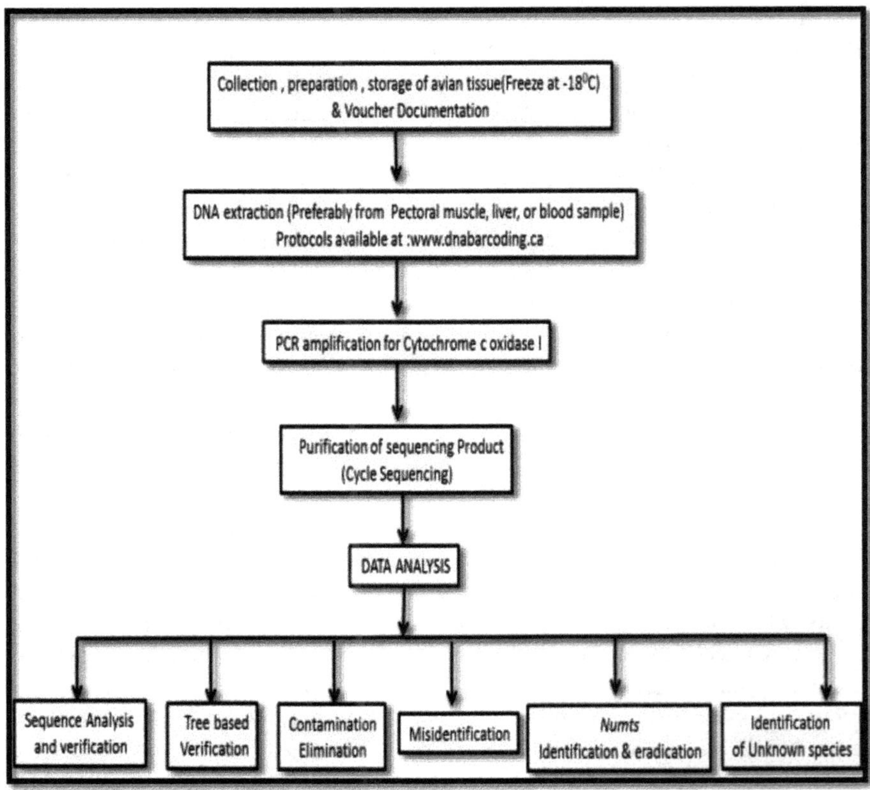

Fig. 1 Modus operand—sampling in avian species for DNA barcoding

isotope analysis, Craig and his team (2013) confirmed that two birds are of the same species. Further DNA barcode studies are underway.

Dove et al. (2008) used the technique of using mitochondrial DNA barcodes (*cytochrome c oxidase subunit 1* [CO1]) for identifying the birds, which are the major cause of air collision. The existing completion of the air space between aircrafts and birds is well described by Sodhi et al. 2002. As the sample available after collision is very small, molecular analysis is not possible; therefore, DNA barcoding has emerged as an immensely useful technique. A total of 821 samples were collected from September to December 2006 from the United States for DNA analysis. From the total sample collected, 554(67.5%) of the sample was successfully amplified for DNA barcoding, whereas 267(32.5%) could not be amplified. Age of the samples, which was almost 6 months, did not affect the DNA efficiency of the sample. It was observed that the primary state of the sample and its collection method greatly influence the DNA viability. Related studies were performed by many groups to prove this fact (Blackwell and Wright 2006; Christidis et al. 2006; Dolbeer 2006; Doran et al. 1990; Hebert et al. 2003; Hermans et al. 1996; Laybourne and Dove 1994; Linnell et al. 1996; Yoo et al. 2006).

5 Limitations of DNA Barcoding

Despite all the milestones that DNA barcoding has achieved, there are certain issues or limitations which must be clearly recognized and resolved. Some of these mentioned are not only related to biological aspects but also affect statistical status.

A. *Elimination of 'Barcode Gaps'*
   This may be due to shortage of sampling or sampling quality (Meyer and Paulay 2005). Careful assessment of sampling during database assembly phase is a must for elimination of barcode gaps (Wiemer and Fiedler 2007). Meyer and Paulay (2005) also observed the inaccuracies for some specimens in well-defined phylogenies as well as in moderately identified groups.
B. *Intrinsic Risks Due to Mitochondrial Lineage*
   As it is a well-known fact that the mitochondrial DNA (mtDNA) is strongly related to maternal inheritance, the usage of mtloci many times leads to pseudo-estimation of sample divergence, thereby leading to an ambiguous identification. Some processes like interspecific hybridization and endosymbiont infections may lead to the relocation of the mitochondrial gene group (Dasmahapatra and Mallet 2006). In all these conditions, nuclear loci analysis is required to clear the phylogenetic relationships. These should be clearly marked in BOLD during its database entry.
C. *NUMTs - Nuclear Copies of COI*
   This is another issue to be cautiously addressed in DNA barcoding. Nuclear mitochondrial DNAs (NUMTs) are described as nuclear copies of mitochondrial DNA sequences, which are translocated into the nuclear genome (Williams and Knowlton 2001). The range of NUMTs varies with species, e.g. none or few in Anopheles to more than 500 in humans (Richly and Leister 2004). NUMTs can be predicted during sequence or amino acid alignment. They can be traced by the sequence checking method available in BOLD, namely "rejection of inconsistent amino acid alignment".
D. *Rate of Genomic Evolution*
   The rate at which species evolve is not the same for all species. There have been cases where due to lack of COI sequence resolving power has misled taxa from primary single gene method to multiple region barcoding system (especially when COI sequence is not species specific or in taxa with low mitochondrial evolutionary rate). This is termed as NON-COI Barcode (Bakker 2007).
E. *Geographical structure*
   Especially in the case of avian species, it can smudge species delineation, as elevated rates of intraspecific divergence are obtained from a geographically remote or isolated population (Hebert et al. 2003); therefore, there must be a major consideration in DNA barcoding. To resolve a prominent question of boundary between population and species, we need to incorporate a broad range of intraspecific samples in the reference databases. Furthermore, there is a need to conduct extensive studies in tropical taxa and those with limited dispersal.

These studies must be undertaken with broader taxonomic lineage and wider geographic boundaries to discover all sister taxa as well.

6 Future Prospects

Birds are a better example for the study on the taxonomy and phylogenetic level than any other organism. Birds are probably one of the best groups of vertebrates in frozen-tissue collections, with more than 300,000 tissue samples covering nearly 75% of known bird species (Stoeckle and Winker 2009). This specificity of birds makes them ideal for the analysis of the effectiveness of any optimized genetic method for species identification, or, in simple words, DNA barcoding. Aves are the first group on which the large-scale DNA barcoding studies were conducted (Lijtmaer et al. 2012). Birds are found to be present in almost all the habitats and show sensitivity to the change in the environmental conditions. Therefore, they play an important role in environmental monitoring as indicator species. (Wormworth and Sekercioglu 2011). DNA barcoding is one of the simple and reliable methods to identify the different species of the aves of past, present, and future and their correlation with the environmental changes. In spite of its simplicity and reliability, there are few disadvantages of DNA barcoding as the technique relies only on mtDNA a single gene analysis. DNA barcoding can be more reliable and useful if it can be supplemented with other barcodes, especially from the nuclear barcodes (Dasmahapatra) and the carbon dating for the fossils. There are many recent advances in barcoding techniques; one of them is metabarcoding. Metabarcoding is a fast and quick method to assess biodiversity. The method is the combination of DNA barcoding (where the universal PCR primers are used to mass-amplify DNA Barcodes) and high-throughput DNA sequencing.

As the proverb stands 'and miles to go before I sleep', DNA barcode has still to solve many more mysteries by systematizing millions of sequences being generated every day in laboratories all over the world.

7 Conclusion

To determine the time of diversification of modern birds is a difficult task. Many methods are required to determine and the different DNA sequencing studies reveal that the ancestors of modern birds were inhabiting South America 95 million years ago (Claramunt and Cracraft 2015). DNA barcoding is one of the powerful tools to determine the age and the origin of a bird species. It is efficient in displaying the avian realms and demarcating their taxonomic boundaries. DNA barcoding is also helpful in clearing the ambiguity for point of origin of an avian species from their point of existing habitat. Although there are few drawbacks of using DNA barcoding, its success in distinguishing species in a wide range of taxa and in identifying the cryptic species is remarkable (Schirtzinger et al. 2012). It can be performed on a small quantity of samples collected from the fossils, which means the fossil's DNA barcoding can give insight into the origin of bird species when it combines with other techniques.

Acknowledgements The authors would like to acknowledge the University of Tabuk, Tabuk, Saudi Arabia.

The author would also like to thank the Department of Biology, Faculty of Sciences, Saudi Digital Library and University Library for providing the facility for Literature survey and collection.

References

Bakker (2007) Second International Barcode of Life Conference, TAIPEI

Blackwell BF, Wright SE (2006) Collisions of red-tailed hawks (*Buteo jamaicensis*), Turkey vultures (*Cathartes aura*), and black vultures (*Coragyps atratus*) with aircraft: implications for bird strike reduction. Journal of Raptor Research 40:76–80. BioOne

Christidis L, Norman JA, Johnson RN, Lindsay S (2006) Forensic identification of aviation bird strikes in Australia. In: Australian government, Australian transport safety bureau research and analysis report, aviation safety research Grant B2005/0017. Canberra, Australia, Australian Museum

Claramunt S, Cracraft J (2015) A new time tree reveals earth history's imprint on the evolution of modern birds. Sci Adv 1(11):e1501005

Dalton DL, Vermaak E, Smit-Robinson HA, Kotze A (2016) Lack of diversity at innate immunity Toll-like receptor genes in the Critically Endangered White-winged Flufftail (Sarothrura ayresi). Scientific Reports, *6*

Dasmahapatra KK, Mallet J (2006) DNA barcodes: recent successes and future prospects. Heredity 97:254–255

Dolbeer RA (2006) Height distribution of birds recorded by collisions with civil aircraft. J Wildl Manag 70:1345–1350. BioOne

Doran HJ, Cross TF, Kelly TC (1990) Electrophoretic identification of bird species involved in collisions with aircraft. Comp Biochem Physiol 97:171–175

Dove CJ, Rotzel NC, Heacker M, Weigt LA (2008) Using DNA barcodes to identify bird species involved in bird strikes. J Wildl Manag 72(5):1231–1236. https://doi.org/10.2193/2007-272

Ewan Grant-Mackie (2006). http://legacy.biotechlearn.org.nz/themes/barcoding_life/barcoding_new_zealand_swamp_hen

Hebert P, Cywinska A, Ball SL, de Waard JR (2003) Biological identification through DNA barcodes. Proc R Soc Lond B 270:313–321

Hebert PD, Stoeckle MY, Zemlak TS, Francis CM (2004) Identification of Birds through DNA Barcodes. PLoS Biol 2(10):e312

Hermans J, Buurma L, Wattel J (1996) Identification of bird remains after bird–airplane collisions, based on DNA sequences analysis. Proceedings of the International Bird Strike Committee Europe 23:203–207

Lambert DM, Baker A, Huynen L, Haddrath O, Hebert PD, Millar CD (2005) "is a large-scale DNA-based inventory of ancient life possible?" (PDF fulltext). J Hered 96(3):279–284

Laybourne RC, Dove CJ (1994) Preparation of bird-strike remains for identification. Proceedings of the Bird Strike Committee Europe 22:5531–5543

Lijtmaer DA, Kerr KC, Stoeckle MY, Tubaro PL (2012) DNA barcoding birds: from field collection to data analysis. DNA Barcodes, Methods and Protocols, pp 127–152

Linnell MA, Conover MR, Ohashi TJ (1996) Analysis of bird strikes at a tropical airport. J Wildl Manag 60:935–945

Martin Collinson (2013) CSI: Birding . . . barcoding the past, present and future; the conservation. Academic rigour, journalistic flair

Meyer CP, Paulay G (2005) DNA barcoding: error rates based on comprehensive sampling. PLoS Biol 3:2229–2238

Richly E, Leister D (2004) NUMTs in sequenced eukaryotic genomes. Mol Biol Evol 21:10811084

Schirtzinger EE, Tavares ES, Gonzales LA, Eberhard JR, Miyaki CY, Sanchez JJ (2012) Alexis Hernandez et al. multiple independent origins of mitochondrial control region duplications in the order Psittaciformes. Mol Phylogenet Evol 64(2):342–356

Sodhi NS (2002) Competition in the air: birds versus aircraft. Auk 119:587–595. BioOne

Stoeckle M, Winker K (2009) A global snapshot of avian tissue collections: state of the enterprise. Auk 126:684–687

Wiemer M, Fiedler K (2007) Does the DNA barcoding gap exist? – a case study in blue butterflies (Lepidoptera: Lycaenidae). Front Zool 4(8). https://doi.org/10.1186/1742-9994-4-8

Williams ST, Knowlton N (2001) Mitochondrial pseudogenes are pervasive and often insidious in the snapping shrimp genus Alpheus. Mol Biol Evol 18:1484–1493

Wormworth J, Sekercioglu CH (2011) Winged sentinels: birds and climate change. Cambridge University Press, New York

Yoo HS, Eah J, Kim JS, Min YKM, Paek WK, Lee H, Kim C (2006) DNA barcoding Korean birds. Mol Cells 22:323–327

DNA Barcoding: A Potential Tool for Invasive Species Identification

Muniyandi Nagarajan, Akash Nambidi Parambath, and Vandana R. Prabhu

Abstract Invasive alien species epitomize the group of nonindigenous species that invade a geographic area and impose detrimental effects environmentally, ecologically, and economically. The management of invasive species will be less complicated at the primary stage of invasion since their population will be represented by a few numbers of individuals. But the detection of these few individuals will be a strenuous task if they are cryptic and exhibit low detectability. Molecular diagnostic tools bestow favorable support for the precise and brisk detection of morphologically indefinite alien species. The more recent outbreak of DNA barcoding lends inspiration for the assessment of biodiversity in a more accurate and also in an inexpensive manner. This typical procedure stands out as a reliable toolkit for the detection of individual/bulk samples, forensic residues, and environmental DNA and enhances rapid response management. Several DNA barcodes, including mitochondrial COI gene, rbcL, matK, trnH-psbA, and ITS (nuclear internal transcribed spacer), have been extensively used as a global bioidentification system for detecting the alien species that invade different ecosystems. This chapter discusses invasive species and the implication of DNA barcoding in their identification as well as management process.

Keywords Invasive species · DNA barcoding · COI gene · Identification · Management

1 Introduction

Biodiversity, which is an elision of biological diversity, acts as the bedrock of structure and function of an ecosystem and thereby contributes to the balance of nature. The ecosystem balance is disturbed by the small-scale risks associated with

M. Nagarajan (✉) · A. N. Parambath · V. R. Prabhu
Department of Genomic Science, School of Biological Sciences, Central University of Kerala, Trivandrum, Kerala, India
e-mail: nagarajan@cukerala.ac.in

S. Trivedi et al. (eds.), *DNA Barcoding and Molecular Phylogeny*,
https://doi.org/10.1007/978-3-030-50075-7_3

the diverseness that spread rapidly and, thus, the emergence of potent policies like biosecurity comes into sight. These small-scale risks include problems associated with infectious diseases, genetically altered organisms, biological weapons like a number of biological agents that are used as a weapon against humans, plants, or animals, and the ways by which all these can be minimized (Meyerson et al. 2002). It also involves the consequential role of biological invasions that cause a threat to the medley of life forms. According to the Convention on Biological Diversity (CBD), invasive alien species (IAS) is the species whose establishment outside their natural habitat threatens ecosystem, habitat, or species inflicting harm to both environment and economy. IAS pressues ecosystem stability, producer sustenance, and consumer confidence (Cock et al. 2003). It causes unalterable detrimental effects on biodiversity, like loss of habitat, followed by the extinction of native species by replacing them with invasive species. Since agricultural as well as commercial activities have an inevitable dependence on our natural ecosystem, the undesirable impact of invasive species costs billions of dollars to the economy. Breaching of biogeographic barriers and the influx of exotic species all over the world through increased international trade and tourism, climate change, etc. show a substantial part in serious economic impacts. For example, IAS stands as the paramount of environmental damages and the cost of these losses add up to almost $120 billion per year in the United States. Also, about 42% of the endangered species are at risk predominantly because of invasive species (Pimentel et al. 2005).

Over the centuries, the introduction of exotic species has been done intentionally or unintentionally by anthropogenic actions. These exotic species established rapidly in those regions, causing major economic and environmental issues. Thus, in an ecosystem, it is more desirable to prevent the spread of invasive species before they become more established, because that would be more cost effective and secure. To contend this, several management measures have to be taken like forbidding the introduction and establishment of alien species, early detection, and annihilation of exotic species (Simberloff et al. 2005). The most significant part in monitoring and predicting a new specimen is to accurately distinguish the specimen to the species level. It requires the collection of a large amount of accurate data about the species entering non-native habitat. However, this comprehensive identification process is interrupted by the nonavailability of a taxonomic expert. Also, the existence of morphologically undistinguishable invasive species makes the detection procedure tedious and leads to the demand for a rapid and accurate identification technique for their detection. This leads to the unraveling of DNA-based tools for identifying and monitoring the invasive species.

DNA-based molecular tools guarantee to enhance over traditional monitoring approaches by extending the detection sensitivity and faster performance. Among the different DNA-based molecular tools, DNA barcoding has been proposed as the standardized universal method for monitoring and identification of different species (Hebert et al. 2003). Barcoding uses a very short DNA sequence from a standard part of the genome, which is common across all taxa. Thus, it stands out of several other molecular diagnostic tools because it can serve a dual purpose, both acting as a taxonomist's toolbox enriching their knowledge and being an innovative device for nonexperts who need to make a quick identification. The variation between two organisms is known as the barcoding gap (Meyer and Paulay 2005). DNA barcoding enables rapid identification of the specimens at any life stage (Palumbi and Cipriano 1998). This success of DNA barcoding in species identification is dependent mainly on the accuracy of a reference database, which consists of voucher specimens that are morphologically identified using existing methods and compared to the sequence of a given barcode (Ratnasingham and Hebert 2007). The main motive of DNA barcoding is to parent the global BoLD database, which includes the sequence of voucher specimen as well as the sequence imported from other databases.

2 Invasive Species: A General Outlook

The species architecture of an ecosystem depends mainly on environmental conditions, nature of disturbances, and counterbalance of extinction and recruitment. The disturbances caused by invasive species are one among the major type of disturbances that alters the species composition. Invasive species are benefited if they are unaccompanied by their natural enemies. This ecological release helps them to get established in the new habitat to which they are introduced. Each species has distinctive potential for becoming invasive, and to study about this, knowledge about their rare dispersal events along long distances is required. Other factors such as habitat disturbances, distribution frequency, and reproductive maturity also play a crucial role in determining the invasive potential of different species. Among two million species that are characterized, almost 10% (200,000) species have the ability to become cogent invaders. Some examples of IAS that belong to different categories are given in Table 1.

3 Impacts of Invasive Alien Species

The establishment of invasive species can cause harm to the native species, economy, environment, and human health. Consequences of invasion can be dramatic. It can also alter the evolutionary pathway of native species and ultimately destroy the ecological balance by species extinction. Even though the species transported across biogeographic barriers increases, species richness does not show much variation. Only a few of them got established and about 1% of them become pests. But over the years these additions have become significant (Williamson 1996). In New Zealand, alien-established plant species equals the proportion of native plant species. Most of the countries contain 20% or more alien plant species (Vitousek et al. 1996). Vast lands and water capes in different regions are dominated by invasive alien species like Star thistle (*Centaurea solstitalis*) in California and Cheatgrass (*Bromus tectorum*) in the United States (Mack 1985), all of which disturb the balance of local biodiversity.

Table 1 Different types of Invasive Alien Species

Organisms	Descriptions
Avian malaria (*Plasmodium reticulum*)	It was introduced to Hawai'i by means of some exotic birds, which was further spread in 1826 by a vector named southern house mosquito (*Culex quiquefasciatus*) that occupied the water barrels in a sailing ship. Unlike non-native birds, native birds of Hawai'i do not have resistance against avian malaria and their number declined quickly. Birds such as honeycreepers which have evolved into a diverse array of species are mostly affected by this virus. Avian malaria has led to the extinction of at least 10 native bird species and terrorizes many more in the islands of Hawai'i.
Water hyacinth (*Eichhornia crassipes*)	It is native to South America and considered as the most dreadful aquatic weed in the world. It has got large purple- and violet-colored flowers, which makes them visually attractive and popular ornamental plant. It is now commonly seen in almost 50 countries as it is a very fast-growing aquatic weed that doubles its population size within 12 days. It rapidly spreads all over the waterways making it difficult for swimming, fishing, and boat traffic. Moreover, it prevents sunlight and oxygen from reaching water column that results in the dramatic reduction of aquatic life forms and eventually changes the entire aquatic ecosystem.
Miconia (*Miconia calvescens*)	A native of South America and it is a highly ornamental plant. Its red and purple leaf makes it very attractive. It was imported to the botanical garden on the island of Tahiti in 1937. Currently, more than half of the island is invaded by this plant, being spread by fruit-eating birds making it difficult for the survival of endemic species. It was also introduced to islands in Hawai'i as an ornamental plant, where its population is still increasing.
Crazy ant (*Anoplolepis gracilipes*)	They are invaders seen on Christmas Island in the Indian Ocean. They have caused many environmental problems to natural ecosystems, which include the extermination of land crab (*Gecarcoidea natalis*) population. These invaders attack a variety of arthropods, birds, and mammals and are also involved in the protection of sap-sucking scale insects that destroy forest canopy. They have invaded almost 5% of Christmas Island and if this continues it will lead to the extinction of many species on the island.
Bull frog (*Lithobates catesbeianus*)	It was introduced to northern Belgium in 1990s along with fish transports from other European countries. Later it spread to other parts of the country because of the availability of suitable reproductive conditions like nutrient-rich ponds, lack of predators, and presence of lots of algae.
Western mosquito fish (*Gambusia affinis*)	It is a small fish native to southern United States. It was introduced to different parts of the world as a biological control of mosquito mainly by mosquito control agencies without knowing its adverse effects. It has become a pest worldwide because it eats the eggs of rare and economically desirable fish.
The house sparrow (*Passer domesticus*):	It is native to Asia and was introduced to eastern Africa on trading ships long ago. Even though it mainly feeds on insects, it can cause harm to humans by destroying fruit trees, crops, and rooftops by its nesting activity. Introduction of this species also led to a decrease in the number of several native species by taking over its nesting place.

(continued)

Table 1 (continued)

Organisms	Descriptions
Brown tree Snake (*Boiga irregularis*)	It is native to Australia, Indonesia, Papua New Guinea, and the Solomon Islands. It had been introduced to Guam on military aircrafts in the 1940s. Their population rate exploded as it was unaccompanied by their natural predators, which led to serious economic and ecological destruction. It is involved in the complete extermination of Guam's native bird species. It has also entered Hawai'i, the United States, Spain, and many others by concealing itself in boats, aircrafts, and airplane wheel-wells causing a serious threat to the biological diversity.
Feral pig (*Sus scrofa*)	It was once domesticated but has been released or escaped into the wild. Subsequently, it spread to different parts of the world like Australia, Canada, the United States, and many others. It causes a serious threat to native vegetation, including humans by damaging crops and transmitting disease like leptospirosis.

As global trade increases, the rate of exchange of exotic species also increases. Some examples from the past also show dominant nature of invasive species in their non-native habitat. The biota of Red sea and Mediterranean Sea were isolated until the construction of Suez Canal in 1869. Suez Canal opened the way for movement of different species across Red Sea and Mediterranean Sea. Over 250 species moved into Mediterranean Sea from Red Sea (Por 1978). This invasion resulted in the displacement of native fish to the depths as the invaders preferred shallow, warmer water at surface (Golani 1993). This helped the invaders to exploit the new habitat and increase in number. As their population grew greater in size, they had greater negative impact on ecosystem. The impacts of anthropogenic actions like fire or pollution decrease over time, but impacts of IAS tend to increase over time. The impacts of IAS can be mainly categorized as follows.

3.1 Ecological Impacts

Alteration of the local biodiversity of an area or ecological process by invasive species can cause many ecological impacts. Their impacts can be seen at different level starting from individual species to local population that disturbs the function of entire ecosystem. For example, invasive species can hybridize with the native species causing alteration in the gene pool resulting in reduced fitness of native species or alternatively invaders become more stable. Hybridization event eventually leads to the establishment and propagation of invasive species. Invasive species particularly plants are capable of changing the structure of soil, causing erosion problems and altering resource availability like minerals and water. This in turn leads to cascade of negative impacts on ecosystem. For example, the Paper Bark Tree (*Melaleuca quinquenervia*), an invasive species introduced to Florida, resulted in the alteration of soil structure and resource availability. Subsequently the wetlands in this area are subjected to degradation (Porazinska et al. 2007). Likewise invasion of Cheatgrass (*Bromus tectorum*) in the northwestern United States enhanced the fire frequency by increasing the habitat of Cheatgrass (Mack 1985).

3.2 Economic Impacts

Invasive species may cause crucial economic losses to the society either in the form of direct economic impacts like loss of crops due to invasive crop pests, preventing the export of products that are infected by IAS, etc. Also, indirect or secondary impacts such as human health issues, cost of responding and prevention of the problems caused by IAS, damage to infrastructure due to ecosystem changes, etc. stand foremost in the list of economic impacts contributed by IAS.

New Zealand ranks as one of the most highly invaded areas in the world due to its increased number of introduced mammals, birds, and several exotic plant species. Some among those species that had been introduced for prosperity in the field of agriculture, forestry, horticulture, etc. were found to engender threats to native biodiversity. In the last 150 years, these invading species are found to be naturalizing at a fixed rate of 12 per year. The New Zealand economy has a loss of about 400 million NZ Dollar in a year due to exotic species and also about 440 million NZ Dollar costs for the prevention of these losses. The sum of these two costs contributes to 1% of New Zealand's GDP (Williams and Timmins 2002). The estimated annual cost for damages and losses rendered by the domestic cats in the United States was about 17 billion dollars, which only represents the damage done by feral cats to certain bird species. The total loss triggered by both feral and urban (pet) cats together adds up to around 34 billion dollars (Pimentel et al. 2005). It is unfortunate that a single invasive alien species, the water hyacinth (*Eichhornia crassipes*), cost about several millions of US dollars in the United States from 1980 to 1991. For instance, in Florida, more than 43 million US dollars are spent to counteract and another three million dollars for the management of this cryptic species every year. 500,000 US dollars are spent annually in California for the control of this weed (Mullin et al. 2000).

3.3 Impact on Human Health

Apart from creating a loss to the biodiversity, invasive species also are a menace to human health. Invasive pathogens can affect health directly, or otherwise invasive vectors can sometimes alter the transmission cycles of native or non-native pathogens (McMichael and Bouma 2000). Many of the non-native species, including insects, rodents, and birds, can act as reservoirs for disease and can even carry diseases such as yellow fever and malaria.

3.4 Consequences of Species Mixing

It has been studied that invasive species can adapt to their new environmental conditions pretty quickly, which is known as the direct evolutionary consequence of species mixing. Huey et al. (2000) demonstrated the evolution of Fruit fly, which took only 20 years to change its wing size to get adapted to the new habitat, west coast of North America. Even though the developmental basis for the change in wing size was different from that of native European populations, the functional result was the same. On the other hand, there are studies that show the evolution of native species toward introduced species (Carroll and Dingle 1996). In addition to direct responses, there are many indirect responses toward species mixing. The major causes of these responses are hybridization and introgression. Rhymer and Simberloff (1996) summarized that hybridization of invasive species with native species can cause loss of fitness in the native species. This may even lead to the extinction of native species.

4 Traditional Methods and Identification of Invasive Species

Exploration and classification of living organisms became an essential process throughout history. The need for recognition of unknown species varies, and it is spread across different fields like forensic studies, conservational biology, agriculture, and so on. Traditional methods provide fundamental information for the assessment of unknown species. It basically works by collecting necessary information about the uniqueness, locality, and prosperity of alien species. This information is necessary for the management of invasive species, which are one among the growing threats of the ecosystem. Traditional methods mainly rely on morphological characters for the identification process. They are mainly of two types: one that documents all the vital information about invasive species, including the pattern of distribution and abundance, which are necessary for their management process, and the other one involves collecting information about the ecological relationship or correlation between native and invasive species.

Even though traditional methods are widely used, they have some disadvantages. The morphological characters used for the assessment of different species vary greatly between individuals. These methods also fail to identify the species, if the amount of specimen available is too low. Moreover, these methods are very time-consuming and require highly trained taxonomists. The lack of availability of trained taxonomists lengthens the process. Thus, molecular techniques are considered to be a suitable strategy for the identification of invasive species. Variations at the molecular level are exploited using these techniques. They also complete the process in quick time, which makes them more preferable than traditional methods. There are a number of molecular techniques available for species identification (e.g., standalone molecular methods like immunological (Symondson et al. 1999) or protein-based techniques (Soares et al. 2000) and PCR-based molecular diagnostic methods (Dinesh et al. 1993). But the main problems with these techniques are that they are limited to a finite range of taxa and the inconsistency in PCR-based techniques. Thus, DNA barcoding technique emerged as a powerful tool for species identification and has the potential to overcome all these limitations (Hebert et al. 2003).

5 DNA Barcoding and Identification of Invasive Species

DNA barcoding stands out of many other molecular tools for expeditious species identification, as it contributes answers to questions that were previously beyond the reach of traditional disciplines. DNA barcode, in its simplest definition, is a short sequence of DNA that in virtue should be easily generated and characterized for almost all species in our planet (Savolainen et al. 2005). A gene region to perform as a DNA barcode should have some salient properties like: (i) the typical gene segment should possess notable species-level genetic variability, (ii) it should own conserved flanking regions for the generation of universal PCR primers for extended taxonomic applications and above all, (iii) it should be a short sequence so that DNA extraction and amplification can be done flawlessly. The COI gene is generally being used for almost all animal species. It has been proved to be highly effective in identifying birds, butterflies, fishes, flies, etc. In contrast, the efficacy of COI gene is not exhibited in case of plants (Kress and Erickson 2007; Eberhardt et al. 2012). Therefore, the search for effective gene regions for this major group of organism leads to the recognition of a combination of plastid genes rbcL, matK, and trnH-psbA.

Biological invasions remain the biggest peril to the biodiversity after habitat destruction. They put scientists into a tiresome procedure, that is to define the widely divergent criteria of "invasive species." Early detection of dawning non-native species enhances the brisk management actions that ensure their eradication (Simberloff 2003; Vander et al. 2010). If the detection of such alien species is pursued after their firm establishment, then the eradication becomes difficult and more expensive (Simberloff et al. 2005). In case of many alien species, the early detection is contradicted by low detectability, which represents the probability that an organism will be observed if present. DNA barcoding is useful for the early detection, where identifiable material is too damaged or present in a negligible amount (Armstrong and Ball 2005; Darling and Blum 2007).

5.1 Pests

Expansion of invasive pests in the geographic range requires quick detection and response because otherwise their management becomes tedious. Hence, DNA barcoding that can lend insights beyond the data obtained through morphological analysis can act as an aid for rapid detection of pest species and thereby enhance the downstream management procedures. Terebrantia and Tubulifera, two recognized suborders of the order Thysanoptera, are terrific pests that are commonly called as thrips. Chilli thrips belonging to the order Thysanoptera, native to South Asia, invaded U. S via commodity shipments and later triggered a huge economic loss to states like Hawaii (Seal et al. 2010) and Florida (Brown and Osbourne 2008). Also, *Thrips parvispinus* is an insect vector that can transmit Topoviruses effects to a number of plant species in unrelated plant families used in agriculture, horticulture, etc. On the basis of the integrated approach of morphology and DNA barcoding, the invasion of this dreadful pest was reported for the first time in India from papaya plantations (Kaomud et al. 2015). Apart from enhancing the early detection of alien species (Onah et al. 2015), DNA barcoding also helps to find out their source regions (Bellis et al. 2015) as well as their introduction patterns. Mastrangelo et al. (2014) differentiated *Heliothis armigera* (Gram Pod Borer), an invasive pest species from the native *H. zea* (corn earworm) of Brazil, using DNA barcoding and thereby disclosed the spread of former species into Brazil. The identification of *Agrilus ribesi* (buprestid beetle), whose invasion to North America had been overlooked for a century, has been detected recently by the application of DNA barcoding (Jendek et al. 2015).

5.2 Aquatic Species

More than two-thirds of our earth is covered with oceans and other water bodies; thus, the evaluation of aquatic species diversity is a demanding task. Also, the increase in global population engenders the tendency of exploiting marine resources for energy, food, etc. Sequentially, this puts thrust on the aquatic environment and requests for rapid management. DNA barcoding has been used for the assessment of invasive species that are predominant in the aquatic ecosystem. Aquatic alien species are nonindigenous species that cause peril to the diversity of native species and thereby affect the ecological stability of the infested water body. Floating pennywort (*Hyodrocotyle ranunculoides*), member of the Araliaceae plant family, is an invasive aquatic species that are native to North America. This typical invasive species seems to cause serious problems to the waterway management outside of its original allocated area in Western Europe and Australia. Discrimination of *H. ranunculoides* from its related species is a challenging task. Sequencing one of the variable chloroplast loci, namely, trnH-psbA, served to discriminate the invasive plant species, *H.ranunculoides* from its closely related congeners like *H. vulgaris*, *H. verticillata*, etc. (Van De Wiel et al. 2009).

Myriophyllum, Cabomba, and Ludwigia belonging to the family Hydrocharitaceae represent a significant aquatic invasive species, but many of their related species that are morphologically similar in their vegetative state are noninvasive and are commercially traded. Thus, the detection of invasive plants constituting to this family is a demanding task. The study conducted by Ghahramanzadeh et al. (2013) revealed a noncoding spacer (trnH-psbA) as the best-performing DNA barcode for the differentiation of invasive and noninvasive species belonging to Hydrocharitaceae family. *Driessena polymorpha* (zebra mussel) and *D. rostriformis bugensis* (quagga mussel) are found to be the most competitive invaders in the freshwaters of North America and Europe. Though the phenotypic plasticity found in the genus *Dreissena* blocks their reliable discrimination as well as identification, the adult *Dreissena* individuals were differentiated with the help of CO1gene (Jonathan and Karine 2013).

5.3 Plants

The most challenging aspect about invasive plant species is that they scarcely represent a constant, static entity. The knowledge about the introduction of these plant species as well as the actual level of genetic diversity between them is not well explained. Shaik et al. (2016) conducted a study using DNA barcoding to understand the genetics and ecophysiology of plant invaders of Australia belonging to the family Cucurbitaceae. *Cucumis myriocarpus* (prickly paddy melon), *Citrullus lanatus* (camel melon), and *Citrullus colocynthis* are very much morphologically related to each other and were found to cause a great infestation to the natural as well as the agricultural ecosystem by acting as a harmful weed in several places in Australia. The camel melon and prickly paddy melon are annuals, whereas *C. colocynthis* is perennial and hence the management procedures differ among them. The chloroplast gene (ycf6-psbM) and nuclear gene (G3pdh intron region) served as the barcode for interspecific and intraspecific variability among these three cucurbitaceous invasive species (Shaik et al. 2016).

5.4 Mammals

Herpestes auropunctatus, the small Indian mongoose, is a carnivore that feeds on almost all major vertebrate groups, invertebrates, and even plants (Lewis et al. 2011). This typical species, which is believed to be native to India and Myanmar (Veron et al. 2007), is often confused with the Javan or small Asian mongoose (*H. javanicus*). All literature published before 2007 presumes that the mongoose introduced to Hawaiian and Caribbean islands was *H. javanicus*. Bennett et al.

(2011) conducted a study exposing the efficacy of DNA barcoding approaches with mtDNA cytochrome *b* to discriminate between the two species *H. auropunctatus* and *H. javanicus* as well as the other sympatric members of the genus *Herpestes* (*H. naso, H. urva, and H. edwards*). Further, the study enhanced the confirmation of *H. auropunctatus* invasion in the Hawaiian and Caribbean islands.

6 Conclusion

DNA barcoding has rejuvenated the field of species identification by serving as a robust molecular tool for resolving species diversity. It helps in the accurate identification of specimens, while confusion prevails with traditional techniques. In addition, the availability of an open database of DNA barcodes constituting a broad range of taxonomic groups makes the detection of unknown species easy. Barcoding technique that exhibits a significant role in early detection as well as the prevention of biological invasions is now being adopted by biosecurity agencies in order to control the spread of invasive species.

Acknowledgments The authors are grateful to the Science and Engineering Research Board (SERB), Department of Science and Technology, Government of India, New Delhi, for the financial Assistance (SR/S0/AS-84/2012). Vandana R. Prabhu is grateful to the DST-INSPIRE (IF160266) for the financial support in the form of research fellowship.

References

Armstrong KF, Ball SL (2005) DNA barcodes for biosecurity: invasive species identification. Philos Trans R Soc Lond Ser B Biol Sci 360(1462):1813–1823

Bellis GA, Gopurenko D, Cookson B, Postle AC, Halling L, Harris N, Yanase T, Mitchell A (2015) Identification of incursions of Culicoides Latreille species (Diptera: Ceratopogonidae) in Australasia using morphological techniques and DNA barcoding. Austral Entomol 54(3):332–338

Bennett CE, Wilson BS, DeSalle R (2011) DNA barcoding of an invasive mammal species, the small Indian mongoose (Herpestes javanicus; E. Geoffroy saint-Hillaire 1818) in the Caribbean and Hawaiian islands. Mitochondrial DNA 22(1–2):12–18

Brown SH, Osbourne LS (2008) Chilli Thrips (Scirtothrips dorsalis): a Landscaper's guide. University of Florida, IFAS Extension, Lee County Extension

Carroll SP, Dingle H (1996) The biology of post-invasion events. Biol Conserv 78:207–214

Cock MJW, Kenis M, Wittenburg R (2003) Biosecurity and forests: an introduction-with particular emphasis on forest pests. FAO Forest Health and Biosecurity Working Paper FBS/2E

Darling JA, Blum MJ (2007) DNA-based methods for monitoring invasive species: a review and prospectus. Biol Invasions 9:751–765

Dinesh KR, Lim TM, Chan WK, Phang VPE (1993) RAPD analysis: an efficient method of DNA fingerprinting in fishes. Zool Sci 10:849–854

Eberhardt U, Kress WJ, Erickson DL (2012) Methods for DNA barcoding of fungi. In: DNA barcodes: methods and protocols: methods in molecular biology, Humana press. Springer Science+Publishing Media, LLC, pp 183–206

Ghahramanzadeh R, Esselink G, Kodde LP, Duistermaat H, van Valkenburg JL, Marashi SH, Smulders MJ, van de Wiel CC (2013) Efficient distinction of invasive aquatic plant species from non-invasive related species using DNA barcoding. Mol Ecol Resour 13(1):21–31

Golani D (1993) Trophic adaptations of Red Sea fishes to the eastern Mediterranean environment – review and new data. Israel J Zool 39:391–402

Hebert PDN, Cywinska A, Ball SL, de Waard JR (2003) Biological identifications through DNA barcodes. Proc R Soc Lond 270:313–321

Huey RB, Gilchrist GW, Carlson ML, Berrigan D, Serra L (2000) Rapid evolution of a geographic cline in size in an introduced fly. Science 287(5451):308–309

Jendek E, Grebennikov VV, Bocak L (2015) Undetected for a century: Palaearctic *Agrilus ribesi* Schaefer (Coleoptera: Buprestidae) on currant in North America, with adult morphology, larval biology and DNA barcode. Zootaxa 4034:112–126

Jonathan M, Karine VD (2013) Using DNA barcoding to differentiate invasive Dreissena species (Mollusca, Bivalvia). Zookeys 365:235–244

Kaomud T, Vikas K, Devkant S, Rajasree C (2015) Morphological and DNA barcoding evidence for invasive Pest Thrips, Thrips parvispinus (Thripidae: Thysanoptera). Newly recorded from India. J Insect Sci 15(1):105

Kress WJ, Erickson DL (2007) A two-locus global DNA barcode for land plants: the coding *rbcL* gene complements the non-coding *trnH-psbA* spacer region. PLoS One 2(6)

Lewis DS, van Veen R, Wilson BS (2011) Conservation implications of small Indian mongoose (Herpestes auropunctatus) predation in a hotspot within a hotspot: the Hellshire Hills, Jamaica. Biol Invasions 13(1):25–33

Mack RN (1985) Invading plants: their potential contribution to population biology. In Studies on Plant Demography: White J. (Academic, London), pp. 127–142

Mastrangelo T, Paulo DF, Bergamo LW, Morais EGF, Silva M, Bezerra-Silva G, Azeredo-Espin AML (2014) Detection and genetic diversity of a heliothine invader (Lepidoptera: Noctuidae) from north and northeast of Brazil. J Econ Entomol 107:970–980

McMichael AJ, Bouma MJ (2000) Global changes, invasive species and human health. In: Mooney HA, Hobbs RJ (eds) Invasive species in a changing world. Island Press, pp 191–210

Meyer CP, Paulay G (2005) DNA barcoding: error rates based on comprehensive sampling. PLoS Biol 3(12):e422

Meyerson LA, Reaser JK, Chyba CF (2002) A unified definition of biosecurity. Science 295 (5552):44

Mullin B.H. et al, (2000) Invasive plant species. Ames, Iowa: Council for Agricultural Science and Technology. Issue Paper No. 13

Onah IE, Eyo JE, Taylor D (2015) Molecular identification of tephritid fruit flies (Diptera: Tephritidae) infesting sweet oranges in Nsukka agro-ecological zone, Nigeria, based on PCR-RFLP of COI gene and DNA barcoding. Afr Entomol 23(2):342–347

Palumbi SR, Cipriano F (1998) Species identification using genetic tools: the value of nuclear and mitochondrial gene sequences in whale conservation. J Hered 89:459–464

Pimentel D, Zuniga R, Morrison D (2005) Update on the environmental and economic costs associated with alien-invasive species in the United States. Ecol Econom 52:273–288

Por FD (1978) Lessepsian migration. Springer, Berlin Heidelberg New Yark

Porazinska DL, Pratt PD, Glblin-Davis RM (2007) Consequences of *Melaleuca quinquenervia* invasion on soil nematodes in the Florida Everglades. J Nematol 39:305–312

Ratnasingham S, Hebert PDN (2007) BOLD: the barcode of life data system. Mol Ecol Notes 7 (3):355–364. (www.barcodinglife.org)

Rhymer JM, Simberloff D (1996) Extinction by hybridization and introgression. Ann Rev Ecol Syst 27:83–109

Savolainen V, Cowan RS, Vogler AP, Roderick GK, Lane R (2005) Towards writing the encyclopedia of life: an introduction to DNA barcoding. Philos Trans R Soc Lond Ser B Biol Sci 360 (1462):1805–1811

Seal DRW, Klassen W, Kumar V (2010) Biological parameters of Scirtothrips dorsalis (Thysanoptera: Thripidae) on selected hosts. Environ Entomol 39(5):1389–1398

Shaik RS, Zhu X, Clements DR, Weston LA. (2016) Understanding invasion history and predicting invasive niches using genetic sequencing technology in Australia: case studies from Cucurbitaceae and Boraginaceae. Conservation Physiology, Volume 4, Society for Experimental Biology, https://doi.org/10.1093/conphys/cow030

Simberloff D (2003) How much information on population biology is needed to manage introduced species? Conserv Biol 17(1):83–92

Simberloff D, Parker IM, Windle PN (2005) Introduced species policy, management, and future research needs. Front Ecol Environ 3:12–20

Soares RP, Santanna MR, Gontijo NF, Romanha AJ, Diotaiuti L, Pereira MH (2000) Identification of morphologically similar Rhodnius species (Hemiptera: Reduviidae: Triatominae) by electrophoresis of salivary heme proteins. Am J Trop Med Hyg 62(1):157–161

Symondson WOC, Gasull T, Liddell JE (1999) Rapid identification of adult whiteflies in plant consignments using monoclonal antibodies. Ann Appl Biol 34(3):271–276

Van De Wiel CCM, Van Der Schoot J, Van Valkenburg JL, Duistermaat H, Smulders MJM (2009) DNA barcoding discriminates the noxious invasive plant species, floating pennywort (Hydrocotyle ranunculoides L.f.), from non-invasive relatives. Mol Ecol Resour 9:1086–1091

Vander ZMJ, Hansen GJ, Higgins SN, Kornis MS (2010) A pound of prevention, plus a pound of cure: early detection and eradication of invasive species in the Laurentian Great Lakes. J Great Lakes Res 36(1):199–205

Veron G, Patou ML, Pothet G, Simberloff D, Jennings AP (2007) Systematic status and biogeography of the Javan and small Indian mongooses (Herpestidae, Carnivora). Zool Scripta 36 (1):1–10

Vitousek PM, D'Antonio CM, Loope LL, Westbrooks R (1996) Biological invasions as global environmental change. Am Sci 84(5):468–478

Williams PA, Timmins S (2002) Economic impacts of weeds in New Zealand. Biol Invasions:273–282

Williamson M (1996) Biological Invasions. Chapman and Hall, New York, 244 p

Part II
DNA Barcoding of Microbes

Microbial DNA Barcoding: Prospects for Discovery and Identification

Anand Mohan, Bableen Flora, Madhuri Girdhar, and S. M. Bhatt

Abstract DNA barcoding is a technology used for the identification of any biological species by sequencing obtained through amplification of DNA. Here, we seek a standardized protocol along with the choice of markers used for different microbial species: bacteria, fungi, algae, plants, and animals. Various projects, consorts, tools, and databases have been discussed. With the advancement of technology, the future scope and applications have also been identified with an argument on the limitations along with the concept of universality on the DNA barcoding.

Keywords Microbial · DNA barcoding · Databases · Species identification · Protocol for barcoding

1 Introduction

DNA barcoding is a technology using gene sequences to differentiate species, like the retail stores in which barcodes are used to sell and differentiate the different items. Hebert et al. (2003) state that DNA barcoding is a technique in which species identification is performed by using DNA sequences from the small fragment of the genome and resulted in ecological studies in which traditional taxonomic identification is not possible. The short DNA sequences of 400–800 bp long are called DNA barcodes. Hebert and Gregory (2005) stated that DNA barcode is not just any DNA sequence, but it is a standardized sequence of a minimum length and quality from an agreed-upon gene that is deposited in a major sequence database and attached to a

A. Mohan (✉) · B. Flora · M. Girdhar
School of Bioengineering and Biosciences, Lovely Professional University, Phagwara, Punjab, India

S. M. Bhatt
Sant Baba Bhag Singh University, Jalandhar, Punjab, India

S. Trivedi et al. (eds.), *DNA Barcoding and Molecular Phylogeny*,
https://doi.org/10.1007/978-3-030-50075-7_4

voucher specimen whose origins and current status are recorded. The ecological imbalances occur due to global warming, leading to loss of diversity of species; therefore, there is a need to identify and classify organisms and study their evolutionary relationships so as to conserve the threatened species. Casiraghi et al. (2010) described that one of the best advantages of a DNA barcode is to associate all life-history stages and genders or to identify organisms from part like parasites or to segregate a matrix containing a mixture of biological species. Phylogenetic analysis and searching for the clade along with taxonomy is the chief function of DNA barcoding (Hajibabaei et al. 2007).

Earlier Carolus Linnaeus 'father of taxonomy' classifies species that we still use today. In his publication, Systema Naturae, Linnaeus classified species into hierarchy. He proposed broad groups, called kingdoms to classify microbes, animals, and plants. These kingdoms were further divided into phylum, classes, orders, genera (genus is singular), and then species. But as the variations are increasing, it leads to the evolution of new species that broaden our biodiversity. So, we need a refined and fast method of classification of living organisms. DNA barcoding brought a revolution not just in classifying the species but also in obtaining many more applications in day-to-day life. It also leads to the discovery of new species (Hebert and Gregory Ryan 2005). Raja et al. (2017) demonstrated the utility of DNA barcoding for the quality of food by identifying fungi commonly present in dietary supplements. They have studied mushroom samples used in dietary supplements via their fungal barcoding. In DNA barcoding, barcode gap analysis or phylogenetic tree is built. Das and Deb (2015) analyzed that a barcode gap exists if the minimum interspecific variation is bigger than the maximum intraspecific variation.

DNA barcoding can be done from small, damaged, or industrially processed material, and the DNA barcodes can be efficiently processed and analyzed through different databases.

The most relevant tool used for DNA barcoding is BOLD, the Barcode of Life Data Systems, composed of a set of integrated databases. It includes Public Data Portal and BIN database acting as primary data sources, whereas publication and primer databases support it. One can easily analyze their data in BOLD before submitting it to GenBank, EMBL, and DDBJ, which are known as the International Nucleotide Sequence Databases. They are the permanent public repository for barcode data records.

2 Origin of Barcoding

Paul Hebert: The Father of DNA Barcoding.

Paul D. N. Hebert is a molecular biologist and director of the New Biodiversity Institute of Ontario at the University of Guelph in Canada. Hebert is an avid insect collector and admirer of biodiversity. Hebert was working on gene sequencing, and most of the people approached him for the sequencing of their soft corals, which gave him the idea of barcoding. Governments, museums, and universities were

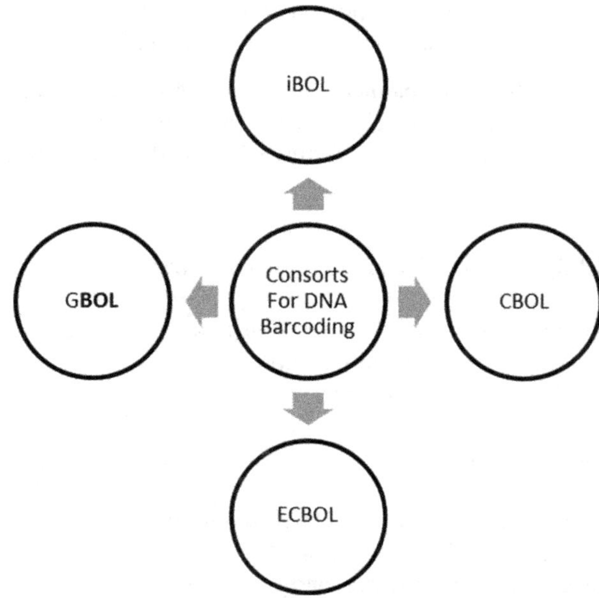

Fig. 1 Different consorts for DNA barcoding of life

focusing to track the invasive species, which motivated him. Hebert started his effort at

The Biodiversity Institute of Ontario—the world's first barcode factory—and at present, different consortia, databases, labs, networks, and project diversity have been added and showed a successful endeavor. Different consortia for barcode of life are shown in Fig. 1.

2.1 Consorts

2.1.1 iBOL

iBOL (International Barcode of Life)

The biodiversity department at the University of Guelph, Canada, initiated an international life project of barcoding, expanding its nodes to 27 nations in order to maintain the database system for barcoding, namely, BOLD (Barcode of Life Data System). The International Barcode of life is aiming to develop a worldwide network of taxonomists, biologists, and geneticists to scrutinize various species including plants, animals, and fungi so as to boost barcode reference library (BOLD) (http://www.barcodinglife.org). BOLD is a blend of bioinformatics and analytical analysis of collected barcode data. It includes the collection, analysis, and quality assurance under malleable protection (http://ibol.org) (Ratnasingham and Hebert 2007). iBOL diversified some specialized groups for the repository of different species including

FormicidaeBOL, TrichopteraBOL, MammaliaBOL, FishBoL, BeeBOL, and MarBOL (Dhawan et al. 2013; Trivedi et al. 2016).

The iBOL global partnership currently involves 25 nations spreading it to Central, National, and Regional nodes. CCDB—the Canadian Center for DNA Barcoding—is one of the first largest factories launched at the University of Guelph.

2.1.2 CBOL (Consortium for the Barcode of Life)

Another international inventiveness aims to establish a global method for the identification of a variety of flora and fauna through DNA barcoding. Presently, it has developed its roots to more than 130 organizations from 40 countries (http://ibol.org/cbol/). Diversity in the development of software has been observed as for barcoding of spider (SPecies IDentity and Evolution in R), BRONX (Barcode Recognition Obtained with Nucleotide eXposés), CLOTU, TaxonGap, CAOS (Characteristic Attributes Organization System), and others (Bhargava and Sharma 2013). CBOL is a leading organization as it improvises barcoding through conferences, outreach activities, and workshops.

2.1.3 QBOL

QBOL is an international consort of 20 partners involving universities, research institutes, and various organizations and stimulated by Plant research international institute of Netherlands. QBOL has an ambition of collecting, diagnosing, and storage of DNA barcodes of plant pathogenic groups. In addition, QBOL supports the collaboration between European and NonEuropean diagnostic laboratories (Bonants et al. 2010).

Various international barcoding activities and tools have been established to expand the horizon of DNA barcoding inclusive of iBarcode (www.ibarcode.org), Barcode Blog (http://phe.rockefeller.edu/barcode/blog), Canadian Barcode of Life Network (www.bolnet.ca), Polar BOLPolar barcode of Life (www.polarbarcoding.org), N-BOL, The Netherlands Barcode of Life (www.dnabarcoding.nl), G-BOL, German Consortium for barcode of life (www.g-bol.de), ECBOL, European Consortium for Barcode of Life (www.ecbol.org), and many more.

2.1.4 ECBOL (European Consortium for Barcode of Life)

The European Consortium for Barcode of Life is site-specific workbench which includes collecting and scrutinizing taxonomic types niched in the European area. ECBOL is funded by EuroBioFund in order to widen identification of a variety of biodiversity in Europe (www.ecbol.org).

2.1.5 GBOL (Germany Consortium for Barcode of Life)

Another regional specific consortium for barcoding is the German consortium, which is funded by the Federal Ministry of Education and Research, Germany. Diversified flora and fauna from botanical gardens contribute to enhancement of barcode library of GBOL as well as strengthen transboundary communication. The deposited data have been analyzed and stored using the BOLD system. At present, nearly 4.6 million DNA barcodes belonging to ca. have been recorded in the International BOLD database system in different categories of plants, animals, and fungi (https://www.bolgermany.de) (Geiger et al. 2016).

2.2 Organization for DNA Barcoding

The Biodiversity Institute at the University of Guelph, Ontario, is the largest hub for DNA barcoding. Among various boards of Directors, Paul Herbert is the main initiator of barcoding of Life. Along with the collaboration and support from the Science Advisory Board and Action Committee from Planetary Biodiversity Mission, the organization has been diversified into various subdepartments that are facilitated with teamwork of the associate director and faculty advisor for the planning and implementation of the designated program. The University of Waterloo and York University also participated as advisor faculty in different subgroups. The organization setup of DNA barcoding is shown in Fig. 2.

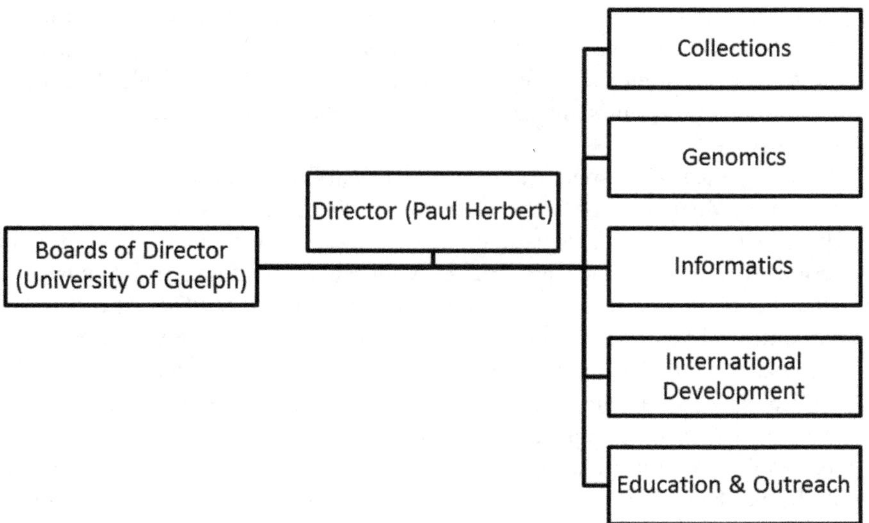

Fig. 2 Organization setup for Barcode of Life (http://biodiversitygenomics.net/about/organization/)

3 Protocol

The various steps for DNA barcoding are

1. Specimen collection—The collection of a specific species from a diversified area is the initial step for the DNA barcoding. An area rich in biodiversity is searched for a particular species that needs to be sequenced and analyzed. Samples can be collected from the field, natural history museums, zoos, botanical gardens, seed banks, etc.
2. Tissue sampling—After collecting the specimen, species is identified and the sample needs to be taken from the particular species, for instance, if fungal species need to be sequenced and the particular colony or cell should be taken for the extraction of DNA.
3. Isolation of DNA—The extraction of DNA for the amplification and sequencing is done in this step. Various methods for extraction and purification of DNA are applied.
4. Polymeric chain reaction (PCR)—The small amount of DNA needs to be amplified with the help of polymeric chain reaction (PCR). In this, the extracted DNA is denatured, annealed, and renatured. Specific primers are required for the amplification of DNA. The amount of DNA becomes large and can easily be analyzed.
5. Analysis and sequencing—A number of copies have been made using PCR for the extracted DNA of specific species. Now, the sequencing is done. Different methods used for sequencing are Sanger method and the new-generation sequencing method. The sequence is represented by a series of letters, CATG representing the nucleic acids—cytosine, adenine, thymine, and guanine.
6. Managing available data—A sequence is now obtained, and we need to create DNA barcodes. Various tools can be used for detecting barcodes. After generating the barcodes, either it needs to get matched with the available data in the barcode library or submit the new data to the databases such as BOLD by creating an Excel file.
7. Publishing data—The submitted data are now published and hosted in the various databases, which help for further study of a particular species or its conservation.
8. Using barcode data—The various databases act as a barcode library for the various species. This data can be used by different users for their own purposes. There are three types of users present for such data usage. The general public needs to read the available data like in institutions for the study of a particular species. The application user usually gets benefit from available data. The application users use the sequence or taxonomic identification for the purpose of conservation of endangered species and maintaining the record of extinct species for ecological balance. This data leads to the discovery of new species as the third user, researchers, utilized this data for the study of evolutionary lineages and relationships among different species. The flow chart of steps involved in the DNA barcoding is shown in Fig. 3.

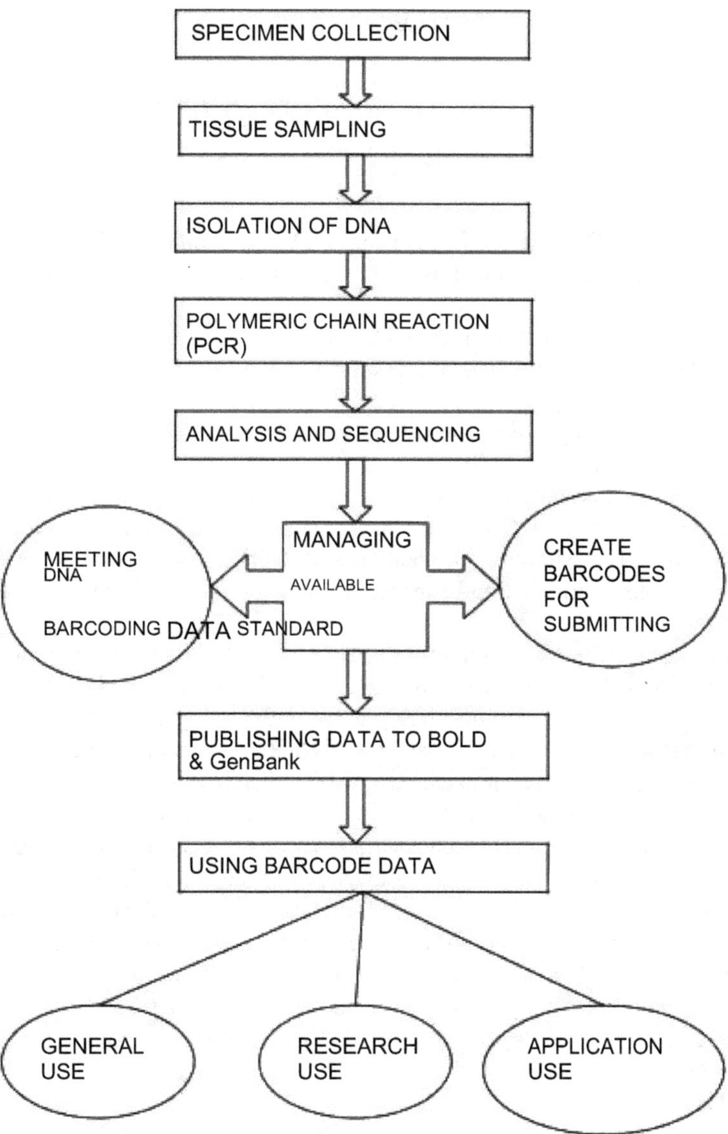

Fig. 3 Flow chart of protocol and application of DNA barcode data

4 Choice for Barcode Marker

It is one of the important parameters for DNA barcoding as different species show different markers best suited for the barcoding. The following are some of the markers used till now, which are arranged as per species:

1. Animals
   In animals, the COI marker is used for DNA sequencing. COI stands for cytochrome C oxidase subunit 1. It is 650 bp long and has single mtDNA.
2. Plants
   Plants show low mitochondrial DNA constitution of nucleotides, and so, COI cannot be used in plants. Plastid markers have been used for barcoding in plants (Fazekas et al. 2008).
   Different coding and noncoding regions have standardized the DNA barcoding system. Few coding loci such as rpoB, rpoC1, rbcL, matK, and 23S rDNA and noncoding loci such as trnH-psbA, atpF-atpH, and psbK-psbI were identified and used as DNA barcoding for plants. Therefore, development of markers for DNA barcoding of plants is still a difficult task as it is hard to find a particular finest combination of marker for a large variety of plant species.
3. Bacteria
   According to BBC News, scientists have estimated the total number of bacteria on earth to be five million trillion. Such diverse species need to be identified and should be analyzed morphologically and phylogenetically. Researchers found that the best marker for bacteria is 16sRNA, which can be easily amplified and sequenced for making DNA barcodes. QBOL is a consortium of 20 partners including research institutes, universities, etc. from all over the world and shares their research in the field of bacteria, fungi, viruses, nematodes, etc. QBOL is focusing more on the Q-species including genera Xylella, Xanthomonas, Clavibacter, and Ralstonia affecting the plant health. Since the 1990s, the most common housekeeping genetic marker being used as a marker gene in bacterial identification is 16sRNA. Another gene chaperonin-60 (cpn60), which is 555 bp long, was also used as a bacterial barcode marker gene. Chaperonin-60 (cpn60) is also known as GroEL and Hsp60 used for bacterial identification.
4. Viruses
   Viruses are estimated to be about 10 times more than the total number of cells present on earth. Viruses can be sequenced using next-generation sequence technology. In next-generation sequencing, micro- and nanotechnologies are employed to reduce the sample size, lower the cost, and enable the parallel sequencing reactions. Arracacha virus, potato black ringspot virus, tomato chocolate virus, and others have been sequenced by taking their whole genome. In 2012, watercress and maize lethal necrosis cause severe damage to Kenya maize crop. The watercress white vein has been discovered.

5. Algae
   Several molecular markers were suggested from time to time to identify the algal species. Various markers used for algae are COI for brown and red algae, tufA for green algae, rbcL for diatoms, and many more. For carrageenophytes, various molecular markers have been introduced for the molecular taxonomy such as cox1 and cox2–3 spacer, ITS, LSU, rbcL, RuBisCO, and the 23S UPA (Conklin et al. 2009).
6. Fungi
   Fungus is a member of the eukaryotic organism that includes microorganisms such as yeasts and molds. Fungi play a major role in maintaining the nutrient cycle by breaking down dead organic material. Fungi can grow in a symbiotic manner in plants called mycorrhizae. Moreover, many fungi can be used in the formation of drugs, used as food like mushrooms, truffles, and morels, and the bubbles in champagne, beer, and bread. Fungi also cause various plant and animal diseases. Rusts, smuts and stem, and leaf rots are some of the plant diseases causing severe damage to crops. Ringworms and athlete's foot are some of the animal diseases. QBOL focusses on identification of fungi, Viruses, nematodes, and phytoplasmas, submitted through DNA barcoding. QBOL has listed 19 Q-species for barcoding. Fungi are the second largest kingdom of eukaryotic life highly diverse and widespread. Fungi help in the major production of enzymes, foods, and antibiotics industrially for which it contributes economically. There are different kinds of markers analyzed and used for the barcoding. Many researchers are trying to find the basic best marker suitable for all the fungi.

   Various molecular methods are being used for the phylogenetic analysis and for taxonomical identification. DNA barcoding has provided a better platform for the same but also produced a particular challenge for the researcher. Some markers like COI are fit for mushrooms, whereas they are unsuitable for other fungi. So, there is a need for a unique marker for the identification of fungi. Many researchers found the ITS sequence best for the taxonomical identification for most of the fungi. Fajarningsih (2016) used the internal transcribed spacer (ITS) region of nuclear DNA (rDNA) for the identification of various species of fungi. Das and Deb (2015) also described various ITS databases for fungi. The ITS region is very much effective for DNA barcoding. They have also analyzed various databases for fungi. QBOL, AFTOL, UNITE, MycoBank, etc. are some of the major databases for fungi. A newly developed database for fungi has been reported by Irinyi et al. (2015), which focuses on the majority of human and animal pathogenic fungi (ISHAM-ITS, freely accessible at http://www.isham.org/ or directly from http://its.mycologylab.org).

   Dentinger et al. (2011) have done a comparative study of two different markers, COI and ITS, and found that COI is not suitable for mushrooms and rusts but may be fit for the other fungi. ISHAM also identified that cox1 is not suitable for DNA barcoding and also described the need for universal marker for the taxonomical advances of fungi. Dulla et al. (2016), ISHAM, revealed that the majority of medically important species had a low variability in ITS regions. Therefore, ITS sequencing can be utilized for the identification of most medical relevant fungal species. Cox1 is found to be unsuitable to be used for DNA barcode (Casiraghi et al. 2010). One of the best advantages of a DNA barcode is to associate all life-history stages and genders or to identify organisms from a part like parasites or to segregate a matrix containing a mixture of biological species.

5 Tools and Databases for DNA Barcoding

There are various databases and tools used for DNA barcoding. As we require different markers for barcoding, different databases and tools have been created for different species to ease the access. For the large-scale DNA barcoding, funds have been provided by the National Science Foundation (Division of Biological Infrastructure) to develop effective bioinformatics tools. The goal of this project is to develop standardized bioinformatics tools for large-scale identification of species.

Some of the tools and databases are listed below:

Tool	Organism	Method	Source
AFTOL	Fungi	Structural and biochemical database	https://aftol.umn.edu/
B	Plants	Sequence quality and contig overlap	http://www.nybg.org/files/scientists/dlittle/B.html
BLOG (barcoding with logic)	Animals	Diagnostic and character based	http://dmb.iasi.cnr.it/blog-downloads.php
BOLD	Protists, fungi, plants, and animals	Taxonomy and identification based	http://www.boldsystems.org/
BRONX	Plants	Sequence identification and taxonomic hierarchy	http://www.nybg.org/files/scientists/dlittle/BRONX.html
CBS prediction servers	All	Sequence analysis	http://www.cbs.dtu.dk/services/
CLOTU	Fungi	Taxa and amplicon data	http://www.bioportal.uio.no
DNA-BAR	Bacteria and fungi	Character-based diagnostic	http://dna.engr.uconn.edu/~software/cgi-bin/barcode/barcode.cgi?num_targets¼0
Eco primers	Bacteria and algae	Barcode markers	ecoPrimers.tar.gz
GBIF	All	Open-access database	http://www.gbif.org/
ISHAM	Fungi	Mycological classification	http://www.isham.org/
MEGA software	All	Molecular evolutionary and genetic analysis	http://www.megasoftware.net/
jMOTU		Clustering barcodes	http://www.jmotu.com-about.com/
MycoBank	Fungi	Pairwise sequence alignment and polyphasic identification	http://www.mycobank.org/

(continued)

Tool	Organism	Method	Source
OFBG	Plants	Spp. discrimination using oligonucleotide frequencies	http://www.nbri.res.in/ofbg.php
OTUbase	Microbes	Opertion serversomic based	http://www.bioconductor.org/packages/release/bioc/html/OTUbase.html
P.A.T.H.S	Fungi, plants	Species relationship	http://barcoding-paths.it
Q-Bank	Plant pest	Identification and disease detection database	http://www.q-bank.eu/
TaxI	Plant	Distance based	axel.meyer@uni-konstanz.de
Taxonerator	Microbes	Taxonomy based	http://www.taxonnerator.com-about.com/
UNITE	Ectomycorrhizal fungi	Species hypothesis	http://unite.ut.ee

6 Significance of DNA Barcoding

DNA barcoding has brought a vital revolution in the world of taxonomy of species. It has contributed in three ways as illustrated by Casiraghi et al. (2010):

1. Molecularization—It involves the use of the variable molecular marker acting as a discriminator for various species. This segregates the species at the molecular level.
2. Computerization—Various computer-based databases used to keep the record of various species as well for the phylogenetic analysis.
3. Standardization—Various standards have been made to approach variability among the species.

7 Limitations

1. Lack of Universality
   DNA barcoding also gave rise to several challenges, and a collaborative effort is being made for the amendments in the protocol. There is a need for multiple analyses; Blaxter (2004) found the necessity of core plurality with more than one sequence in one sample. This will improve DNA preservation and taxonomic units. There is no universal gene for DNA barcoding and no single gene that is conserved in all domains of life and exhibits enough sequence divergence for species discrimination (Purty and Chatterjee 2016). New species have been developed by hybridization, and their taxonomical and molecular analyses are required, which can be possible, but a single marker is required for the detection of their variations (Ali et al. 2014).

2. Coamplification
   There is a need for combined amplification to fasten the process as well as for better analyses of the sequences.
3. Lack of Taxonomic Experts
   Taxonomic experts are required for the correct analysis of the submitted data as well as for the discovery of new species, which leads to ecologically and economically advancements.

8 Applications and Future Aspects

1. Controlling the Pest
   DNA barcoding helps in the control of various pests by identifying their taxonomy and prevents the huge amount of farm production. Fruit flies have been controlled by global tephritid barcoding identification.
2. Identification of Disease Vectors
   DNA barcoding helps to identify the vector species that cause infectious diseases to animals and humans. This leads to easy detection of diseases and thus can be cured with effective way. A global mosquito barcoding is initiated for the building of a reference barcode library that can help to diagnose and cure chronic diseases. Abel et al. (2015) and their colleagues used STAMP to trace the growth and decline of V. cholerae infection in rabbits. They have added about 500 different barcodes for the batch of V. cholerae. Animal products legally used in the production of various medicines can be analyzed and separated from those used illegally. Yan et al. (2013) developed a DNA barcoding technique that can be used to discriminate between a wide range of animal species whose horns are used in traditional medicines.
3. Conservation of Natural Resources and Biodiversity
   Natural resources need to be conserved. The illegal products from hardwood trees for the trade purpose and illegal trade of endangered species should be stopped. Using DNA barcoding, the endangered species can be easily identified and conserved. Databases like FISH-BOL act as a reference library to improve the management and conservation of natural resources. About 90% primate population has been reduced due to hunting for bushmeat. DNA barcoding helps to prevent the endangered species. Devloo-Delvaa et al. (2016) identified a biological pollutant, pygmy mussel (Xenostrobus securis), and nearly 130 specimens are analyzed to obtain the genetic variability and new sources of pollution. Soil contains a huge amount of biodiversity. DNA barcoding helps to identify various species and use them in the right way. Various national and international projects have been established to understand the biodiversity of soil focused by Orgiazzi et al. (2015).

4. Enhanced Quality of Water
   DNA barcoding helps in taxonomical advances of organism present in rivers, streams, lakes, and ponds. It enhances the quality of water by detecting the species obtained and helps in cure of water-borne diseases.
5. Improvement of Food Quality
   The packed food quality can be improved by deep study of a particular species. Trivedi et al. (2016) assessed cryptic species in the marine environment. Various projects on the identification of marine organism are MarBOL, CeDAMar, CMarZ, SHARK-BOL, etc. This helps to find the various food-borne species and their effects on the quality of food. Many researchers are trying to find out various edible species of fungi and track the adulterations in the food. A protein for bodybuilding and antibiotics contain a large number of products from the exploitation of animals. Staats et al. (2016) centralized the concept of metabarcoding for tracking various adulterations and analyzing samples from alleged wildlife crime incidents.
6. Generation of New Tools and Software
   Scientists are moving at a faster rate to achieve the identification of a variety of species and to detect cryptic species to conserve the biodiversity. New databases and software have been discovered for the faster access and maintenance of a large amount of data. Generation of new tools and software will enhance the process as well as create a new source of income as a future scope.

9 Conclusion

After so many years of DNA barcoding, the approach is still controversial. The lack of universality makes it a difficult task to be resolved soon. A single marker cannot be used for all the species as different species show different levels of amplification for the different region. In plants, chloroplast DNA is used, whereas in animals, mitochondrial DNA is used. In the case of microbial barcoding, different species show different effects on the choice of markers. For instance, bacterial barcoding can be done using marker 16sRNA, whereas COI is best for algae. Even the classes in a particular species also show the need for a single marker as the classes also have different markers like in the case of fungi. A basic protocol of extracting DNA and sequencing and submitting or analyzing of DNA barcoding has been discussed.

The success of the DNA barcoding can be estimated from the increasing amount of barcoding projects and consorts that include a large number of scientists from all over the world. Various applications in food security and identification of pathogen lead to the easy diagnosis of various diseases. The phylogenetic relationship and the clade information can be analyzed using DNA barcoding.

The BOLD data system is the main axis of the barcoding. BOLD assembles standard information on the voucher specimen and its current status, which allows advancements in the findings of deposited sequences.

Bioinformatics has contributed at a large scale. Different software and tools have been developed for DNA barcoding to ease the process. This leads to an easy comparison between genomes with high-throughput technology as well as an efficient system, which includes a large-scale assay of species. This improves the research that will further have a great impact in the field of medicine, agriculture, ecology development, and biodefense.

References

Abel S, Wiesch PA z, Chang H-H, Davis BM, Lipsitch M, Waldor MK (2015) Sequence tag– based analysis of microbial population dynamics. Nat Methods 12:223–226. https://doi.org/10.1038/nmeth.3253

Bhargava M, Sharma A (2013) DNA barcoding in plants: evolution and applications of in silico approaches and resources. Mol Phylogenet Evol 67:631–641

Blaxter ML (2004) The promise of a DNA taxonomy. Philos Trans R Soc Lond B 359:669–679

Bonants P, Groenewald E, Rasplus JY, Maes M, de Vos P, Frey J, Boonham N, Nicolaisen M, Bertacini A, Robert V, Barker I, Kox L, Ravnikar M, Tomankova K, Caffier D, Li M, Armstrong K, Freitas-Astua J, Stefani E, Cubero J, Mostert L (2010) QBOL: a new EU project focusing on DNA barcoding of Quarantine organisms. EPPO Conference on Diagnostics, organized in cooperation with the Food and Environment Research Agency (Fera), York, GB

Casiraghi M, Labra M, Ferri E, Galimberti A, Mattia FD (2010) DNA barcoding: theoretical aspects and practical applications. In: Nimis PL, Vignes Lebbe R (eds) Tools for identifying biodiversity: progress and problems, pp 269–273

Conklin KY, Kurihara A, Sherwood AR (2009) A molecular method for identification of the morphologically plastic invasive algal genera Eucheuma and Kappaphycus (Rhodophyta, Gigartinales) in Hawaii. J Appl Phycol 21:691

Das S, Deb B (2015) DNA barcoding of fungi using ribosomal ITS marker for genetic diversity analysis: a review. Int J Pure Appl Biosci 3(3):160–167

Dentinger BTM, Didukh MY, Moncalvo J-M (2011) Comparing COI and ITS as DNA barcode markers for mushrooms and allies (Agaricomycotina). PLoS One 6(9):e25081. https://doi.org/10.1371/journal.pone.0025081

Devloo-Delvaa F, Mirallesa L, Ardura A, Borrell YJ, Pejovic I, Tsartsianidou V, Garcia-Vazquez E (2016) Detection and characterization of the biopollutant Xenostrobus securis (Lamarck 1819) Asturian population from DNA barcoding and eBarcoding. Mar Pollut Bull 105(1):23–29

Dhawan AK, Balwinder S, Manmeet BB, Arora R (2013) Integrated Pest Management. Scientific publishers, Jodhpur, India, pp 78–95

Dulla EL, Kathera C, Gurijala HK, Mallakuntla TR, Srinivasan P, Prasad V, Mopati RD, Jasti PK (2016) Highlights of DNA barcoding in identification of salient microorganisms like fungi. J Mycol Med 26(4):291–297. https://doi.org/10.1016/j.mycmed.2016.05.006

Fajarningsih DN (2016) Internal transcribed spacer (ITS) as DNA barcoding to identify fungal species: a review. Squalen Bull Mar Fish Postharvest Biotech 11(2):37–44

Fazekas AJ, Burgess KS, Kesanakurti PR, Graham SW, Newmaster SG et al (2008) Multiple multilocus DNA barcodes from the plastid genome discriminate plant species equally well. PLoS One 3(7):e2802

Geiger MF, Astrin JJ, Borsch T, Burkhardt U, Grobe P, Hand R, Hausmann A, Hohberg K, Krogmann L, Lutz M, Monje C, Misof B, Morinière J, Müller K, Pietsch S, Quandt D, Rulik B, Scholler M, Traunspurger W, Haszprunar G, Wägele W (2016) Genome 59:661–670. dx.doi.org. https://doi.org/10.1139/gen-2015-0185

Hajibabaei M, Singer CAG, Hebert NDP, Hickey AD (2007) DNA barcoding: how it complements taxonomy, molecular phylogenetics and population genetics. Trends Genet 23(4):167–172

Hebert NDP, Cywinske A, Ball LS, Dewaard RJ (2003) Biological identifications through DNA barcodes. Proc R Soc Lond B 270:313–321

Hebert DNP, Gregory RT (2005) The promise of DNA barcoding for taxonomy. Syst Biol 54 (5):852–859

Irinyi L, Lackner M, de Hoog S, Meyer W (2015) DNA barcoding of fungi causing infections in humans and animals. Fungal Biol 120:125–136. https://doi.org/10.1016/j.funbio.2015.04.007

Orgiazzi A, Dunbar BM, Panagos P, Groot DAG, Lemanceau P (2015) Soil biodiversity and DNA barcodes: opportunities and challenges. Soil Biol Biochem 80:244–250

Purty RS, Chatterjee S (2016) DNA barcoding: an effective technique in molecular taxonomy. Austin J Biotechnol Bioeng 3(1):1059

Raja AH, Baker RT, Little GJ, Oberlies HN (2017) DNA barcoding for identification of consumer-relevant mushrooms: a partial solution for product certification? Food Chem 214:383–392

Ratnasingham S, Hebert PDN (2007) BOLD: the barcode of life datasystem. (www.barcodinglife.org). Mol Ecol Notes 7:355–364

Staats M, Arulandhy JA, Gravendeel B, Holst-Jensen A, Scholtens I, Peelen T, Prins WT, Kok E (2016) Advances in DNA metabarcoding for food and wildlife forensic species identification. Anal Bioanal Chem 408:4615–4630

Trivedi S, Aloufi AA, Rehman H, Saggu S, Ghosh SK (2016) DNA barcoding: tool for assessing species identification in reptilia. J Entomol Zool Stud 4(1):332–337

Yan D, Luo JY, Han YM, Peng C, Dong XP et al (2013) Forensic DNA barcoding and bio-response studies of animal horn products used in traditional medicine. PLoS One 8(2):e55854. https://doi.org/10.1371/journal.pone.0055854

DNA Barcoding on Bacteria and its Application in Infection Management

Mohammad Zubair, Farha Fatima, Shamina Begum, and Zahid Hameed Siddiqui

Abstract The development of bacterial DNA barcoding with error-independent and time-saving technique has increased dramatically, which helps in identifying the microbes, understanding the microbial biodiversity, and analyzing the diseases, related to these pathogens. The study has discussed the developmental stages of DNA barcode research and compared it with the gold standard methods. It has evaluated the possible application of magnetic- and nanoparticle-related applications in bacterial DNA barcoding. These innovative techniques help in the identification of a number of infectious pathogens, which are present in trace amount. Bioimaging detection of an infectious microorganism has proved to be effective for the development of fluorescent nanoparticle, super-paramagnetic nanoparticle, and metallic nanoparticle. The living organisms, present with functional materials, have a vast application in clinical research of water-soluble conjugated polymers. However, the performance of DNA barcoding on a laboratory scale is easy, and it tends to give maximum benefit to the community if they developed a global system to share their findings and interpret the sequenced data by genome.

Keywords Bacteria · DNA Barcoding · Infection Management

Abbreviations

Bg	Bacillus globigii
BLAST	Basic local alignment search tool
BOLD	The barcode of life data system

M. Zubair (✉) · S. Begum
Department of Medical Microbiology, Faculty of Medicine, University of Tabuk, Tabuk, Kingdom of Saudi Arabia

F. Fatima
Department of Zoology, Faculty of Life Science, Aligarh Muslim University, Aligarh, India

Z. H. Siddiqui
Department of Biology, Faculty of Science, University of Tabuk, Tabuk, Kingdom of Saudi Arabia

S. Trivedi et al. (eds.), *DNA Barcoding and Molecular Phylogeny*,
https://doi.org/10.1007/978-3-030-50075-7_5

CPNs	Conjugated polymer nanoparticles
CTX	Cefotaximases
E. coli	Escherichia coli
FRET	Fluorescence resonance energy transfer
KDa	Kilo daltons
cpn60	Chaperonine 60
MLST	Multilocus sequence typing
MOTU	Molecular operational taxonomic unit
MTB	Mycobacterium tuberculosis
NCBI	National Center for Biotechnology Information
MEC-A	novel penicillin-binding protein, PBP-2a
NPs	Nanoparticles
NTM	Nontuberculosis Mycobacteria
PAIs	Potential pathogenicity islands
PCR	Polymerase chain reaction
QBOL	The quarantine barcoding of life
RIF	Replication initiation factor
S. aureus	*Staphylococcus aureus*
SHV	Sulfhydryl variable enzymes
Tuf	Elongation factor Tu

1 Introduction

Recently, the extraction of genomic information from the samples of patients has been considered as a major development among researchers and healthcare professionals. The development in research areas has emerged from the advancement in ultra-high-throughput sequencing (UHTS) technologies. However, this genomic analysis gives a boost in scientific and diagnostic research, when the genome size was particularly small. The impressive increase in the study of genomic data and molecular research in clinical microbiology has been evolved as an important tool in diagnostic laboratories (Figs. 1 and 2).

Short DNA sequences are sequentially used for identifying at the species level as they have emerged as an accurate, standardized, and fast method. In 2003, Hebert et al. (2003) have proposed a new DNA barcoding technique (Hebert et al. 2003). The core intention of this study is the species identification based on a short segment of DNA, extracted from a standardized area of the specified genome. The validation of the DNA sequence is practically observed identifying different species. A major advantage of DNA barcoding is the ability to enrich the biodiversity studies such as discovery and species identification. The relationships between the genetic and evolutionary aspects are studied effectively by researchers using the DNA barcoding. These relationships are extracted from cross-reading distributional, molecular, and morphological data (Savolainen et al. 2005). The identification by

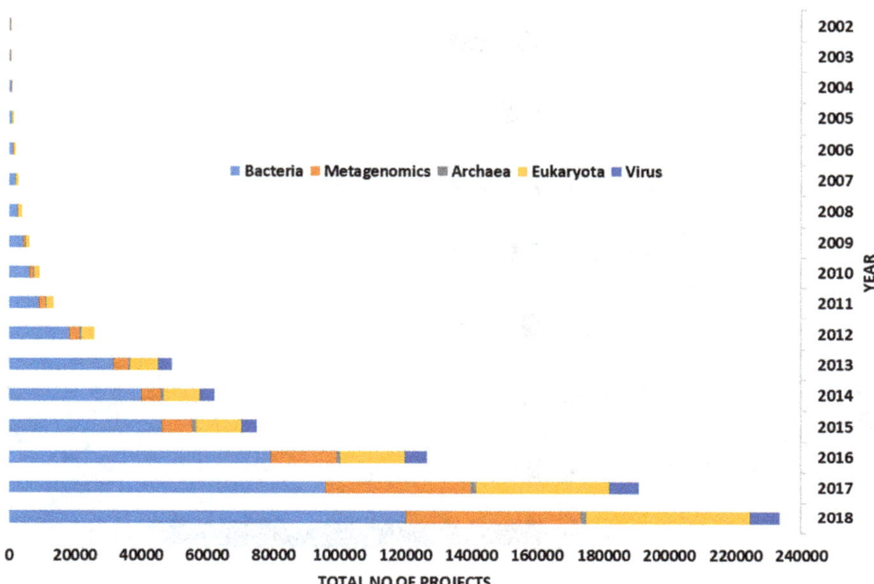

Fig. 1 Timeline of the project completed by domain from the year 2002–2018. The completed genome projects are deposited in public databases from 2002 to 2018 extracted from Mukherjee et al. 2017

DNA barcoding at the species level is performed by using a short DNA sequence from a standard part of the genome (Hebert et al. 2003). Hajibabaei et al. (2007) have compared a library of reference barcode sequences with barcode sequence extracted from specified identity individuals obtained and unidentified specimen, respectively. In contrast, standardized identification method implies an adverse impact on the species mapping on the entire earth. The emergence of particular adverse impact is shown when DNA sequencing technology is economically extractable. DNA barcode recommends that taxa are identified as a 11-digit universal product code by the standardized DNA sequences, identifying the market's retail products (Neigel et al. 2007).

The possession analysis, publication, and storage of DNA barcode records are assisted from a database of informatics worktop, The Barcode of Life Data System (BOLD). The association of BOLD is particularly seen with traditional bioinformatics by obtaining distributional, morphological, and molecular data. The researchers having firm knowledge about DNA barcoding can accessibly use BOLD. Ratnasingham and Hebert (2007) show that affordable specialized services are used linked to BOLD, helping the researchers to acquire barcode designation in the sequence databases internationally. BOLD serves as the universal starting point, which is ultimately referred to specialized databases for the identification of species, for example, pathogenic strains, endangered species, and disease vector species (Ball

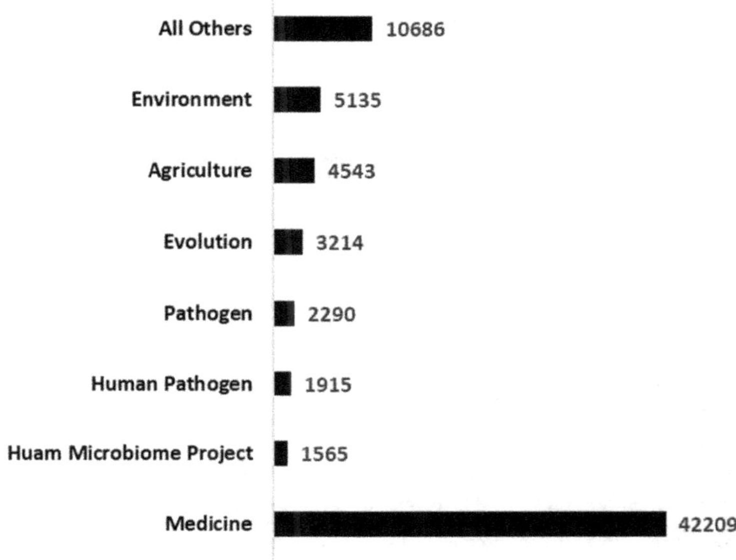

Fig. 2 Bacterial DNA Barcoding project distribution. Distribution of bacterial projects according to their subjects, Medicine (n = 42,209), Environment (n = 5135), Agriculture (n = 4543), Evolution (n = 3214), Human Pathogen (n = 1915), Human Microbiome Project (n = 1565), and All others (n = 10,686) extracted from Mukherjee et al. 2017

and Armstrong 2006). BOLD assists in DNA barcoding study, which is an online resource available to the scientific community.

The DNA barcode data of bacteria were acquired by QBOL (quarantine barcoding of life) to quarantine. Species quantification occurs by using total DNA barcode, which determines the composition of an insect's bacterial symbiotic association and how they alter in time. It also inspects novel bacterial pathogens of insect pests and assesses the hidden biodiversity of soil samples. More than 170 member organizations are included from 50 countries after the launching of the Consortium for the Barcode of Life (CBOL) in 2004. DNA barcoding sensu stricto is promoted by CBOL as a global standard to identify the biological specimens (Miller 2005). In clinical bacteriology, it is very important to identify the pathogen present in a clinical sample rapidly. This help in improving patient care reduces the economic burden. Moreover, it is essential to recognize the presence of a pathogen in a clinical sample, specifically in clinical bacteriology. The importance of testing is revealed from the management of infections and antibiotic treatment for species-level identification and antibiotic susceptibility performance.

A major uprising has been identified in clinical microbiology, the matrix-assisted laser desorption ionization time-of-flight (MALDI-TOF) mass spectrometry (MS), implying positively in identification, typing, and toxin detection (Croxatto et al. 2012). The lacuna in this method rapidly approaches characterization of any bacterial strain. In this regard, multiple analyses were involved under such types of

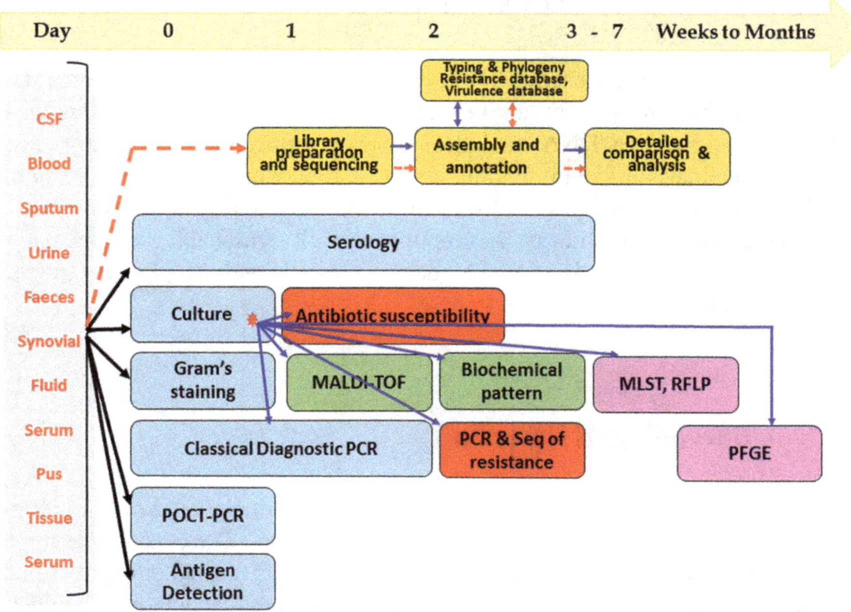

Fig. 3 Timeline provided for schematic representation for clinical sample processing with classical pathogens. The submission of clinical samples is directly based on Gram stain or serology, PCRs, and antigen detection, whereas the bacterium is cultured. The second panel is performed once a pure bacterial culture is extracted. The "time to result" (red dashed arrow) is shortened through direct sequencing on clinical samples specifically for slow-growing bacteria or fastidious bacteria. Techniques are colored according to their application for bacterial detection (blue), species identification (green), antibiotic susceptibility testing (red), and strain typing (pink). Purple color indicates the main steps taken toward genome sequencing. These steps include restriction fragment length polymorphism, point-of-care tests, multilocus sequence typing, MLST, and PFGE (Source: reproduced with permission from Bertelli and Greub 2013)

explicit categorizations; therefore, research laboratories execute a given pathogen specifically. Depending on bacterium type and its complexity of the question, it takes days to months for identifying the bacterium. The chances of minimizing the extent of steps required the pathogen's full characterization and the "time to result" optimization (Fig. 3). The bacterial genomic data of the whole sequence can be obtained from a pure culture, directly from a sample of clinical origin (pus, sputum, body fluids, urine, stool, tissue, blood, serum, etc.) or from a single bacterium present in a clinical sample. In addition, the most costly affair is the genome finishing as it consumes a major extent of time in each phase of a clinical process.

2 Great Criticism on DNA Barcoding

DNA barcoding has been excessively argued by people after its launching. The perception of these people reveals that the traditional methods and taxonomists might vanish through a DNA barcode species identification. In contrast, high throughput facilitates vacuum funding for affording barcode species (Rubinoff et al. 2006). Even though the establishment of DNA barcoding is determined as a global bioidentification and its assumption is still reasonably argumentative (Hickerson et al. 2006), the barcode data have emerged as phylogeographic (Beheregaray 2008).

3 DNA Barcoding of Bacteria

Bacterial pathogens give a clear view of how bacterial population word in groups, but this bacteria-to-bacteria signaling cannot be viewed. There is a paucity of investigating the minimal extent, initiating the infection in a specific organ region. Thereby, the differentiation of clones in a mixed bacterial population should be assessed by developing a marker throughout a prolonged infection. The specific offerings of tagged mutagenesis were assured for the bacteria pools' nature and preferably involved individual mutants through transposon insertions. According to Mecsas et al. (2001), a unique oligonucleotide barcode is identified in each insertion, which offers the sequential path for individual mutants' fate. This method has been widely used for studying virulence factors that are essential for the identification if numerous bacterial pathogens with their stages of infection are present. This specific approach has allowed the researchers to examine the individual bacteria's fate during infection.

Phytoplasmas of bacteria cause a huge loss to agriculture production. The universal DNA barcode-based is utilized as an elongation factor Tu (tuf) gene for phytoplasma identification. New primers amplify 420–444 bp fragment of tuf from all 91 phytoplasma-tested strain (16S rRNA groups -I through -VII, -IX through -XII, -XV, and -XX) (Makarova et al. 2012). *Xanthomonas*, a genus with many important phytopathogens, differentiates bacteria at the species level or below, which is initially associated with plants by using DNA markers. These markers have portions of DNA replication initiation factor (RIF). DNA is a single copy gene consisting of a huge majority of sequences of a bacterial genome and gene-specific primers needed for RIF amplification. The RIF marker system should be expandable to most bacterial genera, including *Pseudomonas* and *Xylella* (Schneider et al. 2011).

Ecology of the microbial community is an area of the fast-growing discipline of microbiology. Color-coded enterobacteria taxa are used along with broad-host-range plasmids for creating an artificial microbial community in the laboratory. Moreover, green fluorescent protein color variants are encoded and microbial fitness is developed by these plasmids. The possibilities have been created for community members to respond differently to the external events by understanding the community composition. In addition, fitness can be measured by understanding the use of community members' responses. The results obtained by pre- and post-test helped to understand about the microbial communities. Barcoding of selected entero-bacterial species using fluorescent proteins provides a simple and speedy method for distinguishing species identity (Le et al. 2013).

4 Applications of DNA Barcoding in Clinical Microbiology

In clinical bacteriology, the process to study the clinical samples has changed minutely over the years. However, the majority of analyses still depend on their gold standard methods known as isolation of a viable microorganism (Figs. 3 and 4). The PCR detection method has developed strongly in recent years. The effectiveness of PCR detection method has allowed researchers to ensure the identification of

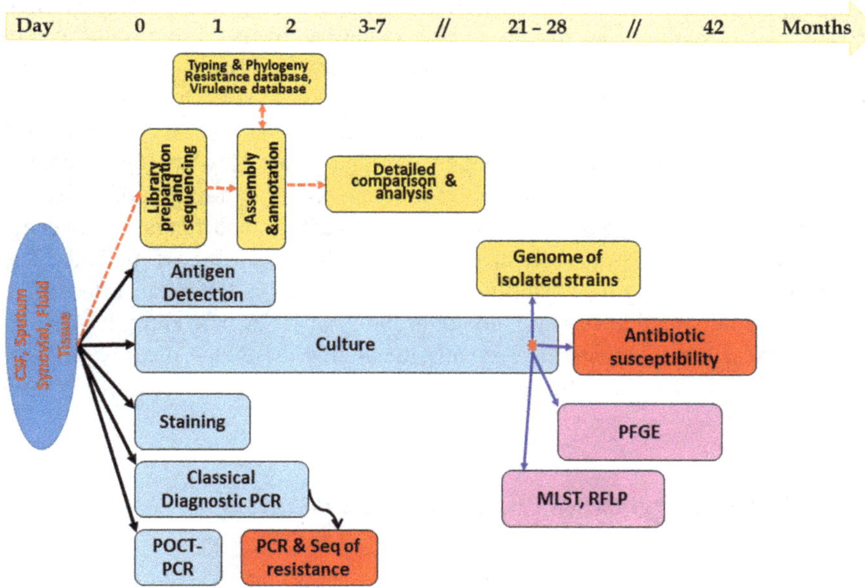

Fig. 4 Timeline provided for the schematic illustration of slow-growing bacterial samples (Mycobacterium tuberculosis). The submission of clinical samples is directly placed for PCRs, Gram stain, and antigen detection, whereas the bacterium is cultured. The second panel is performed once a pure bacterial culture is extracted. The "time to result" (red dashed arrow) is shortened through direct sequencing on clinical samples specifically for slow-growing bacteria or fastidious bacteria. Techniques are colored according to their application for bacterial detection (blue), species identification (green), antibiotic susceptibility testing (red), and strain typing (pink). Purple color indicates the main steps taken toward genome sequencing. These steps include restriction fragment length polymorphism, point-of-care tests, multilocus sequence typing, MLST, and PFGE (Source: reproduced with permission from Bertelli and Greub 2013)

viruses, fastidious organisms, and intracellular bacteria and faster diagnosis. The division of genomics' applications is classified into two types. The determination of biological properties, pure bacterial isolation, and outbreak monitoring are required in Type 1, while clinical sample, including community profiling and metagenomics, is directly applied in Type 2.

The future of clinical microbiology depends on the advancement for the development of methods, which will be helpful in obtaining full genomic data of each and every microbe preferable from clinical samples. The sequence of Plasmodium falciparum was obtained from the cell-depleted samples (Auburn et al. 2011) and the bioterrorism agent *Francisella tularensis* from abscess pus (Kuroda et al. 2012). Samples with high bacterial concentrations or physiological sterile are identified by direct sequencing. If the bacterial concentration in the clinical sample was less, it will be increased by using an initial antibody-based bacterial purification step. Major advantages are provided from the direct clinical sample sequencing, which offers additional time compared to the time required for bacterial culture. It also allows the detection of unculturable or bacteria, which are very difficult to culture and require a longer time to culture.

The significant contributions to assorted regions in virology are enabled by high-throughput sequencing. The specific areas include pathogenesis, molecular epidemiology, virus discovery, host immune system, antiviral pressures, and metagenomics. According to Quinones-Mateu et al. (2014), hepatitis C virus (HCV), influenza virus, and human immunodeficiency virus (HIV) are studied excessively in the DNA barcoding studies. During the last 15 years, the barcoding revealed a new series of enhancements for fungal research in the form of molecular phylogeny. Moreover, a positive consequence on ecology and fungal biodiversity research has emerged sequentially through the development of these methods. The progression of acquiring high-quality sequence databases is the reliance factor for fungi identification and technology-driven usability of DNA barcoding. It is examined that taxonomists contribute effectively in curating these high-quality sequence databases (Begerow et al. 2010).

4.1 Clinical Diagnostics & Species Identification

One of the most crucial steps in developing accurate clinical decisions is the identification of species. The effectiveness of this approach is obtained from the provision of direct information of pathogen strength. Gram staining, biochemical tests, sugar assimilation/fermentation, and colony growth time and morphology are the reliance factors of the gold standard method for identifying bacteria. MALDI-TOF MS is a very low-cost method introduced successfully for routine use (Croxatto et al. 2012). In contrast, approximately unusual species in 50% of cases were undetected by MALDI-TOF (Bizzini et al. 2011). In addition, the computational methods of DNA barcoding were proposed to identify these species, explaining the bacterial/archeal 16S barcode locus benchmark. These loci can be utilized to compare the performance of existing and new methods. The benchmark results indicated that the taxon coverage of reference sequences, which were used as a reference, is incomplete for identifying up to genus or species level. The registration of reference barcode sequences needs to be stepped up as it would help in the identification of species-level and high-throughput DNA barcoding research (Tanabe and Toju 2013).

MOTU (molecular operational taxonomic unit) is considered as fastest and cost-effective identification method. It is helpful for those that are not encountered before in analyzing the evolutionary pattern. MOTU approach possesses problems, particularly in the area of choosing what level of molecular difference defines a biologically relevant taxon (Blaxter 2004). According to Balasingham et al. (2009), a new diagnostic test for the known pathogens has been developed by genome sequences. Moreover, mutant variants such as *Chlamydia trachomatis* harbor a deletion on its cryptic plasmid (Fenollar et al. 2007) or emerging pathogens such as *Tropheryma whipplei* (Ripa and Nilsson 2007). Raw genome sequence may be helpful to identify the new targets for diagnostic PCR or for the development of the serological test (Greub et al. 2009).

Both genetic and phenotypic attributes have defined the species and are fundamentally illustrated by type strains consumed in two international strain centers. It has been proposed that a reference standard should be constituted by reference sequences (Didelot et al. 2012). By using phylogenetic analysis, the classification of the newly sequenced microorganisms should be identified among sequenced reference genomes. The available 1900 bacterial genomes are classified by using multilocus sequence typing (MLST) based on ribosomal protein-encoding genes (Jolley et al. 2012). The presence of bacterial essential genes is considered as an immense advantage in all bacteria, offering additional resolution as compared to 16S rRNA and strongly corroborated for functional preservation. The current classification could cover the newly identified genome by investigating the 53 conserved ribosomal genes. Larsen et al. (2012) developed a web-based method for MLST on preassembled genomes or directly on short sequence reads. Currently, only 66 bacterial species have been identified based on the various alleles by MLST schemes, but it covers the most important clinical pathogens.

4.2 Monitoring of Pathogens during an Outbreak

The outbreak of the virulent *E. coli* O104: H4 is an excellent example of the speed of data generation and implementation-controlled measures obtained by Whole Genome Sequence (WGS). Another outbreak in Germany is diarrhea and hemolytic–uremic syndrome (HUS) caused by Shiga toxin produced by *E coli* strain in May 2011. Within 2 months, more than 3100 nonHUS cases were reported, causing 53 deaths. Many independent research centers rushed to sequence some outbreak and historical reference isolate to combine different sequencing technologies (Mellmann et al. 2011; Rasko et al. 2011; Rhode et al. 2012). The first genome was sequenced and deposited in public databases with <2 weeks from the identified onset of the outbreak and in <62 h from strain isolation (Mellmann et al. 2011). The execution of an open source is considered as an interesting option for the genomics program and to identify the strain (Rhode et al. 2012). This occurrence was due to the unusual E. coli serotype O104: H4, comprising genes from both enterohemorrhagic E. coli (EHEC) and enteroaggregative E. coli (EAEC) (Mellmann et al. 2011; Rasko et al. 2011; Rhode et al. 2012). The gain and loss of plasmid-encoded and chromosomal factors can emerge as a highly pathogenic hybrid of EHEC and EAEC (Mellmann et al. 2011). The formation of new virulent pathogens occurs from the plasticity of bacterial genomes to genetic exchanges. The usefulness and feasibility of rapid draft genome sequencing are demonstrated by this multicenter work.

4.3 Epidemiology Study

Research on epidemiology dynamics of microbes is performed by using two genes (16S rRNA and 60 kDa chaperonin proteins cpn60) of gene-centric metagenomics study to characterize and identify the unknown bacteria. The outbreak efforts of the International Barcode of Life allow us to evaluate the protein coding cpn60 gene and bacteria from the 16S rRNA gene through DNA barcodes (Links et al. 2012). In their findings, this cpn60 universal target has larger barcode gap as compared to 16S rRNA, inferring that cpn60 has been used as a preferred barcode for Bacteria. A cohesive target for identifying the characterization of species-level data is revealed through this large gap. According to Links et al. (2012), DNA barcode is an authentic technique for evaluating novel microbes in epidemiological studies.

The DNA barcode method facilitates entire community phylogenic study of bacteria by using a primer, which was available universally (Barberan and Casamayor 2010; Wang et al. 2012). Small subunit of 16 s rRNA was commonly used as a gene barcode for the study of bacterial and archaeal communities. The majority of research used this marker as a barcode to quantify the microbial load in the community from the known environmental samples based on DNA sequences (Hugenholtz et al. 1998). We have only studied the scrape of the microbial community where millions of bacteria and billions of bacterial species exist (Curtis et al. 2002). Only a little part is sequenced scientifically and submitted in the sequence databases (Wu et al. 2009).

4.4 Virulence Study

The bacterial virulence factor provides information that is important for sequence methods. Opening an information bank for large-scale research based on genomic-phenotypic associations helps to predict the resistance to a specific antibiotic or general antibiotic susceptibility pattern of particular genera, for instance, mecA that confers methicillin resistance to *S. aureus* (Wolk et al. 2009), CTX-M, TEM, and SHV by Enterobacteriaceae (Zubair et al. 2012), or drug targets such as rpoB and others for *Mycobacterium tuberculosis* (Hillemann et al. 2005). It is examined that if the entire genome is accessible for investigating the mechanism of resistance, the susceptibility is revealed as a fast and an economical for rapid-growing bacteria. Moreover, gene mutations are not associated directly with the rapid detection of resistance. The bacterial resistance is a complex matter, and this was not studied extensively by barcoding methods (Iacono et al. 2008; Lieberman et al. 2011).

A study has indicated that diverse severe human infections might be induced by different subtypes of Mycobacterium tuberculosis (MTB) as few symptoms are similar to other pathogens, e.g., Nontuberculosis mycobacteria (NTM) (Liu et al. 2014). A novel DNA barcoding visualization method has been exploited by this study to detect molecular typing of 17 mycobacteria genomes referred from the database of NCBI prokaryotic genome database. The effective interstrain pathogenic variations are represented from the detection of 3 mycobacterium genes (Rv3508, Rv3514, and Rv0279c) obtained from the MT Band's PE/PPE family (Liu et al. 2014). Similarly, many strains that were resistant after laboratory selection may provide in-depth knowledge of evolutionary mechanism for their new resistance gene adaptation from the environment or from other microbes (such as *Bacillus anthracis*) (Serizawa et al. 2010). Many bacterial codes are well-characterized toxins, which are known to cause severe diseases, e.g., HUS by EHEC (Kaper et al. 2004), streptococcus causes toxic shock syndrome (Lappin and Ferguson 2009), and *Corynebacterium diphtheria* causing diphtheria (Holmes 2000).

5 Nanomaterial-Based DNA Barcoding

These innovative techniques identify a number of infectious pathogens quickly, which are present in trace amount. These pathogens, with high sensitivity and specificity, are in invariable demand to prevent global morbidity and mortality. The spread of disease needs to be controlled for reducing the economic burden. Advancement in developing the fluorescent nanoparticle, super-paramagnetic nanoparticle, and metallic nanoparticle is very effective for bioimaging detection of an infectious microorganism. This procedure is helpful in capturing the bacteria and viruses in solutions or biological samples (in vivo and in vitro) (Tallury et al. 2010). Fluorescent-barcoded DNA assay is based on two nanoparticles (NPs): magnetic nanoparticles (MNPs) and gold nanoparticles (Au-NP) (Zhang et al. 2009). This helps in the rapid detection of the *Salmonella enteritidis* gene.

Recently, *Bacillus globigii* (Bg) spores simulate *B. anthracis* and other bacterial species. DNA barcoded with nanowires Gold/silver/nickel (Au/Ag/Ni) is used for multiplexed detection of three antigens (Tok et al. 2006). However, large-scale synthesis of high-quality barcode nanowires is challenging. With the advancement of molecular imaging and multiplexed bioassays, drug discovery, gene profiling, and clinical diagnostics are made easy and cost effective. This ultimately reduces the global burden of mortality rate and endemic spread (Lee et al. 2010). Intracellular bacteria are biodegradable and biocompatible in nature as they act as a good nonviral transfer of plasmid DNA into mammalian cells and surfaces to study the antigen-antibody recognition (Akin et al. 2007).

Another new technique called Multicolor Conjugated Polymer Nanoparticle (CPN) was developed by Feng et al. (2012). The bacteria & CPNs were self-assembled using a simple and time-saving manner. By tuning florescence resonance energy transfer (FRET) efficiencies, the CPN has shown production of assorted colors under single excitation wavelength among CPNs. Different excitation sources of flow cytometry and florescence microscopy have been matched by the wavelength showing up to 170 nm large stokes shifts. It has also been applied for optical barcoding and cell imaging successfully. According to Feng et al. (2012), various barcodes with color-coded microparticles were prepared by mixing *E. coli* and the CPNs together under one excitation. Four colors were used to prepare 4 different CPNs; green, yellow, blue, and red. It is observed that these colors show low toxicity toward cell and were implemented for cell imaging and optical barcoding, instigating new notion for presenting multifunctional structures and to self-assemble living organisms. These living organisms are present with functional materials and have a vast application in clinical research of water-soluble conjugated polymers in biomedical fields (Feng et al. 2012).

Effective control of *Mycobacterium tuberculosis* (MTB) is required to determine the subtype of this organism because their severe human infection symptoms will be induced by other MTB, for example, Nontuberculosis Mycobacteria (NTM). This was the fastest method recommended on the basis of 16S rRNA gene and the combination of three biomarker genes applied to other pathogens. Mycobacterium's three genes were presented by the entire genetic information, comprising a differentiating ability associated with 16S rRNA gene. An accurate molecular typing is obtained by 16S rRNA gene to bring trust for overcoming the disease in the future cases occurred by Mycobacterium and to corroborate the genetic potentials (Liu et al. 2014).

6 Magnetic Barcoding

The detection of pathogenic microorganism on nucleic acid-based magnetic barcoding was introduced by Liong et al. (2013). A magnet was labeled over nanoprobe on microsphere that was specifically sequenced by using nuclear magnetic resonance technique. All the components were mixed up into single and small fluidic cartridges that were first streamlined to be used as a chip. This technology was used to detect drug-resistance strains of Mycobacterium tuberculosis in sputum sample within 2.5 h. The versatility of cost-effective, advanced, and high-throughput technology along with magnetic barcode can be implemented to other diseases and research. This technology could effectively indicate the biodefense and nosocomial infection by modifying the probe sequence (Liong et al. 2013). The effective genetic mutations in chronic diseases, including diabetes, heart diseases, and cancer, are studied by a bedside tool. It may also be used to shed some light for detecting epidermal growth factor receptor causing lung cancer by a single-point mutation in exon 21 (Dufort et al. 2011). The accessibility of particular technology is not restricted to magnetic readout, but it can be broadly used for plasmonic and luminescence studies based on gold nanoprobes and quantum dots, respectively (Hill and Mirkin 2006).

Wang et al. (2012) studied the pathogenicity pool of *Helicobacter pylori* by using this novel integrated technique. To apparently differentiate origin-specific genomic regions, the imaging method transforms frequency matrices to grey-scale levels to investigate H. pylori-barcoded genome. In the National Center for Biotechnological Information (NCBI) on prokaryotic genome database, six strains of H. pylori were compared with barcodes of *E. coli* O157: H7 strain. Wang et al. (2012) submitted the following criteria to be adopted for recognizing the pathogenic potential of *H. pylori* (PAIs);

- Length is greater than 10,000 continuous base pairs.
- Barcode distance is distinct from that of the general background.
- containing genes with known virulence-related functions (as determined by PfamScan and Blast2GO).

Very large DNA fragments (obtained through horizontal transfer) bearing multiple genes that encode virulence factors of bacteria are called PAIs (Schneider et al. 2011). The heterogeneity is associated with great quality of biological systems. The main perceptive of the system is to develop a method which uniquely identifies individual components and has interactions with each other. Peikon et al. (2014) developed a new method that tagged individual cell in vivo with genetic "barcode," which can be recovered by DNA sequencing. The viability of this method has been within the bacterial cells. Peikon et al. (2014) were the first to report an in vivo barcoding scheme with the perspective to label all of the individual cells of an entire tissue or organism. A general DNA barcode data processing system called "Bio-Barcode" was designed by Lim et al. (2009). This system encouraged quick gain in species DNA sequence database to meet global standards by providing specialized services. This makes barcoding inexpensive for Asian researchers.

7 Conclusion

The data of bacterial DNA barcoding are very limited and are not stored in one database to be easily accessible by the scientific community worldwide. Furthermore, the global initiative is required to facilitate the greeted potential of DNA barcoding tool to direct day-to-day infection control in hospitals and the community. The usefulness of DNA barcoding should be acknowledged, which not only is efficient in the identification of unnamed sequences but it may also play an important role in the surveillance and control of microbial endemic diseases. For the designing of unspecified or unidentified bacterial diseases, treatment will be easier in the near future if the research is united and shared in a common platform. It can be done by developing a database, and all the researchers would deposit their sequence data for future reference. Therefore, this area of research is needed globally to counter hazardous bacterial infections for future developments to control their spread.

The data available on bacterial DNA barcoding are limited and used for global bioidentification despite all the criticism on DNA barcoding by the classical researchers (Rubinoff et al. 2006; Will et al. 2005). If a global system is developed for sharing the findings and interpreting the sequenced information, DNA barcoding becomes accessible on a laboratory scale for maximizing benefit to the community. The comparison of assorted strains and the implementation of knowledge are based on the sharing and interpretation of data. In the first stage of global collaboration, all the information related to pathogens will be stored in detail such as resistance profiles and disease details especially for those that will be used as agents of bioterrorism. This is a great challenge for the scientist to form a global database of all the bacteria's sequences, which constitute hundreds and thousands of sequenced genomes annually. Moreover, the associated data, called metadata, will be available to the whole scientific community. It has been concluded that the comparison and sharing of results extracted in different laboratories can be facilitated by the harmonization of procedures such as data sequencing and data analysis.

Acknowledgment The authors are very thankful to all the associated personnel in any reference that contributed in/for the purpose of this research.

Conflict of Interest The authors declare no conflict of interest.

References

Akin D, Sturgis J, Ragheb K, Sherman D, Burkholder K, Robinson JP, Bhunia AK, Mohammed S, Bashir R (2007) Bacteria-mediated delivery of nanoparticles and cargo into cells. Nat Nanotechnol 2(7):441–449. https://doi.org/10.1038/nnano.2007.149

Auburn S, Campino S, Clark TG, Djimde AA, Zongo I, Pinches R, Manske M, Mangano V, Alcock D, Anastasi E, Maslen G (2011) An effective method to purify Plasmodium falciparum DNA directly from clinical blood samples for whole genome high-throughput sequencing. PLoS One 6(7):e22213. https://doi.org/10.1371/journal.pone.0022213

Balasingham SV, Davidsen T, Szpinda I, Frye SA, Tonjum T (2009) Molecular diagnostics in tuberculosis. Mol Diagn Ther 13(3):137–151. https://doi.org/10.1007/bf03256322

Ball SL, Armstrong KF (2006) DNA barcodes for insect pest identification: a test case with tussock moths (Lepidoptera: Lymantriidae). Can J For Res 36(2):337–350. https://doi.org/10.1139/x05-276

Barberan A, Casamayor EO (2010) Global phylogenetic community structure and β-diversity patterns in surface bacterioplankton metacommunities. Aquat Microb Ecol 59(1):1–10. https://doi.org/10.3354/ame01389

Begerow D, Nilsson H, Unterseher M, Maier W (2010) Current state and perspectives of fungal DNA barcoding and rapid identification procedures. Appl Microbiol Biotechnol 87(1):99–108. https://doi.org/10.1007/s00253-010-2585-4

Beheregaray LB (2008) Twenty years of phylogeography: the state of the field and the challenges for the southern hemisphere. Mol Ecol 17(17):3754–3774. https://doi.org/10.1111/j.1365-294x.2008.03857.x

Bertelli C, Greub G (2013) Rapid bacterial genome sequencing: methods and applications in clinical microbiology. Clin Microbiol Infect 19(9):803–813

Bizzini A, Jaton K, Romo D, Bille J, Prod'hom G, Greub G (2011) Matrix-assisted laser desorption ionization–time of flight mass spectrometry as an alternative to 16S rRNA gene sequencing for identification of difficult-to-identify bacterial strains. J Clin Microbiol 49(2):693–696. https://doi.org/10.1128/jcm.01463-10

Blaxter ML (2004) The promise of a DNA taxonomy. Philos Trans R Soc Lond B Biol Sci 359 (1444):669–679. https://doi.org/10.1098/rstb.2003.1447

Croxatto A, Prod'hom G, Greub G (2012) Applications of MALDI-TOF mass spectrometry in clinical diagnostic microbiology. FEMS Microbiol Rev 36(2):380–407. https://doi.org/10.1111/j.1574-6976.2011.00298.x

Curtis TP, Sloan WT, Scannell JW (2002) Estimating prokaryotic diversity and its limits. Proc Natl Acad Sci 99(16):10494–10499. https://doi.org/10.1073/pnas.142680199

Didelot X, Bowden R, Wilson DJ, Peto TE, Crook DW (2012) Transforming clinical microbiology with bacterial genome sequencing. Nat Rev Genet 13(9):601–612. https://doi.org/10.1038/nrg3226

Dufort S, Richard MJ, Lantuejoul S, De Fraipont F (2011) Pyrosequencing, a method approved to detect the two major EGFR mutations for anti EGFR therapy in NSCLC. J Exp Clin Cancer Res 30(1):57. https://doi.org/10.1186/1756-9966-30-57

Feng X, Yang G, Liu L, Lv F, Yang Q, Wang S, Zhu D (2012) A convenient preparation of multi-spectral microparticles by bacteria-mediated assemblies of conjugated polymer nanoparticles for cell imaging and barcoding. Adv Mater 24(5):637–641. https://doi.org/10.1002/adma.201102026

Fenollar F, Puechal X, Raoult D (2007) Whipple's disease. N Engl J Med 356(1):55–66. https://doi.org/10.1056/nejmra062477

Greub G, Kebbi-Beghdadi C, Bertelli C, Collyn F, Riederer BM, Yersin C, Croxatto A, Raoult D (2009) High throughput sequencing and proteomics to identify immunogenic proteins of a new pathogen: the dirty genome approach. PLoS One 4(12):e8423. https://doi.org/10.1371/journal.pone.0008423

Hajibabaei M, Singer GA, Hebert PD, Hickey DA (2007) DNA barcoding: how it complements taxonomy, molecular phylogenetics and population genetics. Trends Genet 23(4):167–172. https://doi.org/10.1016/j.tig.2007.02.001

Hebert PD, Cywinska A, Ball SL (2003) Biological identifications through DNA barcodes. Proc R Soc Lond B Biol Sci 270(1512):313–321. https://doi.org/10.1098/rspb.2002.2218

Hickerson MJ, Meyer CP, Moritz C (2006) DNA barcoding will often fail to discover new animal species over broad parameter space. Syst Biol 55(5):729–739. https://doi.org/10.1080/10635150600969898

Hill HD, Mirkin CA (2006) The bio-barcode assay for the detection of protein and nucleic acid targets using DTT-induced ligand exchange. Nature Protocols 1(1):324. https://doi.org/10.1038/nprot.2006.51

Hillemann D, Weizenegger M, Kubica T, Richter E, Niemann S (2005) Use of the genotype MTBDR assay for rapid detection of rifampin and isoniazid resistance in Mycobacterium tuberculosis complex isolates. J Clin Microbiol 43(8):3699–3703. https://doi.org/10.1128/jcm.43.8.3699-3703.2005

Holmes RK (2000) Biology and molecular epidemiology of diphtheria toxin and the tox gene. J Infect Dis 181(Supplement_1):S156–S167. https://doi.org/10.1086/315554

Hugenholtz P, Goebel BM, Pace NR (1998) Impact of culture-independent studies on the emerging phylogenetic view of bacterial diversity. J Bacteriol 180(18):4765–4774. https://doi.org/10.1016/0167-7799(96)10025-1

Iacono M, Villa L, Fortini D, Bordoni R, Imperi F, Bonnal RJ, Sicheritz-Ponten T, De Bellis G, Visca P, Cassone A, Carattoli A (2008) Whole-genome pyrosequencing of an epidemic multidrug-resistant Acinetobacter baumannii strain belonging to the European clone II group. Antimicrob Agents Chemother 52(7):2616–2625. https://doi.org/10.1128/aac.01643-07

Jolley KA, Bliss CM, Bennett JS, Bratcher HB, Brehony C, Colles FM, Wimalarathna H, Harrison OB, Sheppard SK, Cody AJ, Maiden MC (2012) Ribosomal multilocus sequence typing: universal characterization of bacteria from domain to strain. Microbiology 158(4):1005–1015. https://doi.org/10.1099/mic.0.055459-0

Kaper JB, Nataro JP, Mobley HL (2004) Pathogenic escherichia coli. Nat Rev Microbiol 2 (2):123–140. https://doi.org/10.1038/nrmicro818

Kuroda M, Sekizuka T, Shinya F, Takeuchi F, Kanno T, Sata T, Asano S (2012) Detection of a possible bioterrorism agent, Francisella sp., in a clinical specimen by use of next-generation direct DNA sequencing. J Clin Microbiol 50(5):1810–1812. https://doi.org/10.1128/jcm.06715-11

Lappin E, Ferguson AJ (2009) Gram-positive toxic shock syndromes. Lancet Infect Dis 9 (5):281–290. https://doi.org/10.1016/s1473-3099(09)70066-0

Larsen MV, Cosentino S, Rasmussen S, Friis S, Hasman H, Marvig RL, Jelsbak L, Pontén TS, Ussery DW, Aarestrup FM, Lund O (2012) Multilocus sequence typing of total-genome-sequenced bacteria. J Clin Microbiol 50(4):1355–1361. https://doi.org/10.1128/JCM.06094-11

Le T, Trexel J, Brugler A, Tiebel K, Maezato Y, Robertson B, Blum P (2013) Bar-coded Enterobacteria: an undergraduate microbial ecology laboratory module. American J Edu Res 1(1):26–30. https://doi.org/10.12691/education-1-1-6

Lee H, Kim J, Kim H, Kim J, Kwon S (2010) Color-barcoded magnetic microparticles for multiplexed bioassays. Nat Mater 9(9):745–749. https://doi.org/10.1038/nmat2815

Lieberman TD, Michel JB, Aingaran M, Potter-Bynoe G, Roux D, Davis MR Jr, Skurnik D, Leiby N, LiPuma JJ, Goldberg JB, McAdam AJ (2011) Parallel bacterial evolution within multiple patients identifies candidate pathogenicity genes. Nat Genet 43(12):1275–1280. https://doi.org/10.1038/ng.997

Lim J, Kim SY, Kim S, Eo HS, Kim CB, Paek WK, Kim W, Bhak J (2009) Bio barcode: a general DNA barcoding database and server platform for Asian biodiversity resources. BMC Genomics 10(3):S8. https://doi.org/10.1186/1471-2164-10-s3-s8

Links MG, Dumonceaux TJ, Hemmingsen SM, Hill JE (2012) The chaperonin-60 universal target is a barcode for bacteria that enables de novo assembly of metagenomic sequence data. PloS One 7(11):e49755. https://doi.org/10.1371/journal.pone.0049755

Liong M, Hoang AN, Chung J, Gural N, Ford CB, Min C, Shah RR, Ahmad R, Fernandez-Suarez M, Fortune SM, Toner M (2013) Magnetic barcode assay for genetic detection of pathogens. Nat Commun 4:1752. https://doi.org/10.1038/ncomms2745

Liu B, Zhang X, Huang H, Zhang Y, Zhou F, Wang G (2014) A novel molecular typing method of mycobacteria based on DNA barcoding visualization. J clin bioinforma 4(1):4. https://doi.org/10.1186/2043-9113-4-4

Makarova O, Contaldo N, Paltrinieri S, Kawube G, Bertaccini A, Nicolaisen M (2012) DNA barcoding for identification of 'Candidatus Phytoplasmas' using a fragment of the elongation factor Tu gene. PLoS One 7(12):e52092. https://doi.org/10.1371/journal.pone.0052092

Mecsas J, Bilis I, Falkow S (2001) Identification of attenuated Yersinia pseudotuberculosis strains and characterization of an orogastric infection in BALB/c mice on day 5 postinfection by signature-tagged mutagenesis. Infect Immun 69(5):2779–2787. https://doi.org/10.1128/iai.67.5.2779-2787.2001

Mellmann A, Harmsen D, Cummings CA, Zentz EB, Leopold SR, Rico A, Prior K, Szczepanowski R, Ji Y, Zhang W, McLaughlin SF (2011) Prospective genomic characterization of the German enterohemorrhagic Escherichia coli O104: H4 outbreak by rapid next generation sequencing technology. PloS One 6(7):e22751. https://doi.org/10.1371/journal.pone.0022751

Miller SE, (2005) Proposed standards for BARCODE records in INSDC (BRIs). Request document for continuation of support by the Alfred P. Sloan Foundation submitted by the Smithsonian Institution on behalf of Consortium for the barcode of Life: 22 January 2006, pp. 36–38

Mukherjee S, Stamatis D, Bertsch J, Ovchinnikova G, Verezemska O, Isbandi M, Thomas AD, Ali R, Sharma K, Kyrpides NC, Reddy TBK (2017) Genomes OnLine database (GOLD) v. 6: data updates and feature enhancements. Nucleic Acids Res 45:D446–D456

Neigel J, Domingo A, Stake J (2007) DNA barcoding as a tool for coral reef conservationx. PloS One 6(7):e22751. https://doi.org/10.1007/s00338-007-0248-4

Peikon ID, Gizatullina DI, Zador AM (2014) In vivo generation of DNA sequence diversity for cellular barcoding. Nucleic Acids Res 42(16):e127–e127. https://doi.org/10.1093/nar/gku604

Quinones-Mateu ME, Avila S, Reyes-Teran G, Martinez MA (2014) Deep sequencing: becoming a critical tool in clinical virology. J Clin Virol 61(1):9–19. https://doi.org/10.1016/j.jcv.2014.06.013

Rasko DA, Webster DR, Sahl JW, Bashir A, Boisen N, Scheutz F, Paxinos EE, Sebra R, Chin CS, Iliopoulos D, Klammer A (2011) Origins of the E. coli strain causing an outbreak of hemolytic–uremic syndrome in Germany. N Engl J Med 365(8):709–717. https://doi.org/10.1056/nejmoa1106920

Ratnasingham S, Hebert PD (2007) BOLD: the barcode of life data system. Mol Ecol Resour 7 (3):355–364. https://doi.org/10.1111/j.1471-8286.2007.01678.x. http://www.barcodinglife.org

Rhode H, Qin J, Cui Y, Li D, Loman NJ, Hentschke M, Chen W, Pu F, Peng Y, Li J, Xi F, E. coli O104: H4 Genome Analysis Crowd-Sourcing Consortium (2012) Open-source genomic analysis of Shiga-toxin-producing E. coli O104: H4. N Engl J Med 365(8):718–724

Ripa T, Nilsson PA (2007) A chlamydia trachomatis strain with a 377-bp deletion in the cryptic plasmid causing false-negative nucleic acid amplification tests. Sex Transm Dis 34(5):255–256. https://doi.org/10.2807/esw.11.45.03076-en

Rubinoff D, Cameron S, Will K (2006) A genomic perspective on the shortcomings of mitochondrial DNA for "barcoding" identification. J Hered 97(6):581–594. https://doi.org/10.1093/jhered/esl036

Savolainen V, Cowan RS, Vogler AP, Roderick GK, Lane R (2005) Toward writing the encyclopedia of life: an introduction to DNA barcoding. Philos Trans R Soc Lond B Biol Sci 360 (1462):1805–1811. https://doi.org/10.1098/rstb.2005.1730

Schneider G, Dobrindt U, Middendorf B, Hochhut B, Szijarto V, Emődy L, Hacker J (2011) Mobilization and remobilization of a large archetypal pathogenicity island of uropathogenic Escherichia coli in vitro support the role of conjugation for horizontal transfer of genomic islands. BMC microbiology 11(1):210. https://doi.org/10.1186/1471-2180-11-210

Schneider KL, Marrero G, Alvarez AM, Presting GG (2011) Classification of plant associated bacteria using RIF, a computationally derived DNA marker. PloS One 6(4):e18496. https://doi.org/10.1371/journal.pone.0018496

Serizawa M, Sekizuka T, Okutani A, Banno S, Sata T, Inoue S, Kuroda M (2010) Genomewide screening for novel genetic variations associated with ciprofloxacin resistance in Bacillus anthracis. Antimicrob Agents Chemother 54(7):2787–2792. https://doi.org/10.1128/aac.01405-09

Tallury P, Malhotra A, Byrne LM, Santra S (2010) Nanobioimaging and sensing of infectious diseases. Adv Drug Deliv Rev 62(4):424–437. https://doi.org/10.1016/j.addr.2009.11.014

Tanabe AS, Toju H (2013) Two new computational methods for universal DNA barcoding: a benchmark using barcode sequences of bacteria, archaea, animals, fungi, and land plants. PloS One 8(10):e76910. https://doi.org/10.1371/journal.pone.0076910

Tok JBH, Chuang F, Kao MC, Rose KA, Pannu SS, Sha MY, Chakarova G, Penn SG, Dougherty GM (2006) Metallic striped nanowires as multiplexed immunoassay platforms for pathogen detection. Angew Chem Int Ed 45(41):6900–6904. https://doi.org/10.1002/anie.200601104

Wang J, Soininen J, He J, Shen J (2012) Phylogenetic clustering increases with elevation for microbes. Environ Microbiol Rep 4(2):217–226. https://doi.org/10.1111/j.1758-2229.2011.00324.x

Will KW, Mishler BD, Wheeler QD (2005) The perils of DNA barcoding and the need for integrative taxonomy. Syst Biol 54(5):844–851. https://doi.org/10.1080/10635150500354878

Wolk DM, Struelens MJ, Pancholi P, Davis T, Della-Latta P, Fuller D, Picton E, Dickenson R, Denis O, Johnson D, Chapin K (2009) Rapid detection of Staphylococcus aureus and methicillin-resistant S. aureus (MRSA) in wound specimens and blood cultures: multicenter preclinical evaluation of the Cepheid Xpert MRSA/SA skin and soft tissue and blood culture assays. J Clin Microbiol 47(3):823–826. https://doi.org/10.1128/jcm.01884-08

Wu D, Hugenholtz P, Mavromatis K, Pukall R, Dalin E, Ivanova NN, Kunin V, Goodwin L, Wu M, Tindall BJ, Hooper SD, D'haeseleer P, Zemla A, Singer M, Lapidus A, Nolan M, Copeland A, Han C, Chen F, Cheng JF, Lucas S, Kerfeld C, Lang E, Gronow S, Chain P, Bruce D, Rubin EM, Kyrpides NC, Klenk HP, Eisen JA (2009) A phylogeny-driven genomic encyclopedia of Bacteria and Archaea. Nature 462:1056–1060. https://doi.org/10.1038/nature08656

Zhang D, Carr DJ, Alocilja EC (2009) Fluorescent bio-barcode DNA assay for the detection of Salmonella enterica serovar Enteritidis. Biosens Bioelectron 24(5):1377–1381. https://doi.org/10.1016/j.bios.2008.07.081

Zubair M, Malik A, Ahmad J (2012) Study of plasmid-mediated extended-spectrum β-lactamase-producing strains of enterobacteriaceae, isolated from diabetic foot infections in a north Indian tertiary-care hospital. Diabetes Technol Ther 14(4):315–324. https://doi.org/10.1089/dia.2011.0197

Part III
DNA Barcoding in Plants

DNA Barcoding: Implications in Plant–Animal Interactions

Muniyandi Nagarajan, Vandana R. Prabhu, Ranganathan Kamalakkannan, and Palatty Allesh Sinu

Abstract Trophic interactions between plants and animals occur through various ecological processes such as pollination, frugivory, and herbivory. Among the three, pollination and frugivory represent the mutualistic interactions and herbivory represents a negative antagonistic interaction. Interaction with pollinators and frugivores are statutory for many plants for the dispersal of gametes and offspring, respectively. On the contrary, herbivory plays a substantial role in maintaining the stability of floral communities. Disruption in these natural species interactions can have a detrimental effect on both the interacting species and other species of higher trophic levels, which eventually destabilizes the terrestrial ecosystem. Thus, a thorough perception of the interacting species is imperative for the restoration of biocommunities and conservation of ecosystems. Although plant–animal interactions had been identified traditionally through field observation and tracking the movement of plant visitors, they are inadequate in accurately determining the interacting species as interaction occurs in the vicinity of other co-occurring plants and animals. Similarly, histological and biochemical analyses of gut matter and feces of herbivores/frugivores are not adequate to accurately identify the plant sources that the animals are depending upon. Paucity of an appropriate methodology has been always a limitation in delineating the full range of plant–animal interactions. DNA barcoding technique has revolutionized the field of community ecology. Its unprecedented accuracy and speed have made it ideally the best method of choice for identifying plant–animal interactions at species and population levels. DNA barcoding regions *rbcL, matK, trnH-psbA, trnL*, ITS2 (plant), and COI (animal) are the most common markers used in identifying plant and animal species. These barcodes have been adapted to characterize various plant and animal sources (honey,

M. Nagarajan (✉) · V. R. Prabhu · R. Kamalakkannan
Department of Genomic Science, School of Biological Sciences, Central University of Kerala, Trivandrum, Kerala, India
e-mail: nagarajan@cukerala.ac.in

P. A. Sinu
Department of Animal Science, School of Biological Sciences, Central University of Kerala, Trivandrum, Kerala, India

S. Trivedi et al. (eds.), *DNA Barcoding and Molecular Phylogeny*,
https://doi.org/10.1007/978-3-030-50075-7_6

pollen, gut content, and feces) for unraveling many of the unnoticed trophic associations existing between plants and animals.

Keywords Plant–animal interaction · Pollen · Honey · Seed · Frugivore · DNA barcoding

1 Introduction

Interactions of species build up the food web in any given ecosystem. A stable community has a complex species interaction between and among plants, plants and animals, and animals and animals. The abiotic factors such as water, soil, temperature, energy, etc., facilitate the biotic interactions in a community. Plants and animals occupy a major share of ecological communities. A natural community is understood well through studying almost all possible biotic interactions above and below ground, which includes plant–microbial interactions and plant–animal interactions. While mutualism is a facilitative plant–animal interaction, herbivory is an antagonistic plant–animal interaction. However, these two interactions are vital to maintaining stability in any given terrestrial community and ecosystem.

Pollination networks play a major role in community dynamics in which pollinators interact positively with floral populations. The relationship mainly involves offering food (pollen or nectar) by plants to the pollinators as a reward paid for their service in mediating pollination. Such an important service provided by pollinators serve as the most decisive facet of reproduction in flowering plants and also contributes remarkably to floral evolution and plant adaptations, eventually leading to speciation. Similarly, frugivore–plant relationship remains another classic example of mutualistic interactions. Frugivores (fruit eaters) usually involve in a process of the spatial distribution of the seeds, and thus are vital for the persistence of floral populations. Besides the mutualistic interactions, antagonistic interactions also occupy an important place in maintaining stable ecological communities. A major portion of the plant biomass is a good source of food for a tremendous array of herbivores, comprising the second trophic level of the food chain. Such negative interactions existing between plants and herbivores are considered as one of the chronic causes of plant damage. Though herbivores have a deleterious effect on the plant communities, in an ecological viewpoint, they act as major regulators of ecosystem functions.

Regardless of the ways in which a species interacts with others, most trophic level interactions seen between plants and animals are species specific. For instance, in the pollination networks, a few pollinators manifest foraging preferences toward specific flowering plants. This is substantiative in the case of orchids, where nearly 60% of the orchid families are pollinated by pollen-collecting wasps (Mchatton 2011). A similar cognate relationship has been noted between fruits and frugivores. A number of fruiting plants require the intervention of specific animal vectors for the dispersal of seed and its successful germination. The loss of calvaria forest due to the extinction of dodo bird is a classic example of such species-specific ecological relationship. Likewise, in the case of antagonistic interactions, specifically in the plant–herbivore interactions, most herbivores have a narrow diet range. Though a few plant–animal interactions have been pointed out, the accurate feeding behaviors and foraging preferences of many herbivores, frugivores, and pollinators are still obscure. The organismal interactions between plants and primary consumers, being the fundamental unit on which other higher trophic-level interactions are built, need to be thoroughly studied.

Notwithstanding the ecological importance of plant–animal interactions, their identification has always been a challenge. Several attempts have been made to study the animal and plant species involved in various trophic interactions. Earlier, identification of ecological relationships was based on traditional methods like field observation, microscopic analysis, tracking animals using traps and dyes, etc. Other indirect approaches have included various histological and biochemical procedures that target on analyzing the gut or fecal matter to identify the plant materials. Although researchers detect trophic-level connections between plants and animals through conventional methods, this is not a choice under many situations, such as when they interact in the vicinity of many co-occurring plant and animal species or when the ingested diet meals of primary consumers cannot be easily discerned. Such circumstances require a suitable alternative improved and accurate method of identification processes. In recent years, DNA-based method has received much attention among molecular taxonomists as it allows identification of organisms up to species and the population levels. Specifically, DNA barcoding has proven to be a good candidate for species identification as it uses variable regions in the genome that clearly differs even between closely related species. This chapter focuses on various kinds of plant–animal interactions and the usefulness of DNA barcoding in understanding the trophic relationships between plants and animals.

2 Plant–Animal Interplay: An Ecological Outlook

The first evidence of plant–animal coevolution was reported by Charles Darwin by relating the morphologic specificity between orchids and moths (Darwin 1859; Darwin 1862). The present concept of plant–animal evolutionary associations results from his views. Plant–animal interactions (both mutualistic and antagonistic associations) are the major driving force for plant–animal coevolution (Clare 2014). Exchange of genetic content between the interacting species is assumed to be zero in plant–animal evolutionary interactions (Ehrlich and Raven 1964; Clare 2014). It is through the food chain networks, most of the plant species forge a relationship with pollinators, frugivores, and herbivores (Fig. 1). Therefore, understanding the connections between plants and animals with an ecological perspective is very essential to accurately determine their status in the food chain and to conserve the endangered plant or animal species.

Fig. 1 Different forms of plant–animal interactions. (**a** & **b**) Plant and pollinator. (**c**) Plant and frugivore. (**d**) Plant and herbivore

2.1 Plants and Pollinators

Flowers are the most attractive part of angiosperms that draw the attention of many insects and birds by their color, scent, and shape alone or in combination in order to carry out pollination and set seeds (Ollerton et al. 2011; Schiest and Johnson 2013). In return, the flowers provide the pollinators with food in the form of pollen and/or nectar for their priceless service. The process of pollen transfer is mainly mediated by wind (anemophily) and animals (bees, butterflies, birds etc.). However, besides the anemophilic mode of pollination, a great diversity of flowering plants depends on animals for reproduction (Ollerton et al. 2011). The ecological importance of pollinators, therefore, is much higher in the natural communities as their mutual dependence on the flora impacts persistence of both plant and animal populations (Hegland et al. 2009).

Several studies have been emphasized on the evolution of plant–pollinator interaction, food habits of major pollinators, their role in stabilizing natural communities and conserving the ecosystem. Sargent and Ackerly (2007) described three different processes related to plant–pollinator interactions—habitat filtering, diffuse facilitation, and competitive exclusion to explain which species establish where and persist. A recent study by Robson (2014) postulated the importance of wild insect pollinators on crop yield. Wild insects normally prefer wildflowers as they provide sufficient quantity of nectar. The study ranked 41 wild species that share common pollen vectors with canola (*Brassica napus L.*), a cultivated crop (which does not provide an optimal resource to their visitors), suggesting the usefulness of incorporating selected wild flowering plants in the cropland for attracting important pollinators and increasing crop production. However, it is important to notice that the foraging behavior of the pollinators, the species visited in a given visitation bout, and the diversity of the pollen carrying on the body of the pollinators influence the pollination efficiency of the pollinators.

2.2 *Plants and Frugivores*

Fruit-frugivore interaction is one of the chief components of community ecology that plays a significant role in plant restoration and radiation (Chama et al. 2013). Frugivores (fruit eaters) are natural seed disseminators as they involve in a fundamental spatial process of dispersing propagules to long and short distances. Such spatial patterns of seed dissemination influence the rate of gene migration within and among plant populations and are subjected to post-dispersal processes like selection and predation, eventually affecting the persistence of plant populations (Robledo-Arnuncio and García 2007). Nearly 90% of the plant species bearing fleshy fruits heavily depend on animals to carry out the process of seed dissemination across space and time (Chama et al. 2013). However, it cannot be generalized that frugivores are always in mutualistic association with plants. The seed handling behavior of the animals determines the role played by the animals, whether to be a seed disperser or seed predator (Chama et al. 2013).

Fruit-frugivore interactions are widely studied to get a better understanding of frugivore diet habits, evolution of fruit traits, and plant regeneration dynamics. Encinas-viso et al. (2014) using an individual-based model studied the diversification of fruit traits. The reason for such evolution of trait was predicted as seed dispersion by frugivores, their foraging behavior, and numerical abundance. Another attempt was made by Sankamethawee et al. (2011) to find out the hinge between plants bearing small fleshy fruits and avian frugivores in an evergreen forest of northeast Thailand. Over 53 species of birds, most commonly fairy-bluebirds, barbets, and bulbuls were reported to be interacting with nearly 136 plant species. Majority of these interactions were not specific, except Singkrang (*Saurauia roxburghii*) and figs, which showed a tight link with flowerpeckers and thick-billed pigeons, respectively, for seed dispersal.

2.3 *Plants and Herbivores*

Plant–herbivore interactions are one of the most recognized plant–enemy interactions, which play a central role in regulating the functions of ecosystems (Schallhart et al. 2012). Herbivores are primary consumers that consume plant biomass (a process known as herbivory) and cause damage to plant populations by reducing plant survival and fitness. Effects of such negative interaction in limiting the size of plant population are higher than intra/interspecific competitions existing among plant species (Wirth et al. 2008). Most of the members of class Insecta are herbivores that are adapted to eat plants (Jurado-Rivera et al. 2009). These phytophagous insects illustrate one of the major channels of energy that pass to various trophic levels through the food chain (García-Robledo et al. 2013). Herbivores are classified into generalists and specialists based on what they consume. A generalist feeds on a wide variety of plants whereas the diet of a specialist includes selected plant species. The ungulate vertebrates are the generalist herbivores, while the phytophagous insects are mostly specialist herbivores.

Investigation of herbivores' diet habits is more challenging since their relationship with plants is temporally and spatially dynamic. Feeding choices of herbivores depend on diversity and identity of plants. Schallhart et al. (2012) made an attempt to study the impact of floral identity as well as diversity on food preference of subterranean larvae *Agriotes* by providing different combinations and diversity of food plants such as maize, legume, grass, and forb to them. The results indicated that plant diversity plays a major role in regulating dietary choices of herbivores as the larvae changed their feeding preferences in accordance with increased plant diversity. Though plants are exposed to herbivory, a few plants develop defense mechanisms against this negative interaction. The widespread ant-plant mutualistic association serves as a good example of such defense mechanism. Ants protect their host plants by predation of herbivores and their eggs and in return, plants provide housing and extrafloral nectar for the ants to ensure their perpetual protection against herbivory (Scharmann et al. 2013).

3 DNA Barcoding in Plant–Animal Interactions

Pollinators, frugivores, and herbivores manifest foraging preferences for multiple plant species. Therefore, understanding the food habits of these animals is critical with respect to trophic ecology as they reveal most of the difficult to observe trophic connections between plants and animals in the natural communities. DNA barcoding technique has opened up opportunities to identify plant species from various samples such as pollen, honey, digested plant parts, feces, etc., with utmost precision (Wilson et al. 2010; Nakashima et al. 2010; Gonzalez-Varo et al. 2014; Bruni et al. 2015; Nagarajan et al. 2016; Galimberti et al. 2016). The mitochondrial gene COI, which is extensively being used to differentiate animal species, evolves far too slowly in plants (Hebert et al. 2003; Cowan and Fay 2012; Fazekas et al. 2012; Li et al. 2014; Ali et al. 2015). Because of such lower evolutionary rates, mtDNA markers hold little potential in discriminating plant species (Li et al. 2014). As alternatives, various coding and noncoding regions of the plastid genome have been used as an effective target for barcoding plant species. However, none of these loci work across all the species due to introgressive hybridization and polyploidy events (Li et al. 2014; Ali et al. 2015).

The consortium for the barcode of life-plant working group (CBOL) has suggested a few plastid regions such as *rbcL, matK, trnH-psbA, trnL*, and the nuclear ribosomal internal transcribed spacer (ITS2) for the identification of plant species (Bruni et al. 2015). Though *matK* sequence shows a higher mutation rate, it is less feasible in analyzing complex genome since it requires different primer combinations to amplify various angiosperm families (Hilu and Liang 1997; Dunning and Savolainen 2010). The other noncoding regions such as ITS2 and *trnH-psbA* have been reported as good candidates for discriminating congeneric plant taxa (Hollingsworth et al. 2011). These plant DNA barcoding regions remain as the most successful tool for studying plant–pollinator, plant–frugivore, and plant–herbivore interactions (Boyer et al. 2016). With the aid of such markers, a growing body of research has been conducted to identify plant species using DNA barcoding technique from honey, pollen, and feces for unraveling many of the unnoticed trophic associations existing between plants and animals.

3.1 *Identifying the Provenance of Pollen and Honey*

Pollen and honey are important study subjects for ecologists as it is possible to deduce their botanical constituents. Identification of source constituents of honey and pollen collected from distinct pollinators are fundamental to understand the process of pollen transfer and pollinator-plant interactions. Although traditional methods such as microscopy, direct field observation, and tracking of pollen movements by dyes and traps have been implemented, they lack taxonomic precision in determining the provenance of pollen and honey (Waser and Price 1982; Greenwood 1986; Linhart et al. 1987; Caron and Leblanc 1992; Parra et al. 1993; Richardson et al. 2015). In recent years, the number of studies that have adopted DNA barcoding method (Fig. 2) to identify the floral source of pollinators from the honey and the pollen has increased (Wilson et al. 2010; Schnell et al. 2010; Jain et al. 2013; Bruni et al. 2015; Galimberti et al. 2014; Wong et al. 2015).

In most of the plant–pollinator interactions, pollinators manifest foraging preferences for a vast variety of flowers. However, despite the wide-ranging plant–pollinator interactions, relatively a few pollinators forge an obligatory relationship with specific plant species (Kearns et al. 1998). The specificity of pollinator is conspicuous in the case of *Hylaeus* bees that are endemic to Hawaiian Island. Wilson et al. (2010) delineated that *Hylaeus* bees pollinate only a few native plant species such as *Dubautia menziesii, Leptecophylla tameiameiae*, and *Argyroxiphium*

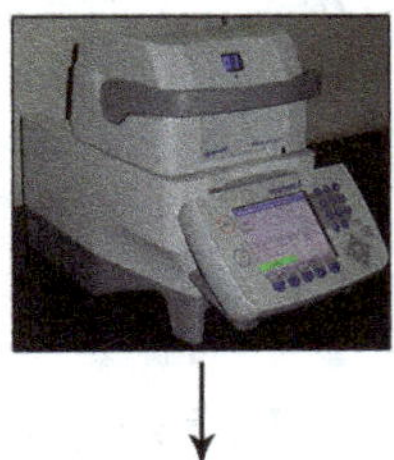

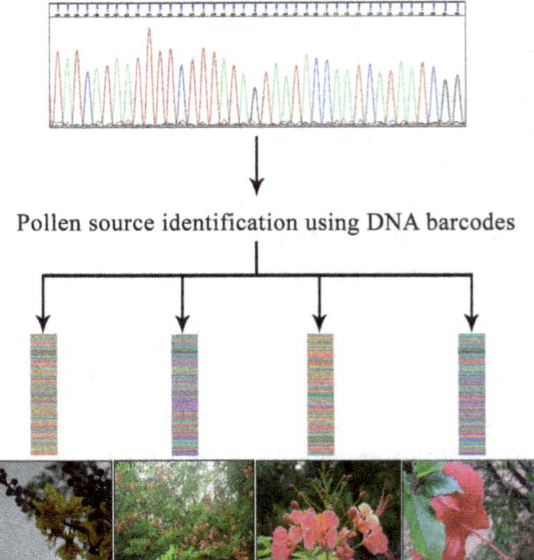

Fig. 2 Schematic view of DNA barcoding approach to study the plant–pollinator interactions

sandwicense in the vicinity of other co-occurring plants at a flowering stage that include both native and a few alien species. The floral choices of Hylaeus bees inhabiting in different geographical locations of the Island were identified by characterizing ITS and 5.8S *rDNA* regions of the pollen samples. All pollen grains collected from the bees belonged only to native plants and among the native pollen, Hylaeus bees carried significantly a large amount of pollen of an endangered species *Argyroxiphium sandwicense,* the commonly known Hawaiian silversword. It indicated that *Hylaeus* bees are important in mediating pollination of native plants and are the key pollinators of endangered silversword species. Recently, a similar study was conducted by Galimberti et al. (2014) using *rbcL* and *trnH-psbA* regions in order to identify the source plants of pollen collected from beehives, which resulted in the identification of 52 source plant species. Though both *rbcL* and *trnH-psbA* have been employed, *rbcL* region did not show much variation between closely related species in comparison with *trnH-psbA*.

Besides, characterizing pollen for unraveling the trophic connections between plants and pollinators, a few studies have identified potential pollinators using mitochondrial markers. For example, Wong et al. (2015) have characterized the pollinators associated with *Persicaria chinensis* using mitochondrial COI gene marker. *Persicaria chinensis,* the commonly known Chinese knotweed, are high-risk invasive species that are of great ecological and economic importance. Despite the economic and ecological value, it serves as a prime example of heterostyly, a structural modification in flowers that turn down self-pollination (Reddy et al. 1977). Such structural modifications of flowers along with strong scent and nectar imply a high chance of cross-pollination by insect vectors in these plants. However, previous studies have reported only a few insect vectors that facilitate pollination in this invasive species (Nishihiro and Washitani 1998; Thomas et al. 2009). Molecular characterization of the floral visitors of Chinese knotweed identified 4 orders and 23 distinct insect species as pollinators of this high-risk invasive species. In general, lack of pollinators does not act as a limiting factor for the propagation of invasive species (Richardson et al. 2000; Bartomeus et al. 2008). The detection of more than twenty distinct pollinator species for Chinese knotweed supported the previous studies. Regardless of the invasive nature of these species, the usefulness of incorporating this invasive species in the cultivation lands would promote pollination of important native plants.

The interactions among nectarivorous insects, specifically honeybees and plants, are easily discerned by characterizing the pollen contained in the honey. Although honeybees store pollen quite separately in their hives, a few infiltrate into the nectar and eventually get deposited in the honey. In many cases, the provenance of honey has been determined from the infiltrate using various genes such as *rbcL, trnH-psbA*, ITS, *adh1*, and *COI* (Schnell et al. 2010; Jain et al. 2013). For instance, Bruni et al. (2015) identified 39 plant species from honey samples.

3.2 Tracking the Seed Dispersal and Disperser

Because plants are sessile, they harness a wide range of animals for dispersing seeds. Therefore, their interaction with fruiting plants needs to be extensively studied to better understand their trophic relationships and seed dispersal events. At many instances, traditional methods face limitations and challenges in resolving fruit-frugivore interactions mainly due to thorough mastication and digestion of ingested plant parts such as seeds, pulp, or skin. DNA barcoding technique has alleviated these limitations and has rendered a new platform to precisely disclose plant–frugivore connections (Hayward 2013; Viana et al. 2015). In recent years, use of DNA barcodes to investigate the feeding ecology of various frugivores has become very popular as they ultimately help to understand plant–frugivore relationships.

The potential of DNA barcodes has been examined by Galimberti et al. (2016) in identifying diet constituents of a frugivorous bird, *Sylvia atricapilla,* in a protected region of northern Italy. By using the plastid DNA markers *rbcL* and *trnH-psbA,* the unscathed seeds and plant parts collected from the bird droppings were analyzed. Comparison of the barcode sequences with the reference DNA sequences of local plants resulted in the identification of 17 plant species in the diets of *Sylvia atricapilla*. Likewise, long distance dispersal (LDD) of seeds has also been identified with the support of plant barcodes. The first empirical report on such possibility of quantifying LDD was provided by Viana et al. (2015). The study analyzed the intact seeds collected from the gizzard of birds that were migrating from the mainland to remote oceanic islands, with the aid of *matK* and *rbcL* markers and seed morphology. Characterization of these barcodes resulted in the identification of 21 different plant species providing evidence that migratory birds are capable of transporting seeds over a long distance during their flight. Further, most of the seeds transported did not appear in the remote Islands, which suggest the significance of post-dispersal filters in impeding the establishment of plant species there. Similar to birds, volant species of mammals are also seen playing an important role in the dispersal of seeds (Muscarella and Fleming 2007; Silveira et al. 2011). An investigation of fruit feeding habits of more distantly related Jamaican bats using *rbcL* markers resulted in the identification of 11 plant taxa from 135 fecal samples (Hayward 2013). But, only 4 taxa were identified when the traditional method was implemented (Giannini and Kalko 2004; Lopez and Vaughan 2007; Teixeira et al. 2009).

On the other hand, a few studies have used mitochondrial DNA markers to reveal animal identities from the fecal or regurgitated samples (Nakashima et al. 2010; Gonzalez-Varo et al. 2014). Common palm civets are widely distributed small mammals of Southeast Asia. Their excessive dependence on fruits and ability to disperse larger seeds over a long distance mark them as important contributors of the ecological process (Davis 1962; Gruezo and Soligam 1990; Joshi et al. 1995; Su and Sale 2007). In general, large seeds appear to be a limiting factor for a number of dispersing agents (Corlett 1998). Civets are one among those few frugivores that disseminate large seeds over a long distance. Formerly, civet species were identified by examining the color, smell, size, shape, etc., of their fecal samples (Sinu et al.

2016). Since distinct civet species have similar body size and hair color, these identification strategies fail to accurately distinguish among several civet species. Nakashima et al. (2010) made an attempt to find out the identity of different civet species from the fecal samples using cytochrome *b* gene. The results indicated that most of the analyzed samples are of common palm civets. Also, seeds present in the feces were measured for determining their size and subsequently allowed to germinate to check its viability as well as to confirm the species. Seeds of fig plant, *Leea aculeata* and *Endospermum diadenum,* were most common in the civet feces and among all the identified plant species; seeds of *Aglaia grandis* were the largest which provided a direct example of the contribution of common palm civets in large seed dispersal. Likewise, Gonzalez-Varo et al. (2014) made an attempt to find out bird-dispersed seeds by characterizing 464 bp COI gene region. The study detected five frugivorous bird species from regurgitated/defecated seeds (belonging to four fleshy fruited plant species) and postulated that the identified bird species are potentially involved in dispersing seeds of a few fleshy fruited plants.

3.3 Identification of Herbivore Host Plants

The plant–herbivore network plays an essential role in regulating the functions of the ecosystem (Schallhart et al. 2012). Insects alone comprise a major group of herbivores and consume a large portion of plant biomass. When taken together, plants and the insects associated with them represent at least 50% of the total known species on earth (Futuyma and Agrawal 2009). Determining the plant–insect trophic link through traditional methods like microscopy, field observation, carbon isotope analysis, etc., is challenging as it is expensive, laborious, and thus provides little reliable information about herbivore trophic dynamics. An increasing number of studies have made use of DNA barcodes, over the past few years, to investigate insect-host plant associations.

Matheson et al. (2008) tested the possibility of identifying the species of plant materials from the insect's gut using 157 bp *rbcL* fragment. Members of eight distinct insect families were collected along with their host plants and they were allowed only to follow a monophagous feeding mode to mitigate taxonomic ambiguity that would arise when multiple plant species are obtained from the guts. The amplification and subsequent sequencing of *rbcL* region identified the plant meal of distinct insects. The study also postulated that plant DNA can be even recovered from insect gut 12 h after ingestion.

Among the diverse families of insects, the beetle family has been reported as one of the most abounding insect families, with much of its diversity having emerged from plant–insect coevolution (Farrell 1998; Oberprieler et al. 2007). Diet habits of beetles and weevils have been widely studied. Jurado-Rivera et al. (2009) studied the host plants-chrysomelid beetle (leaf beetle) interaction by analyzing whole beetle extract using chloroplast *trnL (UAA)* intron region which resulted in the identification of 13 plant families that are consumed by about 76% species belonging to

Chrysomelinae family. The study also reported the preference of beetles for Myrtaceae and Fabaceae families. Likewise, molecular characterization of *trnL* region confirmed 26 plant families, from the guts of weevils of tropical forest feed on (Pinzón-Navarro et al. 2010). However, as an improvement over these studies, Kitson et al. (2013) tested the efficiency of *trnL (UAA)* barcode (by amplifying the region with a different set of primers) in detecting plant species consumed by two closely related species of weevil, *C. murinus* and *C. ovalis.* In general, tropical herbivores (order: Coleoptera) feed on distantly related plant species, which is not concordant with phylogenetic conservatism (Novotny et al. 2002; Ødegaard et al. 2005). The results also revealed that weevil species follow a broad palatable diet, which includes a wide variety of plant species. Quite recently, feeding range of rolled-leaf beetles has been recorded by García-Robledo et al. (2013) with the aid of three loci (*rbcL, trnH-psbA* and ITS2). Among the three loci, good quality sequences were generated only for *rbcL* and ITS2. The *rbcL* locus determined family of a few host plants and in some other cases, it identified the plants up to genus level whereas ITS2 region was successful in the identification host plants up to species level. A couple of years later, Kajtoch (2014) surveyed the ecological interaction between phytophagous beetle and their host plants through amplification of *trnL* and *rbcL* markers, followed by Sanger sequencing and Illumina sequencing for monophagous and polyphagous groups, respectively. The study revealed 224 host plant–beetle interactions, which provided inferences on the feeding habits of a few insect genera. Of these 224 interactions, a few beetles belonging to 7 distinct genera were found feeding on similar plant species. However, a much broad range of feeding interest was observed for the other 4 genera.

Molecular analysis of gut contents of soil-dwelling insects that feed on below-ground parts of the plant has also been researched with the aid of plant-based markers such as *rbcL, trnL,* etc. (Staudacher et al. 2011; Wallinger et al. 2013). Feeding ecology of such subterranean insects would be useful in identifying the insect-plant association in the soil and major soil pests. Herbivory is not limited only with the feeding of plant parts, but using the plant parts for other purposes, such as making a dwell, lining a brood cell, or for cultivating fungal bodies, which become the food of insects (MacIvor 2016; Kambli et al. 2017). Plant–leafcutter bee interaction is an exceptional case of this kind of herbivory. The conventional method to identify the preferred plants of the bees is the direct observation of bees cutting leaves from plants (Kambli et al. 2017). While this might be a tedious part of the fieldwork, the DNA barcoding method can provide an accurate list of preferred plant species through sequencing the DNA extracted from the leaf discs, which are collected from the bee nests (MacIvor 2016).

4 Reconstructing Plant–Animal Link through NGS

DNA barcoding, while being easy, less time consuming, and able to distinguish organisms at species level, has low efficiency when applied to samples containing a mixture of plant species (Wilson et al. 2010; Hawkins et al. 2015). It hugely relies on databases to identify the plant species, by comparing the DNA sequences obtained in the studies with the reference sequences of databases. This also limits the efficiency of the identification process because not all the plant barcode sequences have been deposited in the genetic databases. In addition to these limitations, traditional Sanger sequencing method, which is extensively being used to sequence the DNA barcodes, faces other serious pitfalls such as the requirement of an additional cloning step for multiple amplicon sequencing and failure to generate sequences for all samples at once (Pompanon et al. 2012; Kajtoch 2014). Recent metabarcoding technology has succeeded in dealing with the problems faced by the DNA barcoding method (Harismendy et al. 2009; Shokralla et al. 2012). This expeditious method integrates both the DNA-based identification and next-generation sequencing and allows the users to analyze complex matrices (i.e., honey) containing DNA of different plant species in a single run (McClenaghan et al. 2015).

Diet assessment of several herbivorous animals such as birds, insects, molluscs, ruminants, and other mammals has been carried out using metabarcoding (Soininen et al. 2009; Valentini et al. 2009; Pegard et al. 2009; Kowalczyk et al. 2011; Raye et al. 2011; Pompanon et al. 2012; Kajtoch 2014; Srivathsan et al. 2014; McClenaghan et al. 2015; Keller and Steffan-Dewenter 2014). Kajtoch (2014) provided the first and more accurate report on the feeding habit of polyphagous weevil, *Centricnemus leucogrammus,* by comparing the efficiency of both High throughput sequencing and Sanger sequencing for *rbcL* and *trnL* regions. As expected, next-generation sequencing technique was found extremely suitable in analyzing dietary constituents of the beetle species and resulted in identifying 30 plant genera, whereas Sanger sequencing procedure was successful in delineating only 20 plant genera. Out of the two plant DNA regions used, *rbcL* worked better in NGS. Sequencing success of *trnL* intron was also found slightly higher in NGS than the other. By contrast, in the case of Sanger sequencing method, *trnL* region identified more plant species. Similarly, McClenaghan et al. (2015) studied the dietary constituents of four grasshopper species inhabiting in the same area using *rbcL* marker through Illumina sequencing technology, which provided an explicit idea about the feeding behavior and resource partitioning among these sympatric grasshopper species. Furthermore, the dietary components of two leaf-feeding monkeys from fecal samples using P6 loop of *trnL* region has also been investigated (Srivathsan et al. 2014).

In recent studies, plant–pollinator interactions are also being discovered through metabarcoding of pollen grains and honey (Keller and Steffan-Dewenter 2014). For example, Sickel et al. (2015) identified the source constituents of pooled pollen samples collected from solitary bees using ITS barcodes and Illumina MiSeq, which identified most of the taxa up to species level. Similarly, a parallel study conducted by Richardson et al. (2015), using ITS2, discovered 19 source plant families of pollen grains collected from bee colonies, whereas only 8 families were detected by microscopic melissopalynology (Erdtman 1943; Kearns and Inouye 1993). Hawkins et al. (2015) used *rbcL* barcoding in combination with 454-pyrosequencing to discern the botanical origin of honey sample and identified a broad range of plant taxa that included 46 plant families and 25 orders. Pornon et al. (2016) using *trnL* and ITS1 metabarcoding technique identified plant taxa from pollen mixtures, which were 2.5-fold higher than noted through direct field observation. However, such studies focus only on the foraging preferences of pollinators and on the conservation of endangered plant and pollinator species. A recent study of Gous et al. (2015) has provided insights into the foraging behavior of primitive insect pollinators and the change in their floral preferences over time using ITS1, ITS2, and *rbcL* markers. DNA barcoding followed by Illumina next-generation sequencing detected the source plants of pollen collected from ancient bee specimens. Furthermore, all the three plant barcodes successfully generated sequences even for those specimens dating from 1910 and enabled the plant identification up to the genus level.

5 Conclusion

The history of plant–animal interaction dates back to the cretaceous period when the angiosperm (flowering) plants evolved first. This interaction is considered as one of the major reasons for the overwhelming diversity of angiosperm plants and their associated animals seen today. In an ecological viewpoint, these interactions represent a major conduit of energy that passes through the food cycle to higher trophic levels. Both mutualistic and antagonistic interactions hold equal importance in maintaining biological diversity stable. Most of the plant–animal interactions are obligatory in nature, and knowledge about the interacting species is a prerequisite to update the current state of the interaction, and if required, to develop strategies to revive the weakening-specific interactions. Nevertheless, species-level interactions are much complicated to document in the wild. Since most of the interactions are species specific, ecologists might find DNA barcoding technique as a quick and accurate method to unravel the mysterious interactions existing between plants and animals. In addition to this, it offers a noninvasive procedure to quantify feeding habits of animals from fecal matters, which prevents the euthanization of animals. The growing literature demonstrates the great potential of DNA barcoding technique in redefining the interactions of plants with pollinators, frugivores, and herbivores.

Acknowledgments The authors are grateful to the Science and Engineering Research Board (SERB), Department of Science and Technology, Government of India, New Delhi, for the financial assistance (EMR/2015/000937). Vandana R. Prabhu is grateful to DST-INSPIRE (IF160266) for the financial assistance in the form of research fellowship.

References

Ali MA, Pandey AK, Gyulai G, Park SH, Lee C, Kim SY, Lee J (2015) Introduction to plant DNA barcoding. Plant DNA Barcoding and Phylogenetics:1–14

Bartomeus I, Vilà M, Santamaria L (2008) Contrasting effects of invasive plants in plant-pollinator networks. Oecologia 155:761–770

Boyer S, Benjamin RW, Steve DW (2016) How molecular tools can help understanding species interactions to improve restoration outcomes. PeerJ Preprints. https://doi.org/10.7287/peerj.preprints.2020v1

Bruni I, Galimberti A, Caridi L, Scaccabarozzi D, De Mattia F, Casiraghi M, Labra M (2015) A DNA barcoding approach to identify plant species in multiflower honey. Food Chem 170:308–315

Caron GE, Leblanc R (1992) Pollen contamination in a small black spruce seedling seed orchard for three consecutive years. For Ecol Manag 53(1–4):245–261

Chama L, Berens DG, Downs CT, Farwig N (2013) Habitat characteristics of forest fragments determine specialization of plant-frugivore networks in a mosaic forest landscape. PLoS One 8 (1):e54956

Clare EL (2014) Molecular detection of trophic interactions: emerging trends, distinct advantages, significant considerations and conservation applications. Evol Appl 7(9):1144–1157

Corlett RT (1998) Frugivory and seed dispersal by vertebrates in the oriental (Indomalayan) region. Biol Rev Camb Philos Soc 73(4):413–448

Cowan RS, Fay MF (2012) Challenges in the dna barcoding of plant material. Methods Mol Biol 862:23–33

Darwin CR (1859) On the origin of species by means of natural selection, or the preservation of favoured races in the struggle for life. John Murray, London, UK

Darwin CR (1862) On the various contrivances by which british and foreign orchids are fertilised by insects. John Murray, London, UK

Davis DD (1962) Mammals of the lowland rain-forest of north borneo. Bulletin of the Singapore National Museum 31:3–128

Dunning LT, Savolainen V (2010) Broad-scale amplification of matK for DNA barcoding plants, a technical note. Bot J Linn Soc 164:1–9

Ehrlich PR, Raven PH (1964) Butterflies and plants: a study in coevolution. Evolution 18 (4):586–608

Encinas-Viso F, Revilla TA, Velzen EV, Etienne RS (2014) Frugivores and cheap fruits make fruiting fruitful. J Evol Biol 27(2):313–324

Erdtman G (1943) An introduction to pollen analysis. Chronica Botanica Company, Waltham, Massachusetts, USA

Farrell BD (1998) 'Inordinate fondness' explained: why are there so many beetles? Science 281 (5376):555–559

Fazekas AJ, Kuzmina ML, Newmaster SG, Hollingsworth PM (2012) DNA barcoding methods for land plants. Methods Mol Biol 858:223–252

Futuyma DJ, Agrawal AA (2009) Macroevolution and the biological diversity of plants and herbivores. Proc Natl Acad Sci 106(43):18054–18061

Galimberti A, De Mattia F, Bruni I, Scaccabarozzi D, Sandionigi A, Barbuto M, Casiraghi M, Labra M (2014) A DNA barcoding approach to characterize pollen collected by honeybees. PLoS One 9(10):e109363

Galimberti A, Spinelli S, Bruno A, Mezzasalma V, De Mattia F, Cortis P, Labra M (2016) Evaluating the efficacy of restoration plantings through DNA barcoding of frugivorous bird diets. Conserv Biol 30(4):763–773

García-Robledo C, Erickson DL, Staines CL, Erwin TL, Kress WJ (2013) Tropical plant–herbivore networks: reconstructing species interactions using DNA barcodes. PLoS One 8(1):e52967

Giannini N, Kalko E (2004) Trophic structure in a large assemblage of phyllostomid bats in Panama. Oikos 105:209–220

Gonzalez-Varo JP, Arroyo JM, Jordano P (2014) Who dispersed the seeds? The use of DNA barcoding in frugivory and seed dispersal studies. Methods Ecol Evol 5:806–814

Gous A, de Bruin JJ, Willows-Munro S, Eardley C, and Swanevelder D (2015) The application of next-generation sequencing barcoding in identifying mixed-pollen samples from a historic bee collection. 6th International Barcode of Life Conference, Guelph, Canada. Genome 58: 222

Greenwood MS (1986) Gene exchange in loblolly pine: the relation between pollination mechanism, female receptivity and pollen availability. Am J Bot 73:1443–1451

Gruezo W, Soligam AC (1990) Identity and germination of seeds from feces of the Philippine palm civet (Paradoxurus philippinensis Jourdan). Nat Hist Bull Siam Soc 38:69–82

Harismendy O, Ng PC, Strausberg RL, Wang X, Stockwell TB, Beeson KY, Schork NJ, Murray SS, Topol EJ, Levy S, Frazer KA (2009) Evaluation of next generation sequencing platforms for population targeted sequencing studies. Genome Biol 10(3):R32

Hawkins J, de Vere N, Griffith A, Ford CR, Allainguillaume J, Hegarty MJ, Baillie L, Adams-Groom B (2015) Using DNA metabarcoding to identify the floral composition of honey: a new tool for investigating honey bee foraging preferences. PLoS One 10(8):e0134735

Hayward CE (2013) DNA barcoding expands dietary identification and reveals dietary similarity in Jamaican frugivorous bats. University of Western Ontario–Electronic Thesis and Dissertation Repository. Paper 1697

Hebert PD, Cywinska A, Ball SL, deWaard JR (2003) Biological identifications through DNA barcodes. Proc Biol Sci 270(1512):313–321

Hegland SJ, Nielsen A, Lazaro A, Bjerknes AL, Totland Ø (2009) How does climate warming affect plant-pollinator interactions? Ecol Lett 12(2):184–195

Hilu KW, Liang H (1997) The matK gene: sequence variation and application in plant systematics. Am J Bot 84:830–839

Hollingsworth PM, Graham SW, Little DP (2011) Choosing and using a plant dna barcode. PLoS One 6(5):e19254

Jain SA, De Jesus FT, Marchioro GM, De AraúJo ED (2013) Extraction of DNA from honey and its amplification by PCR for botanical identification. Food Sci Technol, Campinas 33(4):753–756

Joshi AR, Smith JLD, Cuthbert FJ (1995) Influence of food distribution and predation pressure on spacing behavior in palm civets. J Mammal 76:1205–1212

Jurado-Rivera JA, Vogler AP, Reid CA, Petitpierre E, Gómez-Zurita J (2009) DNA barcoding insect-host plant associations. Proc Biol Sci 276(1657):639–648

Kajtoch Ł (2014) A DNA metabarcoding study of polyphagous beetle dietary diversity: the utility of barcodes and sequencing techniques. Folia Biol (Krakow) 62(3):223–234

Kambli S, Aiswarya S, Manoj K, Varma S, Asha G, Rajesh TP, Sinu PA (2017) Leaf foraging sources of leafcutter bees in a tropical environment: implications for conservation. Apidologie. https://doi.org/10.1007/s13592-016-0490-2

Kearns CA, Inouye DW (1993) Techniques for pollination biologists. University Press of Colorado, Boulder, Colorado, USA

Kearns CA, Inouye DW, Waser NM (1998) Endangered mutualisms: the conservation of plant-pollinator interactions. Annu Rev Ecol Evol Syst 29(1):83–112

Keller A, Steffan-Dewenter I (2014) Pollen DNA barcoding using next generation sequencing. https://doi.org/10.1111/plb.12251

Kitson JJN, Warren BH, Florens FBV, Baider C, Strasberg D, Emerson BC (2013) Molecular characterization of trophic ecology within an island radiation of insect herbivores (Curculionidae: Entiminae: Cratopus). Mol Ecol 22(21):5441–5455

Kowalczyk R, Taberlet P, Coissac E, Valentini A, Miquel C, Kamiński T, Wójcik JM (2011) Influence of management practices on large herbivore diet—case of European bison in Białowieza primeval Forest (Poland). For Ecol Manag 261:821–828

Li X, Yang Y, Henry RJ, Rossetto M, Wang Y, Chen S (2014) Plant DNA barcoding: from gene to genome. Biol Rev Camb Philos Soc 90(1):157–166

Linhart YB, Busby WH, Beach JH, Feinsinger P (1987) Forager behavior, pollen dispersal, and inbreeding in two species of hummingbird-pollinated plants. Evolution 41:679–682

Lopez JE, Vaughan C (2007) Food niche overlap among neotropical frugivorous bats in Costa Rica. Rev Biol Trop 55(1):301–313

MacIvor JS (2016) DNA barcoding to identify leaf preference of leafcutting bees. R Soc Open Sci 3 (3):150623

Matheson CD, Muller GC, Junnila A, Vernon K, Hausmann A, Miller MA, Greenblatt C, Schlein Y (2008) A PCR method for detection of plant meals from the guts of insects. Org Divers Evol 7:294–303

McClenaghan B, Gibson JF, Shokralla S, Hajibabaei M (2015) Discrimination of grasshopper (Orthoptera: Acrididae) diet and niche overlap using next-generation sequencing of gut contents. Ecol Evol 5(15):3046–3055

Mchatton R (2011) Orchid pollination: exploring a fascinating world. Amer Orchid Soc:340–349

Muscarella R, Fleming TH (2007) The role of frugivorous bats in tropical forest succession. Biol Rev Camb Philos Soc 82(4):573–590

Nagarajan M, Nimisha K, Nimisha AK, Sinu PA (2016) The promise of genomics in plant-animal interaction studies. In: Sinu PA, Shivanna KR (eds) Mutualistic interaction between flowering plants and animals. Manipal University press, Manipal, India, pp 293–303

Nakashima Y, Inoue E, Inoue-Murayama M, Sukor J (2010) High potential of a disturbance-tolerant frugivore, the common palm civet paradoxurus hermaphroditus (viverridae), as a seed disperser for large-seeded plants. Mammal Study 35(3):209–215

Nishihiro J, Washitani I (1998) Patterns and consequences of self-pollen deposition on stigmas in heterostylous Persicaria japonica (Polygonaceae). Am J Bot 85(3):352

Novotny V, Basset Y, Miller SE, Weiblen GD, Bremer B, Cizek L, Drozd P (2002) Low host specificity of herbivorous insects in a tropical forest. Nature 416(6883):841–844

Oberprieler RG, Marvaldi AE, Anderson RS (2007) Weevils, weevils, weevils everywhere. Zootaxa 1668:491–520

Ødegaard F, Diserud OH, Østbye K (2005) The importance of plant relatedness for host utilization among phytophagous insects. Ecol Lett 8:612–617

Ollerton J, Winfree R, Tarrant S (2011) How many flowering plants are pollinated by animals? Oikos 120(3):321–326

Parra V, Vargas CF, Eguiarte LE (1993) Reproductive biology, pollen and seed dispersal, and neighborhood size in the hummingbird-pollinated Echeveria gibbiflora (Crassulaceae). Am J Bot 80:153–159

Pegard A, Miquel C, Valentini A, Coissac E, Bouvier F, François D, Taberlet P, Engel E, Pompanon F (2009) Universal DNA-based methods for assessing the diet of grazing livestock and wildlife from feces. J Agric Food Chem 57(13):5700–5706

Pinzón-Navarro S, Jurado-Rivera JA, Gomez-Zurita J, Lyal CHC, Vogler AP (2010) DNA profiling of host–herbivore interactions in tropical forests. Syst Entomol 35(1):18–32

Pompanon F, Deagle BE, Symondson WO, Brown DS, Jarman SN, Taberlet P (2012) Who is eating what: diet assessment using next generation sequencing. Mol Ecol 21(8):1931–1950

Pornon A, Escaravage N, Burrus M, Holota H, Khimoun A, Mariette J, Pellizzari C, Iribar A, Etienne R, Taberlet P, Vidal M, Winterton P, Zinger L, Andalo C (2016) Using metabarcoding to reveal and quantify plant-pollinator interactions. Sci Rep 6:27282

Raye G, Miquel C, Coissac E, Redjadj C, Loison A, Taberlet P (2011) New insights on diet variability revealed by DNA barcoding and high-throughput pyrosequencing: chamois diet in autumn as a case study. Ecol Res 26:265–276

Reddy NP, Bahadur B, Kumar PV (1977) Heterostyly in polygonum chinense L. J Genet 63 (2):79–82

Richardson DM, Allsopp N, D'Antonio CM, Milton SJ, Rejmanek M (2000) Plant invasions – the role of mutualisms. Biol Rev 75:65–93

Richardson RT, Lin CH, Sponsler DB, Quijia JO, Goodell K, Johnson RM (2015) Application of ITS2 metabarcoding to determine the provenance of pollen collected by honey bees in an agroecosystem. Appl Plant Sci 3(1):1400066

Robledo-Arnuncio JJ, García C (2007) Estimation of the seed dispersal kernel from exact identification of source plants. Mol Ecol 16(23):5098–5109

Robson DB (2014) Identification of plant species for crop pollinator habitat enhancement in the northern prairies. J Pollinat Ecol 14(21):218–234

Sankamethawee W, Pierce AJ, Gale GA, Hardesty BD (2011) Plant-frugivore interactions in an intact tropical forest in north-East Thailand. Integr Zool 6(3):195–212

Sargent RD, Ackerly DD (2007) Plant–pollinator interactions and the assembly of plant communities. Trends Ecol Evol 23(3):123–130

Schallhart N, Tusch MJ, Wallinger C, Staudacher K, Traugott M (2012) Effects of plant identity and diversity on the dietary choice of a soil-living insect herbivore. Ecology 93(12):2650–2657

Scharmann M, Thornham DG, Grafe TU, Federle W (2013) A novel type of nutritional ant–plant interaction: ant partners of carnivorous pitcher plants prevent nutrient export by dipteran pitcher infauna. PLoS One 8(5):e63556

Schiest FP, Johnson SD (2013) Pollinator-mediated evolution of floral signals. Trends Ecol Evol 28 (5):307–315

Schnell IB, Fraser M, Willerslev E, Thomas M, Gilbert P (2010) Characterisation of insect and plant origins using DNA extracted from small volumes of bee honey. Arthropod Plant Interact 4 (2):107–116

Shokralla S, Spall J, Gibson JF, Hajibabaei M (2012) Next-generation DNA sequencing technologies for environmental DNA research. Mol Ecol 21(8):1794–1805

Sickel W, Ankenbrand MJ, Grimmer G, Holzschuh A, Härtel S, Lanzen J, Steffan-Dewenter I, Keller A (2015) Increased efficiency in identifying mixed pollen samples by meta-barcoding with a dual-indexing approach. BMC Ecol 15:20

Silveira M, Trevelin L, Port-Carvalho M, Godoi S, Mandetta EN, Cruz-Neto AP (2011) Frugivory by phyllostomid bats (Mammalia: Chiroptera) in a restored area in Southeast Brazil. Acta Oecol 37:31–36

Sinu PA, Shibil VK, Manoj K, Rajesh TB (2016) Recent advances on biotic seed dispersal by vertebrates. In: Sinu PA, Shivanna KR (eds) Mutualistic interaction between flowering plants and animals. Manipal University press, Manipal, India, pp 221–237

Soininen EM, Valentini A, Coissac E, Miquel C, Gielly L, Brochmann C, Brysting AK, Sønstebø JH, Ims RA, Yoccoz NG, Taberlet P (2009) Analysing diet of small herbivores: the efficiency of DNA barcoding coupled with high-throughput pyrosequencing for deciphering the composition of complex plant mixtures. Front Zool 6(1):16

Srivathsan A, Sha JC, Vogler AP, Meier R (2014) Comparing the effectiveness of metagenomics and metabarcoding for diet analysis of a leaf-feeding monkey (Pygathrix nemaeus). Mol Ecol Resour 15(2):250–261

Staudacher K, Wallinger C, Schallhart N, Traugott M (2011) Detecting ingested plant DNA in soil-living insect larvae. Soil Biol Biochem 43(2):346–350

Su S, Sale J (2007) Niche differentiation between common palm civet Paradoxurus hermaphroditus and small Indian civet viverricula indica in regeneration degraded forest, Myanmar. Small Carniv Conserv 36:30–34

Teixeira RC, Corrêa CE, Fischer E (2009) Frugivory by artibeus jamaicensis (Phyllostomidae) bats in the Pantanal, Brazil. Stud Neotrop Fauna Environ 44:7–15

Thomas SG, Rehal SM, Varghese A, Davidar P, Potts SG (2009) Social bees and food plant associations in the Nilgiri biosphere reserve, India. Trop Ecol 50:79–88

Valentini A, Miquel C, Nawaz MA, Bellemain E, Coissac E, Pompanon F, Gielly L, Cruaud C, Nascetti G, Wincker P, Swenson JE, Taberlet P (2009) New perspectives in diet analysis based on DNA barcoding and parallel pyrosequencing: the trnL approach. Mol Ecol Resour 9 (1):51–60

Viana DS, Gangoso L, Bouten W, Figuerola J (2015) Overseas seed dispersal by migratory birds. Proc Biol Sci 283(1822):20152406

Wallinger C, Staudacher K, Schallhart N, Peter E, Dresch P, Juen A, Traugott M (2013) The effect of plant identity and the level of plant decay on molecular gut content analysis in a herbivorous soil insect. Mol Ecol Resour 13(1):75–83

Waser NM, Price MV (1982) A comparison of pollen and fluorescent dye carry-over by natural pollinators of Ipomopsis aggregata (Polemoniaceae). Ecology 63:1168–1172

Wilson EE, Sidhu CS, Levan KE, Holway DA (2010) Pollen foraging behaviour of solitary Hawaiian bees revealed through molecular pollen analysis. Mol Ecol 19(21):4823–4829

Wirth R, Meyer ST, Leal IR, Tabarelli M, Lüttge U, Beyschlag W, Murata J (2008) Plant herbivore interactions at the Forest edge. Progress in Botany 69:423–448

Wong MM, Lim CL, Wilson JJ (2015) DNA barcoding implicates 23 species and four orders as potential pollinators of Chinese knotweed (Persicaria chinensis) in peninsular Malaysia. Bull Entomol Res 105(4):515–520

A Molecular Assessment of Red Algae with Reference to the Utility of DNA Barcoding

Zahid Hameed Siddiqui, Zahid Khorshid Abbas, Khalid Rehman Hakeem, Mather Ali Khan, and Md Abdul Ilah

Abstract The ecological and commercial importance of red algae is of high value, the estimated cost seaweed industry produces is US$10 billion. The species which are exploited most are the members of Rhodophyta (*Eucheuma/Kappaphycus*, *Porphyra*, and *Gracilaria*). In order to understand the distribution of seaweed, their identification is necessary which is generally based on morphological characteristics, often resulting in wrong identification of species. DNA barcoding can be used as a contemporary tool for species identification. It can resolve many intrinsic problems of morphological taxonomy, only a small amount of tissue is required for species identification, and the samples can be examined at all stages of development. The application of DNA barcoding can be used in the identification of invasive and endangered species along with conservation biology. In the case of red algae, DNA barcoding proved to be beneficial for the recognition of high-yielding agar strain as well as for cryptic species identification. In this study, several identification-based problems of red algae have been discussed by using different intraspecific markers such as cox1, cox3, and cox2–3 spacer and rbcL and rbcL-rbcS spacer. As per the available data, the mitochondrial marker gene cox1 is more effective than rbcL for the measurement of red algal genetic diversity.

Keywords DNA barcoding · Red algae · cox1 · cox2–3 · rbcL

Z. H. Siddiqui (✉) · Z. K. Abbas
Department of Biology, Faculty of Science, University of Tabuk, Tabuk, Kingdom of Saudi Arabia

K. R. Hakeem
Department of Biological Sciences, Faculty of Science, King Abdulaziz University, Jeddah, Kingdom of Saudi Arabia

M. A. Khan
Bond Life Sciences Center, University of Missouri-Columbia, Columbia, MO, USA

M. A. Ilah
Yaduvanshi Degree College, Mahendragarh, Haryana, India

S. Trivedi et al. (eds.), *DNA Barcoding and Molecular Phylogeny*,
https://doi.org/10.1007/978-3-030-50075-7_7

Abbreviations

AFLP	Amplified Fragment Length Polymorphism
BOLD	Barcode of Life Database
CBOL	Consortium for the Barcode of Life
cox1	Mitochondrial Cytochrome C Oxidase subunit1
cox2	Mitochondrial Cytochrome Oxidase subunit 2
cox3	Mitochondrial Cytochrome Oxidase subunit 3
DNA	Deoxyribonucleic acid
ISSR	Inter-Simple Sequence Repeats
ITS-1 and ITS-2	The internal transcribed spacers
RAPD	Random Amplified Polymorphic DNA
rbcL	large subunit of the ribulose-1,5- bisphosphate carboxylase/ oxygenase (RUBISCO)
rbcL-rbcS	The plastid-encoded RuBisCO spacer
rDNA (SSU rDNA)	The nuclear small subunit
RFLP	Restriction Fragment Length Polymorphism

1 Introduction

In this era of globalization, human activities are causing climate change, pollution, coastal degradation, and introduction of alien and invasive species. During these changes, seaweed populations are also affected severely irrespective of nation and continents. Among the seaweed, the members of Rhodophyta are very important, ecologically as well as commercially. Therefore, in order to understand the distribution of seaweed, identification of organisms is necessary which is generally based on morphological characteristics such as size, shape, and color of the organism's parts. This is the traditional way of identification which is little complex because of its three major limitations (Hebert et al. 2003; Pires and Marinoni 2010). First, the key on the basis of morphological characteristics is not complete or not available for all taxa due to lack of a proper taxonomic description (May 1988) as some groups (flowering plants and vertebrates) are better studied than others (algae, nematodes). Second, the morphological characteristics and their complexity depend on the group under consideration; therefore, sometimes the identification keys need well-trained taxonomists, who are becoming very rare and are not always available for regular identifications: this is referred to as taxonomic impediment. Thirdly the organism to be identified may be too small or at a developmental stage, where even trained taxonomists face difficulty in characterization and identification. In addition to these, the morphology-based methods are time consuming and do not provide classification up to the species level (Rindi et al. 2008; Packer et al. 2009). With the advent of molecular biology, it has been revealed that a large number of biological species show genetic divergence without accompanying morphological disparities and therefore cannot be identified by traditional methods. The recognition of these cryptic species is a major challenge to modern taxonomy (Heinrichs et al. 2011). During the last two decades, small segments of DNA, called DNA barcodes, have been developed for differentiation of species (Yow et al. 2013). In short, DNA barcoding is based on small, single marker sequence similarity comparison to classify species (Hebert et al. 2003). The authors further debated that by incorporating DNA barcoding, the unknown biodiversity can be classified quickly and reliably than traditional taxonomic methods alone (Hebert et al. 2004). The identification of species using DNA barcoding has been effectively documented in animals such as spiders (Barrett and Hebert 2005), fish (Ward et al. 2005), birds (Hebert et al. 2004), and rodents (Robins et al. 2007) as well as in plants (Chase et al. 2005; Fazekas et al. 2012), fungi (Seena et al. 2010), and algae (Saunders 2008).

In animals, with about 600 base pairs, the 5′end portion of the mitochondrial gene cytochrome *c* oxidase 1 (cox1) was selected as the main molecular marker (Hebert et al. 2003). Later on, the same gene was adapted in Rhodophyceae (Saunders 2005) and Phaeophyceae (McDevit and Saunders 2010) for molecular classification. Using cox1, a large number of articles have been published for red as well as brown algae, which indicates the usefulness of this marker for species-level identification, unraveling cryptic diversity and variations within species (Brodie et al. 2008). However, the use of universal plastid amplicon (domain V of the plastid large subunit rDNA) was proposed by Sherwood and Presting (2007) as a DNA barcode in photosynthetic organisms for rapid species identification. In this chapter, we will discuss the ecological and economic importance of red algae and DNA barcoding and its use in red algae as a case study.

2 Ecological and Economic Importance of Red Algae

The Rhodophyta is a separate taxon characterized by the accessory photosynthetic pigments phycocyanin, phycoerythrin, and allophycocyanins arranged in phycobilisomes (Woelkerling 1990). It is commonly known as red algae and encompasses a large number of species with different types of body shape, ranging from unicellular-filamentous (simple) to the blade-pseudoparenchymatous (most complex) (Cole and Sheath 1995). As per Guiry (2012), in AlgaeBase there are around 6131 species of red algae, the total number of estimated species which are described is 7000, and the species which are undescribed are 7000. This information tells us the need for identification and classification of this important taxon (Rhodophyta), which is very important from environmental as well as from an economic perspective. There are several genera of red algae which are found in freshwaters, but the majority of red algae are of marine nature; they are generally reported on tropical and temperate marine shores.

At these shores, they play the role of "keystone species" by building and maintaining coral reefs (Freshwater 2000). This gives banded coral, giant clams, clownfish, shrimps, and other animals a reliable habitat where they can dwell and maintain ecological balance. Further, these algae form flat sheets that fuse and stabilize reef crest and protect reefs from wave damage. There are fossil evidence which indicates that these coralline red algae play this important role from hundreds of millions of years (Freshwater 2000; Graham et al. 2009).

Recently, Rebours et al. (2014) reported the estimated value of the seaweed industry, producing US$10 billion. The species which are exploited most are the members of Rhodophyta (*Eucheuma/Kappaphycus*, *Porphyra*, and *Gracilaria*). Among these species, *Kappaphycus* alone produces US$1.3 billion, while the nori market creates US$1.5 billion. *Porphyra yezoensis*, *Gracilaria*, and some other species are grown in mariculture for use as human food; in fact, most of the algae contain amino acids; proteins; carbohydrates; vitamins B1, B2, B12, and C; and carotenoid that are essential for the normal functioning of the human body. They also contain minerals such as potassium, magnesium, iron, selenium, and a large amount of iodine, and their fat content is very low (around 0.2–2%), which makes them a good source of balanced nutrition (Lee 2008). *Kappaphycus* along with other genera are cultivated for the extraction of gel-forming agar, agarose, and carrageenan. These gelling polysaccharides are commonly used in laboratory cell culture media preparation (Yeong et al. 2008), genomics-/proteomics-based research (Siow et al. 2012), and food processing (Van de Velde et al. 2002).

Red algae are also considered as a source of compounds that can be used against microbial or herbivore attack as well as a potential source for human medicine (Bixler 1996; Villanueva et al. 2008; Graham et al. 2009; Yow et al. 2013). Table 1 lists the medicinal role and industrial uses of some of the important genera of red algae.

3 DNA Barcoding

The existence of huge biodiversity in nature is because of the genetic differences among organisms which cannot be recognized by using traditional morphological-based studies; the identification of such hidden species is a huge challenge to modern taxonomy (Heinrichs et al. 2011). DNA barcoding can resolve many intrinsic problems of morphological taxonomy; we are well aware that the numbers of taxonomists are decreasing day by day, and on the other hand, numbers of species are increasing; therefore, in order to manage this situation, molecular biology helps to solve the taxonomical problems. In molecular biology, only a small amount of tissue is required for species identification, and the samples can be examined at all stages of development (Floyd et al. 2002; Savolainen et al. 2005). Nowadays, barcoding is routinely used for multicellular organisms, such as butterflies (Burns et al. 2008), birds (Hebert et al. 2004), and aquatic hyphomycetes (Seena et al. 2010). Arnot et al. (1993) first used "DNA barcode" while working on *Plasmodium falciparum*, but the use of molecular biology for organism identification was older (McAndrew and Majumdar 1983; Anderson et al. 1985). Allozymes were the first molecules used for species differentiation by Hubby and Lewontin (1966). However,

Table 1 Medicinal Role and Industrial Uses of Some of the Genera of Rhodophyta

S. No.	Algae	Common Name	Product	Uses	References
1	*Gelidium*	Gelidium	Agar	It is used as a laxative in food preparation industry and as an inert carrier for drug in the pharmaceutical industry In bacteriology, mycology and plant tissue culture media preparation It is a constituent of radiology suspending agents and anticoagulants It is a constituent of cosmetics, ointments, and lotions It is also used as a substitute for gelatin, as an antidrying agent in the bakery industry	https://botany.si.edu/projects/algae/economicuses/redalgae.htm
2	*Gracilaria*	–			
3					
4	*Ahnfeltia*	–			
5	*Pterocladia*	–			
6	*Solieria filiformis*	–	Lectin	Antinociceptive and anti-inflammatory activities	Abreu et al. 2016
7	*Pterocladiella capillacea*	–	Lectin	Antinociceptive and anti-inflammatory activities	Silva et al. (2010)
8	*Laurencia papillosa*	Lagot Laki	Carrageenan	Antitumor activities	Murad et al. (2014)
9	*Gracilaria cornea*	–	Carrageenan or Agaran	Neuroprotective effects	Souza et al. (2017)
10	*Tichocarpus crinitus*	–	Carrageenan	Anti-inflammatory	Kalitnik et al. (2017)
11	*Gigartina stellata*	–	Carrageenan	It is used as a stabilizing agent for chocolate, milk, egg- nog, ice cream, puddings, frostings, creamed soups It is used for air freshener gels, tertiary oil treatment, cleaners	https://botany.si.edu/projects/algae/economicuses/redalgae.htm
12	*Chondrus crispus*	Irish moss or Carrageen moss			
13	*Eucheuma*	–		It is also used for enzyme immobilization, electrophoretic and chromatographic media preparation	

(continued)

Table 1 (continued)

S. No.	Algae	Common Name	Product	Uses	References
14	*Gelidium amansii*	Ceylon moss	Bio-fuel	It can be used an alternative bioresource because of high carbohydrate content	Wi et al. (2009)
15	*Mastocarpus stellatus*	False Irish moss	Edible seaweed	It is used for cooking and to make a drink to cure flu and cold	Hardy and Guiry (2006)

During the last 50 years, the algal industry has been increased multifold with a value of over US$6 billion (FAO 2014; Loureiro et al. 2015). Further, the cultivation of algae holds much greater potential ranging from integrated fish farming to biofuel production. Therefore, identification of correct species for differential use is the need of the hour, and DNA barcoding seems to be a promising solution for the same

the recognition of species by DNA barcodes depends upon the target species which should have reasonable genetic differentiation that can be harnessed for separation of species, even where morphological similarities exist. The novelty behind DNA barcode is to find a single segment of DNA which can be used for the identification of all living taxa. Researchers are still trying to find a single segment of DNA which is suitable for the identification of all taxa; in spite of more than two decades of work in this particular area, single DNA marker is yet to be reported (Pečnikar and Buzan 2014). Based on the amount of work in DNA barcoding, Pečnikar and Buzan (2014) laid down certain qualities of DNA barcodes as stated below.

1. The fragment of DNA must be nearly identical in specimens of the same species but different between individuals of different species.
2. The section must be standardized and the same section should be used in different taxonomic groups.
3. The marker must be robust, with conservative primer binding sites that allow it to be readily amplified and sequenced.

3.1 Practical Approaches of DNA Barcoding

The initial idea behind the development of DNA barcodes was for species identification. It is a research tool for taxonomists and helps in species recognition at various stages of growth and development, e.g., in plants from seeds, seedling to plant, whereas in animals from the formation of eggs, larvae, to mature individual (fertile and sterile). It also helps in species identification in case of damaged samples, unisexual species, gut contents, and fecal matter (Kress and Erickson 2012). Further, DNA barcoding has the potential to evaluate the consistency of species with a universal measure of genetic variability based on the barcode sequence data. The application base use of DNA barcoding can be seen in the identification of invasive and endangered species along with conservation biology. It can also be used for checking the identity and purity of materials such as biological products, herbal medicines, dietary supplements, and seafood. DNA barcoding also helps to flag species which are new, especially the cryptic species (Hebert et al. 2004). Nowadays, DNA barcodes are also used to solve basic evolutionary and ecological problems (Kress et al. 2010) and for the recognition of medically important pathogens and their invertebrate vectors. They can also be used for the identification of species which is used for manufacturing of drugs of natural origin (Pečnikar and Buzan 2014). Besides that, DNA barcoding can be done on the herbarium specimens and museum samples, in order to reconfirm the efficacy of this new science. All over the globe, large numbers of natural history museums and herbaria have huge collections of specimens which are classified on the basis of morphology by experienced taxonomists. The application of DNA barcoding in these samples helps in a proper understanding of earlier classification, and simultaneously specimens will be backed up by the DNA barcodes (Ellis 2008; Puillandre et al. 2012).

This method is giving a quick and reliable identity to unrecognized organisms with the use of verified reference material (Erickson and Kress 2012). For this, Consortium for the Barcode of Life (CBOL) has been developed by the efforts of global DNA barcoding. In January 2017, the Barcode of Life Database (BOLD) contained more than 6.9 million specimen records, with 5.29 million having barcodes belonging to over 262,108 species (BOLD Systems 2017a). Among plant database of BOLD, the data for red algae specimen is highest (47,976 specimen records) as per January 2017. Of these, the number of specimens with sequences is 35,231, and the numbers of specimens with barcodes are 24,372, whereas 3062 species have barcodes (BOLD Systems 2017b); these data are self-explanatory about the importance of red algae.

4 Markers for DNA Barcoding in Red Algae

The genetic diversity of seaweed compels the researchers for its measurement; with the advent of molecular tools and advancement in molecular biology, this work is growing at a very fast rate, and a lot of research articles are published on this subject (Conklin et al. 2009). Genetic markers give data which can be used for the analysis of biogeographic studies, population structure, and their phylogenetic relationships, parentage, and relatedness in the gene flow (Féral 2002). A large number of molecular tools are easily available and are widely used to explain the genetic inconsistency of marine macroalgae, such as AFLP (Donaldson et al. 2000; Pang et al. 2010), DNA markers used in allozyme differentiation, RFLP (Kamikawa et al. 2007), ISSR, RAPD (Zhao et al. 2008), and microsatellites (Zhang et al. 2009). Apart from them, a large group of intraspecific markers are also available for measuring genetic diversity of seaweeds.

1. The large subunit of the ribulose-1,5-bisphosphate carboxylase/oxygenase (RuBisCo) plastid gene rbcL (Nam et al. 2000; Gurgel and Fredericq 2004).
2. The plastid-encoded RuBisCO spacer, the rbcL-rbcS (Byrne et al. 2002; Zuccarello et al. 2006).
3. Mitochondrial cytochrome oxidase subunit 3, cox3 (Steel et al. 2000; Coyer et al. 2004; Uwai et al. 2006).
4. Mitochondrial-encoded cox2–3 spacer (Chiasson et al. 2007; Vidal 2008; Vis et al. 2010).
5. Mitochondrial cytochrome oxidase subunit 1 gene, cox1 (Sherwood 2008; Kim et al. 2010; Yow et al. 2011).
6. The nuclear small subunit rDNA (SSU rDNA) and the internal transcribed spacers, ITS-1 and ITS-2 (Cho et al. 2007; Lindstrom 2008; Moniz and Kaczmarska 2010).

For measuring the genetic diversity of red alga, the usefulness of mitochondrial markers, the cox1 (Yang et al. 2007; Yow et al. 2011) and the cox2–3 spacer (Zuccarello and West 2002), has been reported. As per Saunders (2005), the DNA region of mitochondria gives positive results for species identification of red macroalgae, whereas Geraldino et al. (2006) and Kim et al. (2010) reported gene cox1 as an ideal marker for DNA barcoding of red algae. The gene cox1 also helps in elucidating of the cryptic diversity of red algae (Robba et al. 2006), whereas the cox2–3 spacer was proven beneficial for the phytogeographical study of rhodophytes (Andreakis et al. 2007; Vis et al. 2008).

5 DNA Barcoding in Red Algae

Due to its ecological and economic importance, a lot of DNA barcoding work has been done and undergoing in red algae. As stated above, *Porphyra* are commercially edible seaweeds available in different countries (Turner 2003; Nelson and Broom 2005; Ruangchuay and Notoya 2007); therefore, their accurate identification is very necessary before the establishment of commercial cultivation. Because there are only a few morphological characteristics that can be used objectively to identify species, this group of algae has defied taxonomists for a long time. To aggravate the problem, individuals of the same species can be different morphologically (phenotypic plasticity), and individuals of different species can be morphologically similar but highly divergent genetically (Neefus et al. 2002; Milstein and Oliveira 2005; Brodie et al. 2007; Lindstrom 2008).

In Brazil, Milstein et al. (2012) performed an elaborated study on the macroalgal flora of São Paulo state in a biodiversity project. The authors sequenced cox1 and cox2–3 spacer and UPA as three DNA barcode markers for *Porphyra* species from Brazil. The three markers help in the recovery of a cryptic species (*Porphyra* sp. 77) and also clustered two different species (*P. drewiana* and *P. spiralis*) that were synonymized. Further, they reported that varieties such as *P. acanthophora* and

P. spiralis are only differing morphologically and there is no sequence divergence in three studied molecular markers. In *Gelidium elegans*, the plastid gene rbcL and nuclear 18S ITS genes have been analyzed in samples from Japan and Taiwan (Freshwater et al. 1995; Shimada et al. 2000). In another report, sequences of rbcL, photosystem I P700 chlorophyll an apoprotein A (psaA), and cox1 gene (Kim et al. 2011) have shown that *G. elegans* demonstrates high genetic diversity as compared to other red algal species. Recently, Kim et al. (2012) assessed the genetic structure of the Korean species *G. elegans* by analyzing cox1 gene from 272 individuals collected from 36 locations. The authors reported 34 haplotypes, which were unique; the nucleotide and haplotype diversities of cox1 within *G. elegans* were 0.711 ± 0.028 (H) and 0.00736 ± 0.00038 (π), respectively.

In BOLD Systems (2017c), among family Rhodomelaceae, the specimen record of *Polysiphonia* is the highest, i.e., 1456, whereas for the same genus, specimens with sequences and barcodes recorded are 780 and 553, respectively. The phenotypic plasticity in *Polysiphonia* species is very high (Kim et al. 2000); therefore, species identification is very difficult or nearly impossible (Kim and Yang 2006). On the basis of rbcL sequences, *Polysiphonia morrowii* has been reported as an alien species in the Southern Pacific Ocean (Chile and New Zealand) (Kim et al. 2004; Mamoozade and Freshwater 2011). As per the previous records, this species naturally belongs to Northwest Pacific Ocean and has been recorded in China (Segi 1951), Far East Russia (Perestenko 1980), Japan (Kudo and Masuda 1992), and South Korea (Kim et al. 1994). Earlier, the presence of *P. morrowii* was debatable on the basis of its morphology and anatomy in New Zealand (Nelson and Maggs 1996), in the Mediterranean Sea (Erdŭgan et al. 2009), and putatively in the North Sea as *P. senticulosa* Harvey (Maggs and Stegenga 1999). Geoffroy et al. 2012 sampled 110 specimens of *Polysiphonia* and used a DNA barcoding approach for their identification at the species level. A total of 110 rbcL sequences were generated, and the alignment covered 1225 bp with 430 variable sites. The sequences generated in the present study were resolved in the lineage containing the *P. morrowii* sequences from GenBank. Among the 105 individuals of *P. morrowii* collected along the coast of Brittany, three haplotypes were found suggesting several introduction events. *P. stricta* joined *P. pacifica* and were resolved as a sister group of *P. morrowii*. Hence, the authors demonstrated the presence of *P. morrowii* as an alien species in the North-Eastern Atlantic with the help of DNA barcoding.

Gracilaria changii is a very important commercial species in Peninsular Malaysia with high genetic diversity; for better agar yield and high growth rate, it is required to identify the best strain, as these features are species or strain specific. Besides that, for better conservation and management of various species, strain identification is very necessary. Information on intraspecific genetic diversity of this economically important species is scarce (Yow et al. 2011). Keeping these points, Yow et al. (2013) use mitochondrial cox1 gene and cox2–3 spacer for measuring genetic diversity of *G. changii*. The authors reported that cox1 gene is a potential marker for defining intraspecific genetic variation in red algae and cox1 marker is more variable as compared to cox2–3 spacer. In another report, Robba et al. (2006) performed a comparative analysis between plastid RuBisCO spacer and cox1 from numerous samples of red algae and concluded that cox1 is a more sensitive marker for revealing population structure and the hidden diversity of red algal species. Yang et al. (2008) reported similar results in *G. vermiculophylla* and putative relatives and their results suggested that cox1 is a valuable molecular marker within species and for haplotype analyses. Recently, Yang and Kim (2015) performed molecular analyses of Gracilariaceae (Rhodophyta) for identification from the Asia-Pacific region. The authors analyzed mitochondrial cox1 and plastid rbcL genes as a marker; the results revealed 22 species as a distinct entity including five unknown and two new distribution records. On comparative analysis of cox1 and rbcL sequence data, cox1 showed more variation than rbcL data, allowing the perfect discrimination of species.

The Philippines is the leading producer and holds almost 70% of the world's semi-refined carrageenan supply by culturing *Eucheuma denticulatum* (N. L. Burman) F.S. Collins and Hervey and *Kappaphycus sp.* (Llana 1991; Villanueva et al. 2008). The farming of these two species outside the Philippines has been profitable in only a few countries (Hurtado and Agbayani 2002; McHugh 2003). Due to shortage of distinguished morphological characteristics for recognition and high morphological plasticity within the genera, *Eucheuma* and *Kappaphycus* have created misperception about the distributions and spread of three introduced eucheumoid species in Hawaii (Conklin et al. 2009). The authors employed DNA barcoding and used three molecular markers, namely partial nuclear 28S rRNA, partial plastid 23S rRNA, and mitochondrial 5′ cox1, in order to identify *Eucheuma* and *Kappaphycus* samples from Hawaii.

Another red alga *Grateloupia* of family Halymeniaceae has wide geographical distribution, it ranges from tropical to warm temperate regions of the world. There is a very high morphological similarity among various species of *Grateloupia*, especially between *G. elliptica* and *G. lanceolata*. These two species possess blade-like thalli with a leather texture and cruciately divided tetrasporangia; as a result of similar appearance, there is a recurrent confusion which leads to difficulty in the identification. Yang et al. (2013) analyzed the relationships between two species by using plastid rbcL and mitochondrial cox1 gene as a biomarker. The rbcL sequence analyses backed the difference between two species, with an interspecific divergence of 3.7–4.6%. The genetic diversity of cox1 gene within species was estimated to be 0–0.3% in *G. elliptica* and 0–1.0% in *G. lanceolata*, respectively. Finally, the authors reported that cox1 gene is more effective than rbcL. Wolf et al. (2011) reported the presence of exotic *Hypnea flexicaulis* in the Mediterranean Sea by using DNA barcoding employing rbcL and cox1 gene along with morphological analysis.

6 Conclusion

It is a well-known fact that genera of red algae are very important commercially as well as ecologically. Due to human activities, there is the introduction of new species which are exotic and invasive in nature; for ecological conservation, species identification is necessary because of symbiotic nature of these algae which are species specific. Further, for better agar yield and growth, species identification is mandatory. We also know that because of limited characteristics and phenotypic plasticity, identification on the basis of morphology is doubtful by this age-old practice. This chapter deals with resolving taxonomical problem of red algae by means of DNA barcoding; it is a reliable tool when species identification is required in a quick manner. Some cases of red algae are described here; on that basis, it is concluded that cox1 is a more sensitive marker for revealing population structure and the hidden diversity.

Conflict of Interest The authors declare that there is no conflict of interest.

References

Abreu, TM, Ribeiro NA, Chaves HV, Jorge RJ, Bezerra MM, Monteiro H.S, Vasconcelos IM, Mota ÉF, Benevides NM (2016) Antinociceptive and anti-inflammatory activities of the lectin from marine red alga *Solieria filiformis*. Planta Med 6(7), 596–605

Anderson NL, Parish NM, Richardson JP, Pearson TW (1985) Comparison of African trypanosomes of different antigenic phenotypes, subspecies and life cycle stages by two-dimensional gel electrophoresis. Mol Biochem Parasitol 16(3):299–314

Andreakis N, Procaccini G, Maggs C, Kooidtra WHCF (2007) Phylogeography of the invasive seaweed *Asparagopsis* (Bonnemaisoniales, Rhodophyta) reveals cryptic diversity. Mol Ecol 16:2285–2299

Arnot DE, Roper C, Bayoumi RA (1993) Digital codes from hypervariable tandemly repeated DNA sequences in the *Plasmodium falciparum* circumsporozoite gene can genetically barcode isolates. Mol Biochem Parasitol 61(1):15–24

Barrett RD, Hebert PD (2005) Identifying spiders through DNA barcodes. Can J Zool 83 (3):481–491

Bixler HJ (1996) Recent developments in manufacturing and marketing carrageenan. Hydrobiologia 326(327):35–57

BOLD Systems (2017a). http://www.boldsystems.org/index.php/TaxBrowser_Home

BOLD Systems (2017b). http://www.boldsystems.org/index.php/Taxbrowser_Taxonpage?taxid=48327

BOLD Systems (2017c). http://www.boldsystems.org/index.php/Taxbrowser_Taxonpage?taxid=55735

Brodie J, Bartsch I, Neefus C, Orfinidis S, Bray T, Mathieson A (2007) New insights into the cryptic diversity of the North Atlantic–Mediterranean '*Porphyra leucosticta*' complex: *P. olivii* sp. nov. and *P. rosengurttii* (Bangiales, Rhodophyta). Eur J Phycol 42:3–28

Brodie J, Mortensen AM, Ramirez ME, Russell S, Rinkel B (2008) Making the links: towards a global taxonomy for the red algal genus *Porphyra* (Bangiales, Rhodophyta). J Appl Phycol 20:939–949

Burns JM, Janzen DH, Hajibabaei M, Hallwachs W, Hebert PDN (2008) DNA barcodes and cryptic species of skipper butterflies in the genus Perichares in area de Conservacion Guanacaste, Costa Rica. Proc Natl Acad Sci U S A 105:6350–6355

Byrne K, Zuccarello GC, West J, Liao ML, Kraft GT (2002) *Gracilaria species* (Gracilariaceae, Rhodophyta) from southeastern Australia, including a new species, *Gracilaria perplexa* sp. nov.: morphology, molecular relationships and agar content. Phycol Res 50:295–312

Chase MW, Salamin N, Wilkinson M, Dunwell JM, Kesanakurthi RP, Haidar N, Savolainen V (2005) Land plants and DNA barcodes: short-term and long-term goals. Philos Trans R Soc Lond Ser B Biol Sci 360(1462):1889–1895

Chiasson WB, Johanson KG, Sherwood AR, Vis ML (2007) Phylogenetic affinities of the form taxon *Chantransia pygmaea* (Rhodophyta) specimens from the Hawaiian islands. Phycologia 46:257–262

Cho GY, Kogame K, Kawai H, Boo SM (2007) Genetic diversity of *Scytosiphon lomentaria* (Scytosiphonaceae, Phaeophyceae) from the Pacific and Europe based on RuBisCo large subunit and spacer, and ITS nrDNA sequences. Phycologia 46:657–665

Cole KM, Sheath RG (1995) Biology of the Red Sea weeds. Cambridge University Press, New York

Conklin KY, Kurihara AA, Sherwood AR (2009) A molecular method for identification of the morphologically plastic invasive algal genera *Eucheuma* and *Kappaphycus* (Rhodophyta, Gigartinales) in Hawaii. J Appl Phycol 21:691–699

Coyer JA, Hoarau G, Stam WT, Olsen JL (2004) Geographically specific heteroplasmy of mitochondrial DNA in the seaweed, *Fucus serratus* (Heterokontophyta, Phaeophyceae, Fucales). Mol Ecol 13:1323–1326

de Velde V, Knutsen SH, Usov AI, Cerezo AS (2002) 1H and 13C high resolution NMR spectroscopy of carrageenans: application in research and industry. Trends Food Sci Technol 13:73–92

Donaldson SL, Chopin T, Saunders GW (2000) An assessment of the AFLP method for investigating populations structure in the red alga *Chondus crispus*. J Appl Phycol 12:25–35

Ellis R (2008) Rethinking the value of biological specimens: laboratories, museums and the barcoding of life initiative. Mus Soc 6(2):172–191

Erdŭgan H, Akı C, Acar O, Dural B, Aysel V (2009) New record for the East Mediterranean, Dardanelles (Turkey) and its distribution: *Polysiphonia morrowii* Harvey (Ceramiales, Rhodophyta). Turk J Fish Aquat Sci 9:231–232

Erickson DL, Kress WJ (2012) Future directions. In: Kress JW, Erickson DL (eds) DNA barcodes: methods and protocols, methods in molecular biology, vol 858. Humana Press, New York, pp 459–465

Fazekas AJ, Kuzmina ML, Newmaster SG, Hollingsworth PM (2012) DNA barcoding methods for land plants. In: Kress JW, Erickson DL (eds) DNA barcodes: methods and protocols, methods in molecular biology, vol 858. Springer, New York, pp 223–252

Féral JP (2002) How useful are the genetic markers in attempts to understand and manage marine biodiversity? J Exp Mar Biol Ecol 268:121–145

Floyd R, Abebe E, Papert A, Blaxter M (2002) Molecular barcodes for soil nematode identification. Mol Ecol 11:839–850

Food and Agriculture Organization of the United Nations (2014) Fisheries and aquaculture information and statistics services.. http://www.fao.org/figis/. Accessed 2 Feb 2017

Freshwater DW (2000) Rhodophyta. Red algae. Version 24 march 2000 (under construction). http://tolweb.org/Rhodophyta/2381/2000.03.24 *in* the tree of life web project

Freshwater W, Fredericq S, Hommersand MH (1995) A molecular phylogeny of the *Gelidiales* (Rhodophyta) based on analysis of plastid rbcL nucleotide sequences. J Phycol 31:616–632

Geoffroy A, Gallc LL, Destombe C (2012) Cryptic introduction of the red alga *Polysiphonia morrowii* Harvey (Rhodomelaceae, Rhodophyta) in the North Atlantic Ocean highlighted by a DNA barcoding approach. Aquat Bot 100:67–71

Geraldino PJL, Yang EC, Boo SM (2006) Morphology and molecular phylogeny of *Hypnea flexicaulis* (Gigartinales, Rhodophyta) from Korea. Algae 21:417–423

Graham LE, Graham JM, Wilcox JW (2009) Red algae. In: algae, 2nd edn. Pearson Benjamin Cummings, San Francisco, CA

Guiry MD (2012) How many species of algae are there? J Phycol 48:1057–1063

Gurgel CFD, Fredericq S (2004) Systematics of Gracilariaceae (Gracilariales, Rhodophyata): a critical assessment based on rbcL sequence analyses. J Phycol 40:138–159

Hardy FG and Guiry MD (2006) A check-list and atlas of the seaweeds of Britain and Ireland. The British Phycological Society

Hebert PDN, Cywinska A, Ball SL, de Waard JR (2003) Biological identifications through DNA barcodes. Proc Biol Sci 270:313–321

Hebert PDN, Stoeckle MY, Zemlak TS, Francis CM (2004) Identification of birds through DNA barcodes. PLoS Biol 2(10):e312

Heinrichs J, Kreier HP, Feldberg K, Schmidt AR, Zhu RL, Shaw B, Shaw AJ, Wissemann V (2011) Formalizing morphologically cryptic biological entities: new insights from DNA taxonomy, hybridization, and biogeography in the leafy liverwort *Porella platyphylla* (Jungermanniopsida, Porellales). Am J Bot 98(8):1252–1262

https://botany.si.edu/projects/algae/economicuses/redalgae.htm

Hubby JL, Lewontin RC (1966) A molecular approach to the study of genic heterozygosity in natural populations. I. The number of alleles at different loci in *Drosophila pseudoobscura*. Genetics 54:577–594

Hurtado AQ, Agbayani RF (2002) Deep-sea farming of *Kappaphycus* using the multiple raft, longline method. Bot Mar 45:438–444

Kalitnik AA, Karetin YA, Kravchenko AO, Khasina EI, Yermak IM (2017) Influence of carrageenan on cytokine production and cellular activity of mouse peritoneal macrophages and its effect on experimental endotoxemia. J Biomed Mater Res A 105(5):1549–1557

Kamikawa R, Masuda I, Oyawa K, Yoshimatsu S, Sako Y (2007) Genetic variation in mitochondrial genes and intergenic spacer region in harmful algae *Chattonella species*. Fish Sci 73:871–880

Kim KM, Hoarau GG, Boo SM (2012) Genetic structure and distribution of *Gelidium elegans* (Gelidiales, Rhodophyta) in Korea based on mitochondrial cox1 sequence data. Aquat Bot 98:27–33

Kim KM, Hwang IK, Park JK, Boo SM (2011) A new agarophyte species, *Gelidium eucorneum* sp. nov. (Gelidiales, Rhodophyta), based on molecular and morphological data. J Phycol 47:904–910

Kim MS, Lee IK, Boo SM (1994) Morphological studies of the red alga *Polysiphonia morrowii* Harvey on the Korean coast. Korean J Phycol 9:185–192

Kim MS, Maggs C, McIvor L, Guiry M (2000) Reappraisal of the type species of *Polysiphonia* (Rhodomelaceae, Rhodophyta). Eur J Phycol 35:83–92

Kim SY, Weinberger F, Boo SM (2010) Genetic diversity hints at a common donor region of the invasive Atlantic and Pacific populations of *Gracilaria vermiculophylla* (Rhodophyta). J Phycol 46:1346–1349

Kim MS, Yang EC (2006) Taxonomy and phylogeny of *Neosiphonia japonica* (Rhodomelaceae, Rhodophyta) based on rbcL and cpeA/B gene sequences. Algae 2:287–294

Kim MS, Yang EC, Mansilla A, Boo SM (2004) Recent introduction of *Polysiphonia morrowii* (Ceramiales, Rhodophyta) to Punta Arenas, Chile. Bot Mar 47:389–394

Kress WJ, Erickson DL (2012) DNA barcodes: methods and protocols. Methods Mol Biol 858:3–8

Kress WJ, Erickson DL, Swenson NG, Thompson J, Uriarte M, Zimmerman JK (2010) Advances in the use of DNA barcodes to build a community phylogeny for tropical trees in a Puerto Rican forest dynamics plot. PLoS One 5(11):e15409

Kudo T, Masuda M (1992) Taxonomic features of *Polysiphonia morrowii* Harvey (Ceramiales, Rhodophyta). Korean J Phycol 7:13–26

Lee RE (2008) Phycology, vol 107. Cambridge University Press, Cambridge

Lindstrom SC (2008) Cryptic diversity, biogeography and genetic variation in Northeast Pacific species of *Porphyra sensu lato* (Bangiales, Rhodophyta). J Appl Phycol 20:501–512

Llana EG (1991) Production and utilisation of seaweeds in the Philippines. INFOFISH Int 1 (91):12–17

Loureiro R, Gachon CMM, Rebours C (2015) Seaweed cultivation: potential and challenges of crop domestication at an unprecedented pace. New Phytol 206(2):489–492

Maggs CA, Stegenga H (1999) Red algal exotics on North Sea coasts. Helgoländer Meeresun 52:243–258

Mamoozade NR, Freshwater DW (2011) Taxonomic notes on Caribbean *Neosiphonia* and *Polysiphonia* (Ceramiales, Florideophyceae): five species from Florida, USA and Mexico. Bot Mar 54:269–292

May RM (1988) How many species are there on earth? Science 241:1441–1449

McAndrew BJ, Majumdar KC (1983) Tilapia stock identification using electrophoretic markers. Aquaculture 30(1–4):249–261

McDevit DC, Saunders GW (2010) A DNA barcode examination of the *Laminariaceae* (Phaeophyceae) in Canada reveals novel biogeographical and evolutionary insights. Phycologia 49:235–248

McHugh DJ (2003) A guide to the seaweed industry. Technical report 441. Food and agriculture Organization of the United Nations, Rome, Italy

Milstein D, Medeiros AS, Oliveira EC, Oliveira MC (2012) Will a DNA barcoding approach be useful to identify *Porphyra species* (Bangiales, Rhodophyta)? A case study with Brazilian taxa. J Appl Phycol 24:837–845

Milstein D, Oliveira MC (2005) Molecular phylogeny of Bangiales (Rhodophyta) based on small subunit rDNA sequencing: emphasis on Brazilian *Porphyra species*. Phycologia 44:212–221

Moniz MBJ, Kaczmarska I (2010) Barcoding of diatoms: nuclear encoded ITS revisited. Protist 161:7–34

Murad H, Ghannam A, Al-Ktaifani M, Abbas A, Hawat M (2014) Algal sulfated carrageenan inhibits proliferation of MDA-MB-231 cells via apoptosis regulatory genes. Mol Med Rep 11:2153–2158

Nam KW, Maggs CA, McIvor L, Stanhope MJ (2000) Taxonomy and phylogeny of Osmunder (Rhodomelaceae, Rhodophyta) in Atlantic. Eur J Phycol 36:759–772

Neefus CD, Mathieson AC, Klein AS, Teasdale B, Gray T, Yarish C (2002) *Porphyra birdiae* sp. nov. (Bangiales, Rhodophyta): a new species from the Northwest Atlantic. Algae 17:203–216

Nelson WA, Broom JES (2005) Contributions of molecular biology to understanding systematics and phylogeny in the order Bangiales. Nat Hist Res Spec Issue 8:1–12

Nelson WA, Maggs CA (1996) Records of adventive marine algae in New Zealand: *Antithamnionella ternifolia*, *Polysiphonia senticulosa* (Ceramiales, Rhodophyta), and *Striaria attenuate* (Dictyosiphonales, Phaeophyta). NZ J Mar Freshw Res 30:449–453

Packer L, Gibbs J, Sheffield C, Hanner R (2009) DNA barcoding and the mediocrity of morphology. Mol Ecol Resour 9(Suppl S1):42–50

Pang QQ, Sui ZH, Kang KH (2010) Application of SSR and AFLP to the analyses of genetic diversity in *Gracilaria lemaneiformis* (Rhodopyta). J Appl Phycol 22:607–612

Pečnikar ŽF, Buzan EV (2014) 20 years since the introduction of DNA barcoding: from theory to application. J Appl Genet 55:43–52

Perestenko LP (1980) Algae of Peter the Great Bay. Nauka, Leningrad, pp 232–239

Pires AC, Marinoni L (2010) DNA barcoding and traditional taxonomy unified through integrative taxonomy: a view that challenges the debate questioning both methodologies. Biota Neotrop 10:339–346

Puillandre N, Bouchet P, Boisselier-Dubayle M-C, Brisset J, Buge B, Castelin M, Chagnoux S, Christophe T, Corbari L, Lambourdière J, Lozouet P, Marani G, Rivasseau A, Silva N, Terryn Y, Tillier S, Utge J, Samadi S (2012) New taxonomy and old collections: integrating DNA barcoding into the collection curation process. Mol Ecol Resour 12:396–402

Rebours C, Marinho-Soriano E, Zertuche-González JA, Hayashi L, Vásquez JA, Kradolfer P, Soriano G, Ugarte R, Abreu MH, Bay-Larsen I, Hovelsrud G, Rødven R, Robledo D (2014) Seaweeds: an opportunity for wealth and sustainable livelihood for coastal communities. J Appl Phycol 26:1939–1951

Rindi F, Guiry MD, López-Bautista JM (2008) Distribution, morphology, and phylogeny of Klebsormidium (Klebsormidiales, Charophyceae) in urban environments in Europe. J Phycol 44:1529–1540

Robba L, Russell SJ, Barker GL, Brodie J (2006) Assessing the use of the mitochondrial cox1 marker for use in DNA barcoding of red algae (Rhodophyta). Am J Bot 93:1101–1108

Robins JH, Hingston M, Matisoo-Smith E, Ross HA (2007) Identifying Rattus species using mitochondrial DNA. Mol Ecol Notes 7(5):717–729

Ruangchuay R, Notoya M (2007) Reproductive strategy and occurrence of gametophytes of Tha laver Porphyra vietnamensis Tanaka et Pham-Hoang Ho (Bangiales, Rhodophyta) from Songkhla Province. Kasetsart J (Nat Sci) 41:143–152

Saunders GW (2005) Applying DNA barcoding to red macroalgae: a preliminary appraisal holds promise for future application. Philos Trans R Soc B 360:1879–1888

Saunders GW (2008) A DNA barcode examination of the red algal family Dumontiaceae in Canadian waters reveals substantial cryptic species diversity. 1. The foliose Dilsea–Neodilsea complex and Weeksia. Botany 86:773–789

Savolainen V, Cowan RS, Vogler AP, Roderick GK, Lane R (2005) Towards writing the encyclopedia of life: an introduction to DNA barcoding. Philos Trans R Soc Lond Ser B Biol Sci 360 (1462):1805–1811

Seena S, Pascoal C, Marvanová L, Cássio F (2010) DNA barcoding of fungi: a case study using ITS sequences for identifying aquatic hyphomycete species. Fungal Divers 44:77–87

Segi T (1951) Systematic study of the genus *Polysiphonia* from Japan and its vicinity. J Fac Fish Prefect Univ Mie 1:169–272

Sherwood AR (2008) Phylogeography of Asparagopsis taxiformis (Bonnemaisoniales, Rhodophyta) in the Hawaiian islands: two mtDNA markers support three separate introductions. Phycologia 47:79–88

Sherwood AR, Presting GG (2007) Universal primers amplify a 23S rDNA plastid marker in eukaryotic algae and cyanobacteria. J Phycol 43:605–608

Shimada S, Horiguchi T, Masuda M (2000) Two new species of Gelidium (Rhodophyta, Gelidiales), *Gelidium tenuifolium* and *Gelidium koshikianum*, from Japan. Phycol Res 48:37–46

Silva LM, Lima V, Holanda ML, Pinheiro PG, Rodrigues JA, Lima ME, Benevides NM (2010) Antinociceptive and anti-inflammatory activities of lectin from marine red alga *Pterocladiella capillacea*. Biol Pharm Bull 33(5):830–835

Siow RS, Teo SS, Ho WY, Shukor MYA, Phang SM, Ho CL (2012) Molecular cloning and biochemical characterization of galactose-1-phosphate uridylytransferase from *Gracilaria changii* (Rhodophta). J Phycol 48:155–162

Souza RB, Frota AF, Sousa RS, Cezario NA, Santos TB, Souza LM et al (2017) Neuroprotective effects of sulphated agaran from marine alga *Gracilaria cornea*in rat 6-hydroxydopamine Parkinson's disease model: Behavioural, neurochemical and transcriptional alterations. Basic Clin Pharmacol Toxicol 120:159–170

Steel DJ, Trewick SA, Wallis GP (2000) Heteroplasmy of mitochondrial DNA in the ophiuroid *Asterobrachion constricum*. J Hered 91:146–149

Turner NJ (2003) The ethnobotany of edible seaweed (*Porphyra abbottae* and related species; Rhodophyta: Bangiales) and its use by first nations on the Pacific coast of Canada. Can J Bot 81:283–293

Uwai S, Yotsukura N, Serisawa Y, Muraoka D, Hiraoka M, Kogame K (2006) Intraspecific genetic diversity of *Undaria pinnatifida* in Japan, based on the mitochondrial of cox3 gene and ITS1 of nrDNA. Hydrobiologia 553:345–356

Vidal R (2008) Phylogeography of the genus *Spongites* (Corallinales, Rhodophyta) from Chile. J Phycol 44:173–182

Villanueva RD, Montaño MNE, Romero JB (2008) Iota-carrageenan from a newly farmed, rare variety of eucheumoid seaweed–"endong". J Appl Phycol 21:27–30

Vis ML, Feng J, Chiasson WB, Xie SL, Stancheva R, Entwisle TJ, Wang WL (2010) Investigation of the molecular and morphological variability in *Batrachospermum arcuatum*

(Batrachospermales, Rhodophyta) from geographically distant locations. Phycologia 49:545–553

Vis ML, Hodge JC, Necchi OJR (2008) Phylogeography of *Batrachospermum macrosporum* (Batrachospermales, Rhodophyta) from north and South America. J Phycol 44:882–888

Ward RD, Zemlak TS, Innes BH, Last PR, Hebert PD (2005) DNA barcoding Australia's fish species. Philos Trans R Soc Lond Ser B Biol Sci 360(1462):1847–1857

Wi SG, Kim HJ, Mahadevan SA, Yang D-J, Bae H-J (2009) The potential value of the seaweed Ceylon moss (*Gelidium amansii*) as an alternative bioenergy resource. Bioresour Technol 100 (24):6658–6660

Woelkerling WJ (1990) An introduction. In: Cole KM, Sheath RG (eds) Biology of the red algae. Cambridge University Press, Cambridge, pp 1–6

Wolf MA, Sfriso A, Andreolia C (2011) Moroa I (2011) the presence of exotic *Hypnea flexicaulis* (Rhodophyta) in the Mediterranean Sea as indicated by morphology, rbcL and cox1 analyses. Aquat Bot 95:55–58

Yang MY, Han EG, Kim MS (2013) Molecular identification of *Grateloupia elliptica* and *G. lanceolata* (Rhodophyta) inferred from plastid rbcL and mitochondrial COI genes sequence data. Genes Genomics 35:239–246

Yang YM, Kim MS (2015) Molecular analyses for identification of the Gracilariaceae (*Rhodophyta*) from the Asia–Pacific region. Genes Genomics 37:775–787

Yang EC, Kim MS, Geraldino PJL, Sahoo D, Shin JA, Boo SM (2007) Mitochondrial cox1 and plastid rbcL genes of *Gracilaria vermiculophylla* (Gracilariaceae, Rhodophyta). J Appl Phycol 20:161–168

Yang EC, Kim MS, Paul JL, Geraldino PJL, Sahoo D, Shin J-M, Boo SM (2008) Mitochondrial cox1 and plastid rbcL genes of *Gracilaria vermiculophylla* (Gracilariaceae, Rhodophyta). J Appl Phycol 20:161–168

Yeong HY, Khalid N, Phang SM (2008) Protoplast isolation and regeneration from Gracilaria changii (Gracilariales, Rhodophyta). J Appl Phycol 20:641–651

Yow YY, Lim PE, Phang SM (2011) Genetic diversity of *Gracilaria changii* (Gracilariacea, Rhodopyta) from west coast, peninsular Malaysia based on mitochondrial cox1 gene analysis. J Appl Phycol 23:219–226

Yow YY, Lim PE, Phang SM (2013) Assessing the use of mitochondrial cox1 gene and cox 2-3 spacer for genetic diversity study of Malaysian *Gracilaria changii* (Gracilariaceae, Rhodophyta) from peninsular Malaysia. J Appl Phycol 25:831–838

Zhang XC, He Y, Xu D (2009) Screening microsatellite sequences from *Gracilaria lemaneiformis* and its phylogenetic analysis. Period Ocean Univ China 39:259–264

Zhao F, Liu F, Liu J, Ang PO, Duan D (2008) Genetic structure analysis of natural *Sargassum muticum* (Fucales, Phaeophyta) populations using RAPD and ISSR markers. J Appl Phycol 20:191–198

Zuccarello GC, Buchanan J, West JA (2006) Increased sampling for inferring phylogeographic patterns in *Bostrychia radicans/B. moritziana* (Rhodomelaceae, Rhodophyta) in the eastern USA. J Phycol 42:1349–1352

Zuccarello GC, West JA (2002) Phylogeography of the *Bostrychia calliptera–B. pinnata* complex (Rhodomelaceae, Rhodophyta) and divergence rates based on nuclear, mitochondrial and plastid DNA markers. Phycologia 41:49–60

Part IV
DNA Barcoding in Animals

DNA Barcoding of Rays from the South China Sea

B. Akbar John, M. A. Muhamad Asrul, Wahidah Mohd Arshaad, K. C. A. Jalal, and Hassan I. Sheikh

Abstract Fishery management on elasmobranchs is gaining attention in recent years due to their economic value and the key ecological role played by them in their natural habitat. Many species of elasmobranch, especially rays, exhibit overlapping morphological similarities and hence difficult to identify to their species level. As an accurate identification is key for developing a framework for fishery management, we used a molecular approach to identify ray fishes sampled from South China Sea. We used Cytochrome oxidase subunit I (COI) sequencing (~652 bp) to cross-examine field identification of ray fishes. A total of 10 species belonging to 3 families were successfully sequenced/identified from 29 PCR products. BLAST/BOLD analysis was performed and inter- and intra-species genetic distance was calculated. Due to overlapping morphological characters and morphocryptic nature, their accurate field identification is challenging. We also addressed problems in species-level delimitation of ray fishes due to the paucity of information in DNA databanks besides unresolved taxonomic status of available dataset. We further addressed management and action plan for sustainable management of elasmobranch fishery in Malaysia.

Keywords DNA barcoding · COI gene · Ray fish · Batoids

B. Akbar John (✉)
Institute of Oceanography and Maritime Studies (INOCEM), Kulliyyah of Science, International Islamic University Malaysia, Kuantan, Malaysia

M. A. Muhamad Asrul · K. C. A. Jalal
Department of Marine Science, Kulliyyah of Science, International Islamic University Malaysia, Kuantan, Malaysia

W. M. Arshaad
Department of Fisheries, SEFDEC, Kuala Terengganu, Terengganu, Malaysia

H. I. Sheikh
Faculty of Fisheries and Food Science, Universiti Malaysia Terengganu, Kuala Nerus, Terengganu, Malaysia

S. Trivedi et al. (eds.), *DNA Barcoding and Molecular Phylogeny*,
https://doi.org/10.1007/978-3-030-50075-7_8

1 Introduction

In the last century, the number of species extinction from their natural habitat has increased in alarming rate due to the effects of global climate change, habitat destruction and overharvesting. This is due to an increase in the yearly rate of extinction from one species/million to $\geq$1000/million from the estimated two million identified plant and animal species, while the total estimate is ranging from 5 to 50 million animal taxa besides ten million microorganisms (Hebert et al. 2003). Many species are extinct before they are being identified from their natural habitat. This has directed researchers and policymakers to initiate an attempt to preserve species diversity in the face of accelerating habitat destruction having led to the awareness on the importance of accurate species identification.

In aquatic habitat, many closely related species share similar and overlapping morphological characters which could not be differentiated without the assistance from expert taxonomists. For instance, many researchers argued that elasmobranchs exhibit cryptic morphological overlapping characters, species complexity besides ontogenetic colour pattern or variation which are shared among the closely related taxa that makes the accurate identification challenging (Cerutti-Pereyra et al. 2012; Sandoval-Castillo and Rocha-Olivares 2011; Toffoli et al. 2008). It also leads to taxonomic misidentification in the field sampling and thereby effecting conservation-related research and fishery management. Hence, DNA-based species identification was proposed and well established on many cryptic taxonomic forms such as lepidopteron, fishes, spiders, polychaete and many other farms (Blagoev et al. 2016; Burns et al. 2007; Hebert et al. 2003; Hebert and Gregory 2005; Steinke et al. 2016; Ward et al. 2009). The efficiency of DNA barcoding technique is also validated on ray fish identification from Taiwan (Chang et al. 2016), Indonesia (Madduppa et al. 2016), Australia (Ward et al. 2008), India (Bineesh et al. 2016) and other places (Cerutti-Pereyra et al. 2012; Toffoli et al. 2008). Though Malaysian water is well diversified with 84 species of batoids (second largest ray fish diversity in South East Asia), their accurate identification using DNA barcoding technique is still limited while their field identification is highly challenging due to lacking of conventional taxonomist and problems addressed above.

Reports suggested that at least 62.5% of ecology-based papers published in reputed peer journals do not provide supporting documents to justify accurate identification of organism under study (Bortolus 2008). It is also argued that, though checklist on detailed morphological characters of the species under study is available, their field identification still requires expert taxonomists to differentiate species that share overlapping morphological characters, sexual dimorphisms and ontogenetic colour pattern. Interestingly, ray fishes exhibit morphocryptic characters, species complexity and differential ontogenetic variations and hence, in this study, we addressed the efficiency of universal DNA barcoding technique in species identification and field validation of ray fishes.

2 Materials and Methods

2.1 Sample Collection and Preparation

Ray fishes were collected from three different locations from Malaysian waters (Fig. 1). A total of 6 samples from Mukah (west Malaysia), 9 samples from Dungun (Peninsular Malaysia) and 18 samples from Kuantan (Peninsular Malaysia) were collected either by gill nets, hooks or lines. Samples were also confiscated from the local market and identified morphologically using standard references (Ali et al. 2013; Ambak and Terengganu 2012). Tissue samples of 2 × 2 cm were excised from the ventral side of each fish and transferred into 70% ethanol and then at -20 °C prior to analysis. Voucher specimens were photographed for further analysis and reference. Field identification of samples and corresponding geographical location are shown in Table 1.

2.2 DNA Barcoding and Sequencing

Total genomic DNA was extracted using DNeasy Blood and Tissue Kit (Qiagen Inc., Germany) following manufacturer guidelines. Concentration and purity of genomic DNA were analysed using NanoDrop 2000 UV-Vis Spectrophotometer. PCR amplification of target gene (652 base pairs fragments of partial cytochrome

Fig. 1 Location of sampling sites

Table 1 Ray fish samples collected from 3 sampling locations in South China Sea

No.	Sample Code	Field Identification	Sampling Location
1	PMKH 1	*Gymnura japonica*	Mukah, Sarawak, West Malaysia (WM) (2.5958781°N, 112.1824272°E)
2	PMKH 2	*Gymnura poecilura*	
3	PMKH 3	*Gymnura japonica*	
4	PMKH 6	*Himantura uarnacoides*	
5	PMKH 7	*Himantura uarnacoides*	
6	PMKH 8	*Himantura* cf. *uarnacoides*	
7	PDGN 1	*Gymnura poecilura*[a]	Dungun, Terengganu, peninsular Malaysia (PM) (4.9427071°N, 103.1772143°E)
8	PDGN 2	*Gymnura poecilura*[a]	
9	PDGN 4	*Himantura uarnak*[a]	
10	PDGN 5	*Himantura uarnak*[a]	
11	PDGN 6	*Aetobatus narinari*[a]	
12	PDGN 7	*Taeniura lymma*	
13	PDGN 8	*Taeniura lymma*	
14	PDGN 10	*Dasyatis parvonigra*	
15	PDGN 11	*Dasyatis parvonigra*	
16	PKTN 7	*Gymnura japonica*	Kuantan, Pahang, peninsular Malaysia (PM) (3.9800417°N, 103.4098969°E)
17	PKTN 8	*Gymnura japonica*	
18	PKTN 9	*Gymnura japonica*	
19	PKTN 18	*Himantura uarnak*	
20	PKTN 22	*Aetobatus ocellatus*	
21	PKTN 28	*Himantura walga*	
22	PKTN 29	*Himantura walga*	
23	PKTN 30	*Himantura walga*	
24	PKTN 31	*Dasyatis parvonigra*	
25	PKTN 33	*Gymnura poecilura*	
26	PKTN 34	*Gymnura poecilura*	
27	PKTN 35	*Himantura leoparda*[a]	
28	PKTN 41	*Taeniura lymma*	
29	PKTN 52	*Taeniura lymma*	
30	PKTN 62	*Dasyatis parvonigra*	
31	PKTN 87	*Gymnura poecilura*	
32	PKTN 96	*Dasyatis parvonigra*	
33	PKTN 99	*Himantura uarnak*	

Note: ([a]) in scientific names represents unsure field identification by conventional taxonomist due to cryptic morphological characters. PMKH, PDGN and PKTN represent species code used in this paper for samples collected from sampling stations Mukah, Dungun and Kuantan, respectively

oxidase C subunit 1 gene) was carried out using multiple universal primer sets (Table 2).

Each 25 µl of PCR reaction mixture contained 9.5 µl of sterile distilled water, 12.5 µl of 1 × MyTaq™ Mix (BIOLINE), 1.0 µl of each forward and reverse primer

Table 2 PCR primer sets used to amplify the target gene (mitochondrial COX1 gene)

Name	Primer sequence 5′ to 3′	Reference
HCO2198	TAAACTTCAGGGTGACCAAAAAATCA	Folmer et al. (1994)
VF1_t1	TGTAAAACGACGGCCAGTTCTCAACCAACCACAAAGACATTGG	Ivanova et al. (2007)
VR1_t1	CAGGAAACAGCTATGACTAGACTTCTGGGTGGCCAAAGAATCA	Ivanova et al. (2007)
VF2_t1	TGTAAAACGACGGCCAGTCAACCAACCACAAAGACATTGGCAC	Ivanova et al. (2007)
FishF2_t1	TGTAAAACGACGGCCAGTCGACTAATCATAAAGATATCGGCAC	Ivanova et al. (2007)
FishR2_t1	CAGGAAACAGCTATGACACTTCAGGGTGACCGAAGAATCAGAA	Ivanova et al. (2007)
FR1d_t1	CAGGAAACAGCTATGACACCTCAGGGTGTCCGAARAAYCARAA	Ivanova et al. (2007)

(5.0 μM) and 1.0 μl of DNA template (approximately 30–100 ng of DNA). The PCR thermal cycle conditions were set up with an initial of 2 min for denaturation at 94.0 °C, followed by 35 cycles of 30 s for denaturation at 94.0 °C, 40 s for amplification at 52.5 °C, 1 min for annealing at 72.0 °C, finished with final extension for 10 min at 72.0 °C and then held at 4.0 °C. PCR products were gel eluded on 1.5% agarose gel and visualised under AlphaImager HP system (Alpha Innotech, USA). To ensure ample amount of PCR product recovery, optimisation step was performed (if necessary) to ensure quality PCR products were used to derive DNA sequencing. Sequencing was done by outsourcing at Repfon Glamor Sdn. Bhd following Sanger sequencing standard protocol.

2.3 *Data Analysis*

Chromatograms were inspected for ambiguous, noisy base pairs prior to stopping codon analysis. Noisy tails of sequences were trimmed and either one of the strand was reverse complemented and for identifying primer targeted sequence region. Sequences were run in BLAST to identify possible species match and the data were cross-examined with field identification of samples. Sequences with less than 500 bp were excluded for phylogenetic tree construction. Software used for Chromas Lite (Technelysium Pty. Ltd., Australia) 2.1v for sequence editing, ClustalX (University College Dublin, Ireland) 2.1v was used for sequence alignment Mega6: Molecular Evolutionary Genetics Analysis Version 6.0 software used for phylogenetic analysis. *Epinephelus lanceolatus* (genbank accession ID: HQ174835.1) was used as an outgroup to construct NJ phylogenetic tree.

3 Results and Discussion

A total of 29 out of 33 samples were successfully barcoded for a fragment of the COI gene (~652 bp). There were 10 species belonging to 3 families, namely Myliobatidae (*Aetobatus ocellatus*), Gymnuridae (*Gymnura japonica, G. poecilura*) and Dasyatidae (*Dasyatis parvonigra, Neotrygon kuhlii, Taeniura lymma, Himantura leoparda, H. uarnacoides, H. uarnak and H. walga*) were successfully identified either morphologically or at the molecular level (Table 3). Only 16 out of 29 samples showed the best match (similarity percentage) in Genbank BLAST analysis due to insufficient data available in the database. Hence, BOLD data analysis was used to identify the remaining 13 samples and to revalidate BLAST analysed results (Tables 4 & 5). Almost all samples showed high similarity percentage (>99%) with corresponding species data available in the database. However, few samples were misidentified in database similarity analysis. For instance, morphologically identified *Gymnura poecilura* was paired with *G. japonica* in BLAST analysis. Similarly, *Himantura uarnak* paired with *H. leoparda, Aetobatus narinari* showed

Table 3 Taxonomical position and IUCN status of each species of rays sampled and barcoded in this study

Picture	Taxonomy
	Order: Myliobatiformes Family: Myliobatidae Genus: *Aetobatus* Species: *ocellatus* Common name: Eagle ray IUCN status: Vulnerable Current population trend: Decreasing
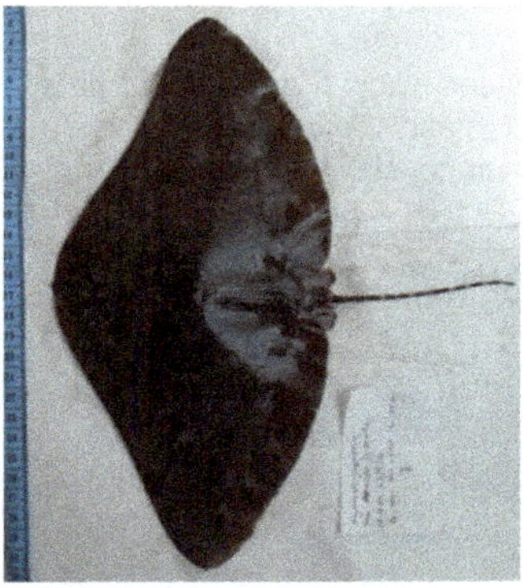	Order: Myliobatiformes Family: Gymnuridae Genus: *Gymnura* Species: *japonica* Common name: Japanese butterfly ray IUCN status: Data deficient Current population trend: Unknown
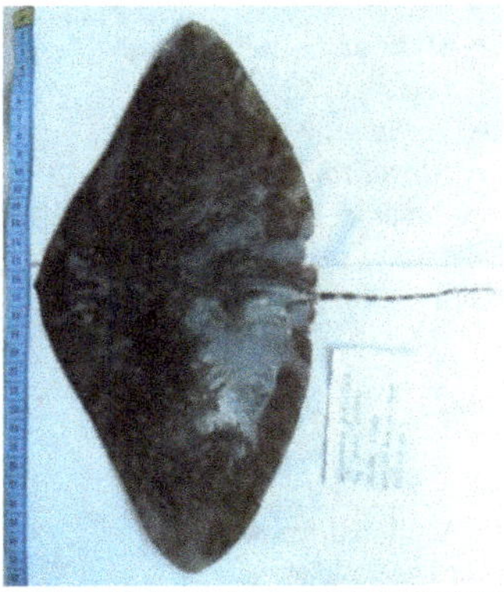	Order: Myliobatiformes Family: Gymnuridae Genus: *Gymnura* Species: *poecilura* Common name: Long tail butterfly ray IUCN status: Near threatened Current population trend: Decreasing

(continued)

Table 3 (continued)

Picture	Taxonomy
	Order: Myliobatiformes Family: Dasyatidae Genus: *Dasyatis* Species: *parvonigra* Common name: Dwarf Black stingray IUCN status: Data deficient Current population trend: Unknown
	Order: Myliobatiformes Family: Dasyatidae Genus: *Neotrygon* Species: *kuhlii* Common name: Blue spotted stingray IUCN status: Data deficient Current population trend: Unknown
	Order: Myliobatiformes Family: Dasyatidae Genus: *Taeniura* Species: *lymma* Common name: Ribbon tailed stingray IUCN status: Near threatened Current population trend: Unknown
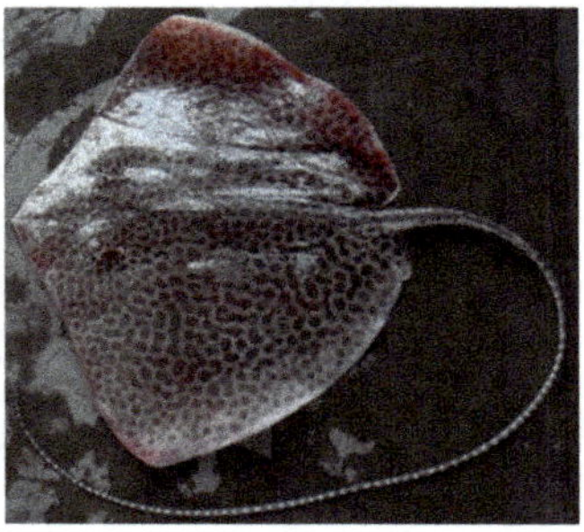	Order: Myliobatiformes Family: Dasyatidae Genus: *Himantura* Species: *leoparda* Common name: Leopard Whipray IUCN status: Vulnerable Current population trend: Decreasing
	Order: Myliobatiformes Family: Dasyatidae Genus: *Himantura* Species: *uarnacoides* Common name: Bleeker's Whipray IUCN status: Vulnerable Current population trend: Decreasing

(continued)

Table 3 (continued)

Picture	Taxonomy
	Order: Myliobatiformes Family: Dasyatidae Genus: *Himantura* Species: *uarnak* Common name: Reticulate Whipray IUCN status: Vulnerable Current population trend: Decreasing
	Order: Myliobatiformes Family: Dasyatidae Genus: *Himantura* Species: *walga* Common name: Dwarf Whipray IUCN status: Near threatened Current population trend: Decreasing

Note: Picture source from SEAFDEC/MFRDMD, Kuala Terengganu. Malaysia; except for *Neotrygon kuhlii* (Photographed by: Lesley Clements; Source http://www.inaturalist.org/observations/2688114)

100% BLAST similarity with *A. ocellatus,* while *Dasyatis parvonigra was* paired with *Neotrygon kuhlii.* This is due to overlapping morphological characters shared by species of the same genera. For instance, *Gymnura poecilura* in Indonesian waters is identical to *G. zonura.* This has been tentatively supported by morphometric character comparisons. In the Philippines, *G. poecilura* is misidentified as *G. japonica.* Hence, the taxonomic review of Gymnuridae in the Indo-West Pacific would be helpful to resolve this problem ("The IUCN Red List of Threatened Species," 2016). Similarly, *A. ocellatus* is usually misidentified as *A. narinari*. White et al. (2010) argued that these two species collected from Western Atlantic region are morphologically similar with minor morphometric differences (tail length, tail tip, background colouration of the dorsal surface) though in some samples they are overlapped. The colouration pattern generally used in field identification is highly controversial due to the direct influence of ambient environmental (water/sediment) conditions having significant influence on skin colouration (Manjaji-Matsumoto and Last 2008). It is also decided by the age, sex, life stages, external stress and behavioural adaptability of fishes (Sandoval-Castillo and Rocha-Olivares 2011; Toffoli et al. 2008; Wynen et al. 2009). Neighbour-joining (NJ) tree showed *Dasyatis parvonigra* to be the sister taxa to *Neotrygon kuhlii*. However, morphological features can clearly distinguish the latter from the former as *D. parvonigra* has no spot on the upper disc in contrast with *N. kuhlii* that has blue spots. In the case of *Himantura sp.,* many researchers have argued that

Table 4 BLAST analysis of generated DNA sequences of ray fishes shows similarity percentage and accurate identification of species (https://blast.ncbi.nlm.nih.gov/Blast.cgi?PAGE_TYPE=BlastSearch)

Sample code	Field identification	Identity ratio (%)	Sequence ID (bp)	Species Identified
PMKH 1	*Gymnura japonica*	652/652 (100%)	EU398804.1 (652)	[a]*Gymnura poecilura*
PMKH 2	*Gymnura poecilura*	652/652 (100%)	EU398804.1 (652)	*Gymnura poecilura*
PMKH 3	*Gymnura japonica*	652/652 (100%)	EU398804.1 (652)	[a]*Gymnura poecilura*
PMKH 6	*Himantura uarnacoides*	651/652 (99%)	KM073009.1 (652)	*Himantura uarnacoides*
PMKH 7	*Himantura uarnacoides*	652/652 (100%)	KM073009.1 (652)	*Himantura uarnacoides*
PMKH 8	*Himantura* cf. *uarnacoides*	651/652 (98%)	KM073009.1 (652)	[a]*Himantura uarnacoides*
PDGN 1	*Gymnura poecilura*[a]	652/652 (100%)	EU398804.1 (652)	*Gymnura poecilura*
PDGN 2	*Gymnura poecilura*[a]	651/652 (99%)	EU398804.1 (652)	*Gymnura poecilura*
PDGN 4	*Himantura uarnak*[a]	651/652 (99%)	KM072997.1 (652)	[a]*Himantura leoparda*
PDGN 5	*Himantura uarnak*[a]	650/652 (99%)	KM072997.1 (652)	[a]*Himantura leoparda*
PDGN 6	*Aetobatus narinari*[a]	652/652 (100%)	EU398508.1 (652)	[a]*Aetobatus ocellatus*
PDGN 7	*Taeniura lymma*	652/652 (100%)	KM073027.1 (652)	*Taeniura lymma*
PDGN 8	*Taeniura lymma*	652/652 (100%)	KM073027.1 (652)	*Taeniura lymma*
PDGN 11	*Dasyatis parvonigra*	652/652 (100%)	EU398733.1 (655)	[a]*Neotrygon kuhlii*
PKTN 7	*Gymnura japonica*	651/652 (99%)	EU398804.1 (652)	[a]*Gymnura poecilura*
PKTN 8	*Gymnura japonica*	649/652 (99%)	EU398804.1 (652)	[a]*Gymnura poecilura*
PKTN 9	*Gymnura japonica*	652/652 (100%)	EU398804.1 (652)	[a]*Gymnura poecilura*
PKTN 18	*Himantura uarnak*	651/652 (99%)	KM072997.1 (652)	[a]*Himantura leoparda*
PKTN 22	*Aetobatus ocellatus*	648/652 (99%)	EU398508.1 (652)	*Aetobatus ocellatus*
PKTN 28	*Himantura walga*	651/652 (99%)	KM072995.1 (652)	*Himantura walga*
PKTN 29	*Himantura walga*	651/652 (99%)	KM072995.1 (652)	*Himantura walga*
PKTN 30	*Himantura walga*	652/652 (100%)	EU398873.1 (655)	*Himantura walga*

(continued)

Table 4 (continued)

Sample code	Field identification	Identity ratio (%)	Sequence ID (bp)	Species Identified
PKTN 31	*Dasyatis parvonigra*	652/652 (100%)	EU398733.1 (655)	[a]*Neotrygon kuhlii*
PKTN 33	*Gymnura poecilura*	652/652 (100%)	EU398804.1 (652)	*Gymnura poecilura*
PKTN 34	*Gymnura poecilura*	649/652 (99%)	EU398804.1 (652)	*Gymnura poecilura*
PKTN 35	*Himantura leoparda*[a]	651/652 (99%)	KM072997.1 (652)	*Himantura leoparda*
PKTN 41	*Taeniura lymma*	652/652 (100%)	KM073027.1 (652)	*Taeniura lymma*
PKTN 52	*Taeniura lymma*	652/652 (100%)	KM073027.1 (652)	*Taeniura lymma*
PKTN 62	*Dasyatis parvonigra*	652/652 (100%)	EU398733.1 (655)	[a]*Neotrygon kuhlii*

Note. Symbol ([a]) represents the species that are not consistent with the initial identification

Table 5 The reconfirmed species after submitting the data on the BOLD system (http://www.boldsystems.org/index.php/IDS_OpenIdEngine)

Sample code	Field identification	Similarity (%)	Species identified by BOLD database
PMKH 1	*Gymnura japonica*	100	[a]*Gymnura poecilura*
PMKH 3	*Gymnura japonica*	100	[a]*Gymnura poecilura*
PMKH 8	*Himantura* cf. *uarnacoides*	99.85	[a]*Himantura uarnacoides*
PDGN 4	*Himantura uarnak*[a]	99.84	[a]*Himantura leoparda*
PDGN 5	*Himantura uarnak*[a]	99.84	[a]*Himantura leoparda*
PDGN 6	*Aetobatus narinari*[a]	100	[a]*A. narinari/ocellatus*
PDGN 11	*Dasyatis parvonigra*	100	[a]*Neotrygon kuhlii*
PKTN 7	*Gymnura japonica*	99.85	[a]*Gymnura poecilura*
PKTN 8	*Gymnura japonica*	99.53	[a]*Gymnura poecilura*
PKTN 9	*Gymnura japonica*	100	[a]*Gymnura poecilura*
PKTN 18	*Himantura uarnak*	100	[a]*Himantura leoparda*
PKTN 31	*Dasyatis parvonigra*	100	[a]*Neotrygon kuhlii*
PKTN 62	*Dasyatis parvonigra*	100	[a]*Neotrygon kuhlii*

existence of species complex and considerable morphological variability within the same forms leads to misidentification in the field and hence urgently require taxonomic revision especially on *H. gerrardi* (Manjaji-Matsumoto and Last 2008; Ward et al. 2009, 2008).

Our primary aim of verifying field identification of ray fish by experts in line with DNA barcoding techniques was successful to some extent due to overlapping morphological characters among the cryptic species besides limitations on the data availability in DNA databases. However, the efficiency of universal barcode gene in species delimitation was apparent in this study. Similarly, previous studies have proposed mitochondrial COI gene and NADH2 gene as reliable molecular markers for identification of elasmobranches (Holmes et al. 2009; Naylor et al. 2012). The efficiency of this method in illegal species trading and confiscation besides consumer preference towards ray products sold in markets are well established (Griffiths et al. 2013). DNA barcoding is also possible when only part of an organism is available (John et al. 2016). In such cases, the method is useful in animal forensics (Holmes et al. 2009). In our study, not all field-identified samples were accurately matched with same species data available in databanks, if Zemlak et al. (2009) rule of thumb for discriminating species (similarity below 96.5%) was used. Thus, expert taxonomical identification of samples in the field would help in ensuring accuracy in species discrimination. It should also be noted that taxonomic decision based on a single molecular marker (for example single gene sequence [COI gene] in this study) that is maternally inherited might not resolve all species identification. Hence, it is argued that COI gene sequencing cannot be used to discriminate recently evolved species as haplotypes are shared between transboundary species (Toffoli et al. 2008).

The paucity of information especially on DNA dataset of cartilaginous fishes in major databases (Genbank, BOLD database) perhaps would be the reason for mismatching of species. Similarly, Cerutti-Pereyra et al. (2012) observed only 19 out of 67 ray fish sequences tested had consistent matches on major databases. They further argued that this might be due to regular anomalies among closely related cryptic species that makes the species identification challenging. Misidentification might also be due to the presence of species complex, miss identification in the field or unavailability of species-specific sequence in databases. On the other hand, limitation associated with the online databases also includes (1) significant number of taxonomically unverified entries on GenBank and BOLD databases and (2) the presence of inconsistent, incorrect, short length duplicated lodged sequences with misleading scientific names.

Phylogenetic and phylogeographical signals are apparent in NJ tree constructed in this study (Fig. 2). All 33 sequences used in this study (including out group ID: HQ174835; *G. japonica, A. narinari* and *H. uarnak*) were clearly segregated into their corresponding congeneric forms. The average intra-species genetic distance was lower (0.25%) than inter-species difference (2.45%). Same species sampled from different sampling zones have segregated into their respective clad. Low genetic variation was observed in the same species collected from a geographically closer population (especially the same species from Kuantan and Terengganu, Peninsular Malaysia) compared to geographically distant population (Sarawak, East Malaysia). A similar observation was noted in an earlier study by Cerutti-Pereyra et al. (2012) who observed higher genetic distance between samples of the same species complex collected from western and eastern coasts.

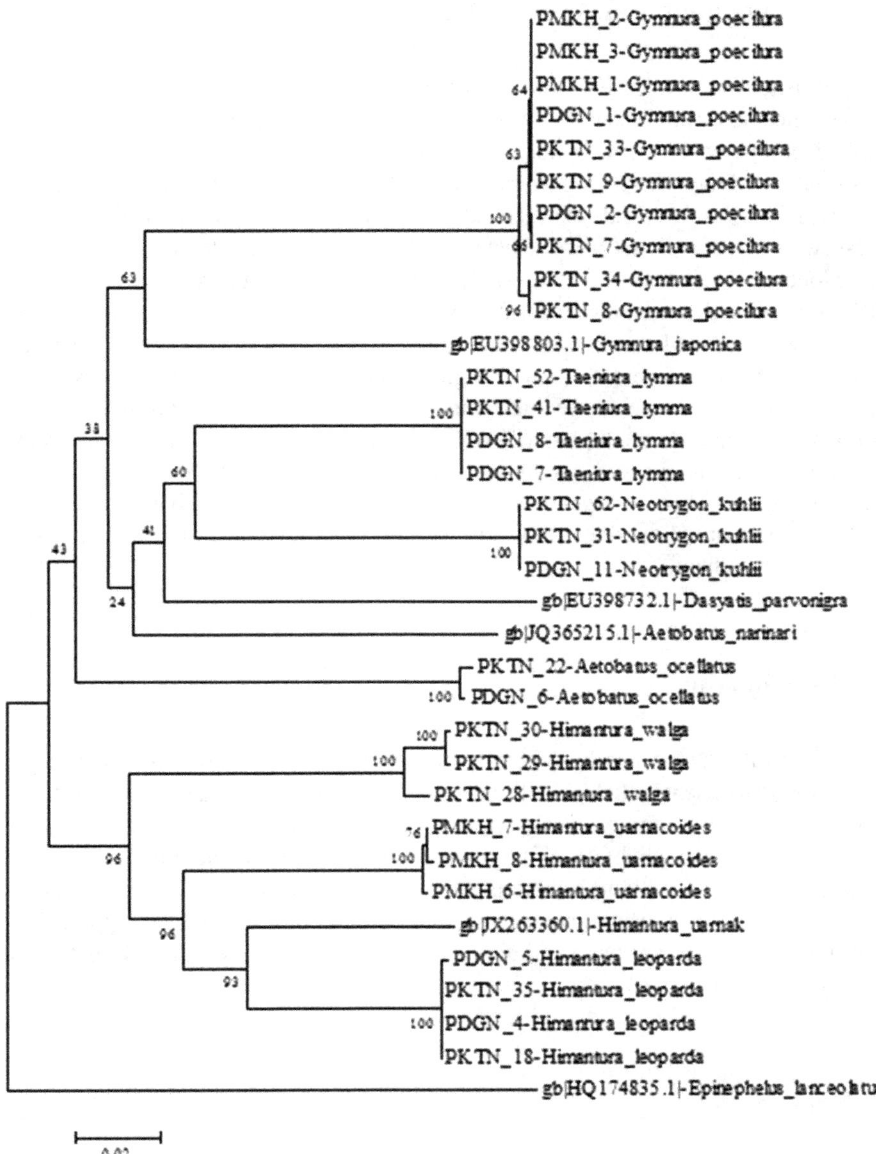

Fig. 2 Neighbour-joining (NJ) phylogenetic tree constructed to determine the evolutionary history of ray fishes. Note: The evolutionary history was inferred using the neighbour-joining method. The optimal tree with the sum of branch length = 1.03083732 is shown. The percentage of replicate trees in which the associated taxa clustered together in the bootstrap test (1000 replicates) is shown next to the branches. The tree is drawn to scale, with branch lengths in the same units as those of the evolutionary distances used to infer the phylogenetic tree. The evolutionary distances were computed using the p-distance method and are in the units of the number of amino acid differences per site. The analysis involved 34 amino acid sequences. All positions containing gaps and missing data were eliminated. There were a total of 620 positions in the final dataset. Evolutionary analyses were conducted in MEGA6

4 Management and Action Plan

As ray fishes are considered to be one of the common table fish in Malaysia, their management and action plan for evaluating their natural stock and fishery pressure is limited compared to shark fishery. Until recently, 84 species of ray fishes were recorded in Malaysian waters compared to the 106 species of rays recorded in Indonesia that makes Malaysia second in South East Asia. Due to high fishery pressure and overexploitation of selected ray fishes (manta ray), their catch is banned in Indonesia since 2012. In Malaysia, the landing of elasmobranch from 1982 to 2012 contributed an average of 1.8% of total demersal fish landing. In general, fishery practice in Malaysia does not target shark and ray fishing. They are being caught by trawls accidentally and brought to the shore for reasonable sales. High consideration was drawn on the management of sharks in Malaysia through national conservation management plan (PLAN 2). The definition of the word '*shark*' includes all cartilaginous or chondrichthyan fishes, comprising sharks, rays, skates and chimaeras. This management plan includes a review on shark population, trading and resource overview, revision, monitoring and evaluation for sustainable fishery management.

5 Conclusion

In conclusion, DNA barcoding is an accurate and rapid technique in a successful validation of consensus field identified samples. There is an urgent need for taxonomic revision on morphologically similar cryptic ray species. Misidentification is perhaps due to the availability of incongruent, inconsistent and ambiguous database in DNA databanks. Although there is a severe criticism on the efficiency of DNA barcoding techniques using single molecular marker sequencing, such approach is still highly useful in species validation and thereby can be used in various fishery management practices.

References

Ali A, Khiok ALP, Fahmi, Dharmadi, Krajangdara T (2013) Field guide to look-alike sharks and rays species of the southeast asian region: Southeast Asians Fisheries Development Center / Marine Fishery Resources Development And Management Department

Ambak MA, Terengganu PUM (2012) Fishes of Malaysia: Penerbit UMT

Bineesh K, Gopalakrishnan A, Akhilesh K, Sajeela K, Abdussamad E, Pillai N et al (2016) DNA barcoding reveals species composition of sharks and rays in the Indian commercial fishery. Mitochondrial DNA A DNA Mapp Seq Anal 28(4):1–15

Blagoev GA, deWaard JR, Ratnasingham S, deWaard SL, Lu L, Robertson J et al (2016) Untangling taxonomy: a DNA barcode reference library for Canadian spiders. Mol Ecol Resour 16(1):325–341

Bortolus A (2008) Error cascades in the biological sciences: the unwanted consequences of using bad taxonomy in ecology. AMBIO J Hum Environ 37(2):114–118

Burns JM, Janzen DH, Hajibabaei M, Hallwachs W, Hebert PD (2007) DNA barcodes of closely related (but morphologically and ecologically distinct) species of skipper butterflies (Hesperiidae) can differ by only one to three nucleotides. J Lepidopterists Soc 61(3):138–153

Cerutti-Pereyra F, Meekan MG, Wei N-WV, O'Shea O, Bradshaw CJ, Austin CM (2012) Identification of rays through DNA barcoding: an application for ecologists. PLoS One 7(6):e36479

Chang CH, Shao KT, Lin HY, Chiu YC, Lee MY, Liu SH, Lin PL (2016) DNA barcodes of the native ray-finned fishes in Taiwan. Molecular Ecology Resources 17(4):796–805

Folmer O, Black M, Hoeh W, Lutz R, Vrijenhoek R (1994) DNA primers for amplification of mitochondrial cytochrome c oxidase subunit I from diverse metazoan invertebrates. Mol Mar Biol Biotechnol 3(5):294–299

Griffiths AM, Miller DD, Egan A, Fox J, Greenfield A, Mariani S (2013) DNA barcoding unveils skate (Chondrichthyes: Rajidae) species diversity in 'ray'products sold across Ireland and the UK. PeerJ 1:e129

Hebert PD, Cywinska A, Ball SL (2003) Biological identifications through DNA barcodes. Proc R Soc Lond B Biol Sci 270(1512):313–321

Hebert PD, Gregory TR (2005) The promise of DNA barcoding for taxonomy. Syst Biol 54 (5):852–859

Holmes BH, Steinke D, Ward RD (2009) Identification of shark and ray fins using DNA barcoding. Fish Res 95(2):280–288

Ivanova NV, Zemlak TS, Hanner RH, Hebert PD (2007) Universal primer cocktails for fish DNA barcoding. Mol Ecol Notes 7(4):544–548

John BA, Sheikh HI, Jalal K, Zaleha K, Kamaruzzaman B (2016) Revised phylogeny of extant xiphosurans (horseshoe crabs). In: DNA Barcoding in Marine Perspectives. Springer, Switzerland, pp 113–130

Madduppa H, Ayuningtyas RU, Subhan B, Arafat D (2016) Exploited but unevaluated: DNA barcoding reveals skates and stingrays (Chordata, Chondrichthyes) species landed in the Indonesian fish market. ILMU KELAUTAN: Indonesian J Marine Sci 21(2):77–84

Manjaji-Matsumoto BM, Last PR (2008) Himantura leoparda sp. nov., a new whipray (Myliobatoidei: Dasyatidae) from the indo–Pacific. In: Descriptions of new Australian Chondrichthyans, vol 22. CSIRO Marine and Atmospheric Research, Hobart, Tas, pp 293–301

Naylor GJ, Caira JN, Jensen K, Rosana K, White WT, Last P (2012) A DNA sequence–based approach to the identification of shark and ray species and its implications for global elasmobranch diversity and parasitology. In: Bulletin of the American Museum of Natural History, pp 1–262

Sandoval-Castillo J, Rocha-Olivares A (2011) Deep mitochondrial divergence in Baja California populations of an aquilopelagic elasmobranch: the golden cownose ray. J Hered 102 (3):269–274

Steinke D, Prosser SW, Hebert PD (2016) DNA barcoding of marine metazoans. In: Marine Genomics: Methods and Protocols. Springer, New York, pp 155–168

The IUCN Red List of Threatened Species (2016) Version 2016–3. Retrieved 28–1-2017, from www.iucnredlist.org

Toffoli D, Hrbek T, Araújo MLG, Almeida MP, Charvet-Almeida P, Farias IP (2008) A test of the utility of DNA barcoding in the radiation of the freshwater stingray genus Potamotrygon (Potamotrygonidae, Myliobatiformes). Genet Mol Biol 31(1):324–336

Ward RD, Hanner R, Hebert PD (2009) The campaign to DNA barcode all fishes, FISH-BOL. J Fish Biol 74(2):329–356

Ward RD, Holmes BH, White WT, Last PR (2008) DNA barcoding Australasian chondrichthyans: results and potential uses in conservation. Mar Freshw Res 59(1):57–71

White WT, Last PR, Naylor GJ, Jensen K, Caira JN (2010) Clarification of Aetobatus ocellatus (Kuhl, 1823) as a valid species, and a comparison with Aetobatus narinari (Euphrasen, 1790)

(Rajiformes: Myliobatidae). In: Descriptions of new sharks and rays from Borneo, vol 32. CSIRO Marine and Atmospheric Research, Hobart, Tas, pp 141–164

Wynen L, Larson H, Thorburn D, Peverell S, Morgan D, Field I, Gibb K (2009) Mitochondrial DNA supports the identification of two endangered river sharks (Glyphis and Glyphis garricki) across northern Australia. Mar Freshw Res 60(6):554–562

Zemlak TS, Ward RD, Connell AD, Holmes BH, Hebert PD (2009) DNA barcoding reveals overlooked marine fishes. Mol Ecol Resour 9(s1):237–242

Molecular Phylogeny of Elasmobranchs

A. Pavan-Kumar, P. Gireesh-Babu, A. K. Jaiswar, S. G. Raje, A. Chaudhari, and G. Krishna

Abstract Elasmobranchs (sharks, rays and skates) are considered as one of the basal and successful lineages in vertebrate evolution. They were originated in lower Devonian period and subsequently radiated in the Carboniferous period with different morphological forms. Elasmobranchs colonized diverse fresh and marine water ecosystems by acquiring various adaptive traits. They play an important role in ecological balance as apex predators and are being used as model organisms for comparative biology and genomics. Before addressing any biological question, it is essential to identify/characterize the species accurately. However, in elasmobranchs, most of the species are yet to be characterized, and still ambiguity persists for some of the species identification. Phylogenetic studies with molecular data would resolve taxonomic ambiguity and provide insights into the evolutionary relationship among elasmobranchs. This chapter summarizes the phylogeny studies reported on elasmobranchs and highlights the significance of molecular phylogeny in resolving taxonomic uncertainties.

Keywords Mitochondrial markers · Nuclear markers · Taxonomy and Management

A. Pavan-Kumar (✉) · P. Gireesh-Babu · A. Chaudhari · G. Krishna
Fish Genetics and Biotechnology Division, ICAR-Central Institute of Fisheries Education, Mumbai, Maharashtra, India

A. K. Jaiswar
Fisheries Resources and Post Harvest Division, ICAR-CIFE, Mumbai, Maharashtra, India

S. G. Raje
ICAR-Central Marine Fisheries Research Institute, Regional Centre, Mumbai, Maharashtra, India

S. Trivedi et al. (eds.), *DNA Barcoding and Molecular Phylogeny*,
https://doi.org/10.1007/978-3-030-50075-7_9

1 Introduction

The elasmobranchs comprising sharks, skates and rays are considered as one of the basal and successful lineages in vertebrate evolution. As per the fossil record, they are reported to be originated during lower "Devonian period" and diversified during the "Carboniferous period" with different morphological and ecological forms (Sorenson et al. 2014; Grogan and Lund 2004). Due to their basal position in the vertebrate evolutionary tree, elasmobranchs have been used as a model organism to study primitive vertebrate physiology, comparative biology and comparative genomics (Venkatesh et al. 2007; Marra et al. 2017). Further, they play an important role in ecological balance as apex predators. Some of the shark fins have high economic value and huge demand in Asia. Due to this high economic value, elasmobranchs have been overexploited, and subsequently the natural stocks are declining at an alarming rate. Consequently, the conservation status of elasmobranchs has been evaluated as critical/endangered (IUCN 2016).

Accurate delineation and classification of biological entities is prerequisite to address any biological question. However, in the case of elasmobranchs, morphological characters often overlap between some of the species due to convergent evolution and cause difficulty in species discrimination. Phylogenetic species concept has been proposed to classify the species and resolve taxonomic ambiguity in various groups (Eldredge and Cracraft 1980; Nelson and Platnick 1981). Compared to the morphological traits, molecular markers could provide more number of characters and are comparable across diverse taxa. Several mitochondrial and nuclear markers were used to reconstruct phylogenetic trees to support/propose hypotheses on elasmobranch phylogeny. This chapter attempts to summarize the reported molecular phylogeny work on elasmobranchs.

Classification of elasmobranchs based on morphology grouped them into two monophyletic groups: batoids and sharks (Regan 1906; White 1937). Later, based on phenetics (similarities), Compagno (1973, 1977) classified elasmobranchs into four separate superorders, namely Galeomorphii (Orders: Heterodontiformes, Carcharhiniformes, Lamniformes, Orectolobiformes), Squalomorphii (Orders: Hexanchiformes, Pristiophoriformes, Squaliformes), Squatinomorphii (Genus *Squatina*) and Batoidea (Orders: Rajiformes, Rhinobatiformes, Myliobatiformes, Torpediniformes, Pristiformes) (Fig. 1a–b). Maisey (1980) merged superorders Squalomorphs and Squatinomorphs as Orbitostylic sharks on the basis of potential Synapomorphic trait (an orbital process that projects from the upper jaw cartilage inside the eye socket) (Fig. 2).

Later, several phylogenetic studies based on morphological characteristics (external, skeletal and muscular) reported that batoids (rays and skates) were derived from sharks and grouped them with the Pristiophoriformes and Squatinomorphii. This clade (Hypnosqualea) along with other Squalomorphs has been termed as "Squalea" and formed as a sister group to galeomorphs (Fig. 3) (Shirai 1992, 1996; de Carvalho 1996; de carvalho and Maisey 1996). The synapomorphic traits considered for this classification were orbital articulation, a basal angle on the suboptimal cranium and widely separated nasal capsules.

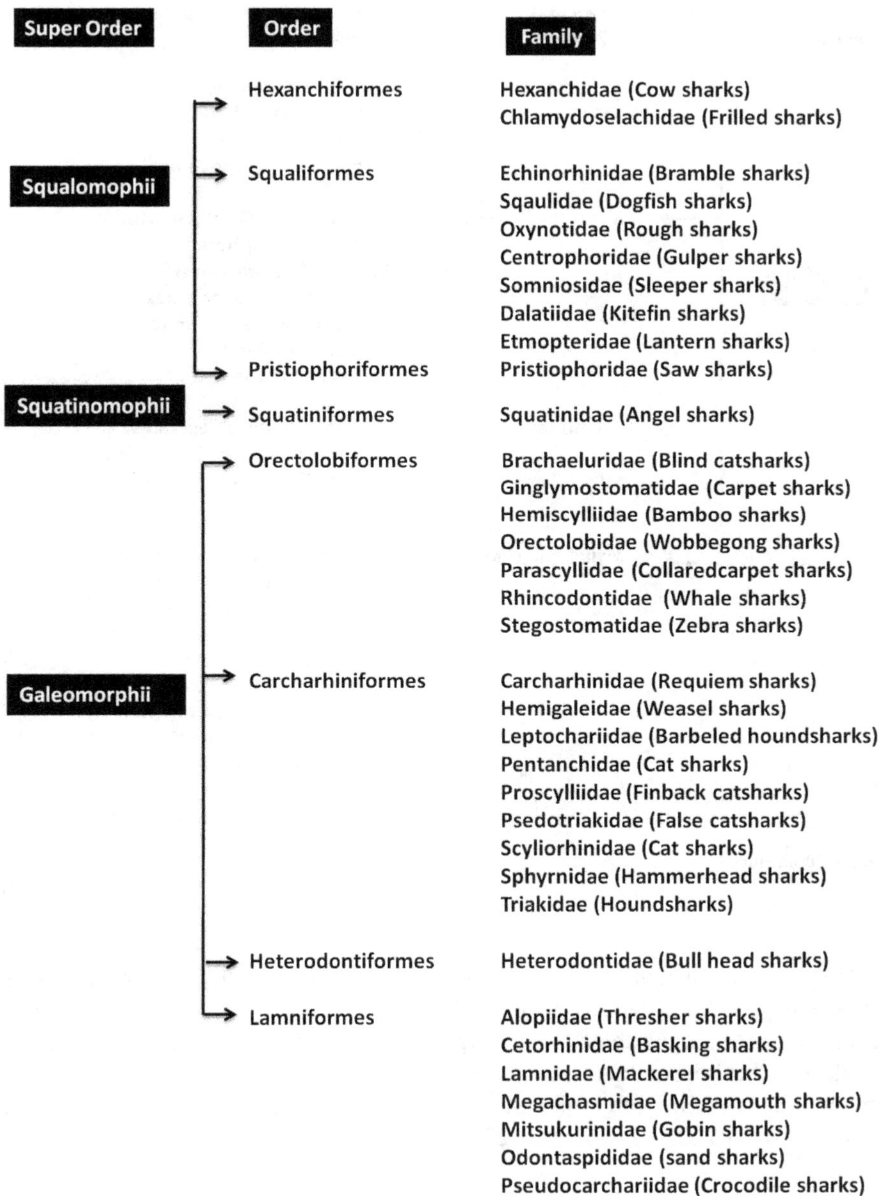

Fig. 1 (**a**) Classification of sharks. (**b**) Classification of rays (Batoidea)

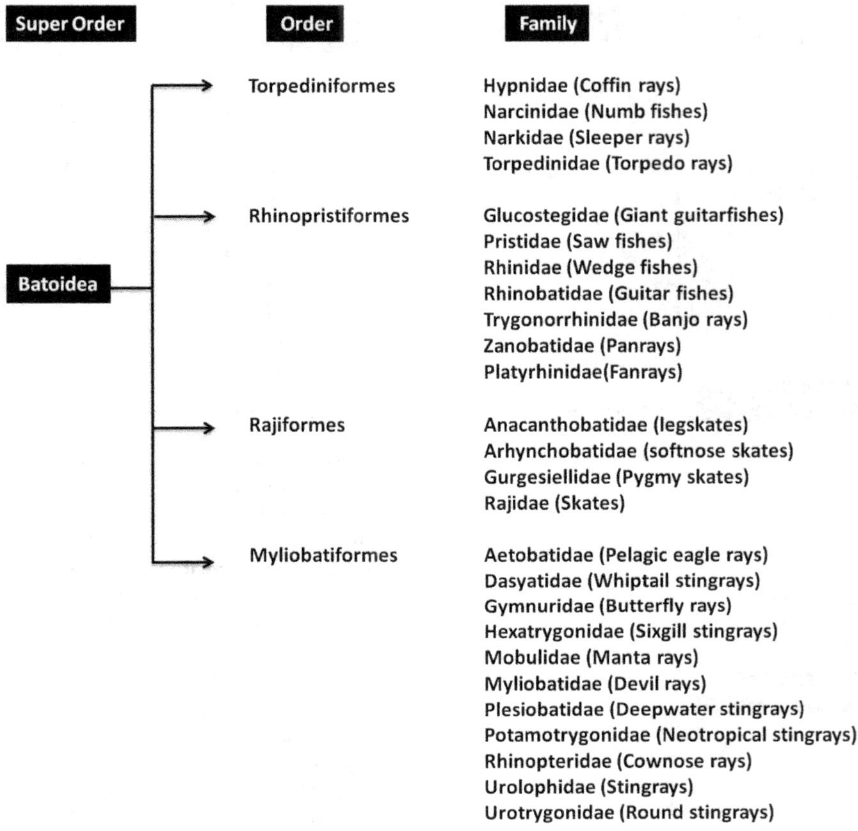

Fig. 1 (continued)

Due to some inherent problems with morphological characteristics such as poor preservation of fossil endoskeleton, convergent evolution of morphological traits and paucity of recognizable synapomorphies for certain groups, molecular markers (nuclear and mitochondrial) have been implied to infer phylogenetic relationship among orders/families of elasmobranchs (Bernardi et al. 1992; Douady et al. 2003, Dosay-Akbulut 2008, Naylor et al. 2012).

2 Interordinal Phylogeny

Several phylogenetic studies based on mitochondrial and nuclear genes supported the major division of sharks into Squalimorphii and Galeomorphii (Douady et al. 2003, Heinicke et al. 2009). Most of the molecular phylogenetic studies rejected

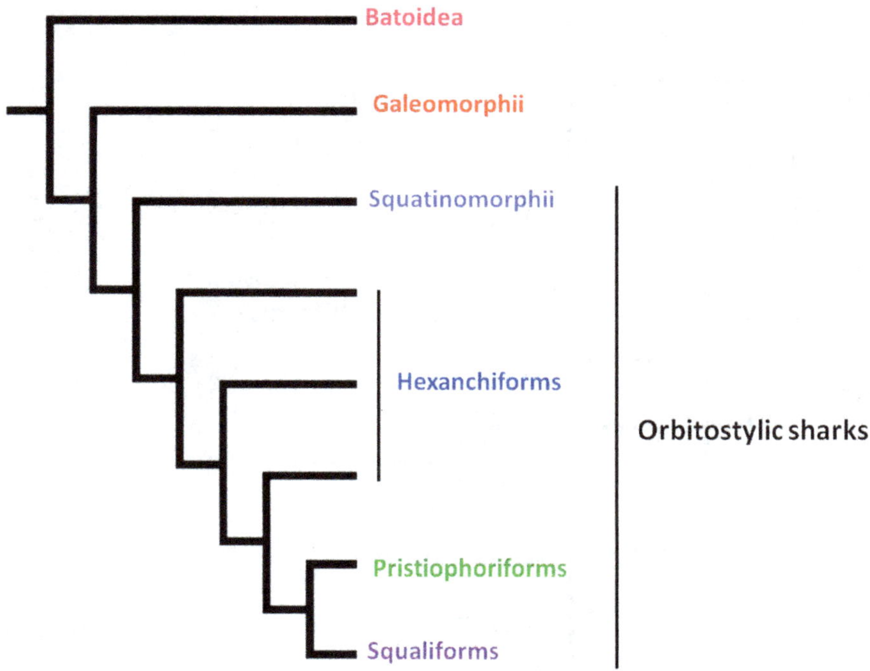

Fig. 2 Hypothesis of elasmobranch relationships proposed by Maisey (1980)

hypnosqualean hypothesis that positioned batoids within sharks (Compagno 1977; de Carvalho and Maisey 1996; Shirai 1992, 1996). As the genomic and species coverage increased, the phylogenetic tree topology and placement of orders changed and more resolved trees have been reported (Vélez-Zuazo and Agnarsson 2011; Naylor et al. 2005, 2012, Pavan-Kumar et al. 2014). Vélez-Zuazo and Agnarsson (2011) reported paraphyletic nature of Orectolobiformes using mito- and nuclear markers. However, their study was based on reported sequences from public databases and not from the original specimens. Later, based on the mitochondrial marker, Naylor et al. (2012) showed the monophyletic nature of all the orders except Squaliformes. The relationship within Squalimorphii was defined as (Hexanchiformes (Squaliformes (Pristiophoriformes, Squatiniformes))). Within Galeomorphii, Pavan-Kumar et al. (2014) showed sister group relation between Lamniformes—Orectolobiformes with limited species coverage (Fig. 4). With a large sample size, Naylor et al. (2012) have reconstructed the phylogenetic tree with 595 elasmobranch species using NADH2 gene and their study stated that "*Heterodontiformes are the immediate sister to a group consisting of the Orectolobiformes and the Lamniformes+Carcharhiniformes*". Further molecular phylogenetic work is needed with both mitochondrial and nuclear markers as the usage of a single marker would lead to soft polytomies (unresolved tree topology).

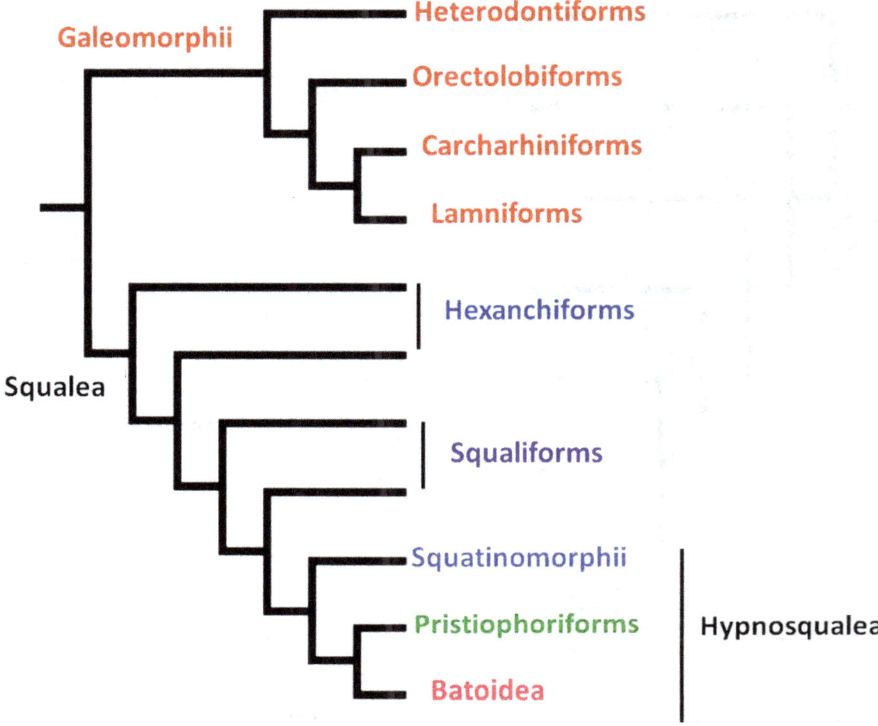

Fig. 3 Hypothesis of elasmobranch relationships proposed by Shirai (1992, 1996)

3 Batoidea

Batoids are characterized by the dorso-ventrally flattened body, greatly expanded pectoral fins attached to the side of the head and ventral mouth. Batoids comprise Myliobatiformes (stingrays), Pristiformes (sawfishes), Rajiformes (skates), Rhiniformes, Rhinobatiformes (guitarfishes) and Torpediniformes (electric rays) (McEachran and Aschliman 2004; Nelson 2006). Several hypotheses based on morphological characters have been proposed and no consensus agreement was observed among all these hypotheses (Compagno 1973, 1977, 1999; Heemstra and Smith 1980; Nishida 1990; Shirai 1992, 1996; McEachran et al. 1996; McEachran and Dunn 1998; Peach and Rouse 2004; Aschliman et al. 2012a, b; Aschliman 2014). A cladistic analysis based on morphological characters supported the derivation of batoids from sharks and placed them within sharks (de Carvalho 1996; Shirai 1996). However, molecular phylogenetic analysis disproved this hypothesis and proposed batoids as distinct group from sharks (Lawson et al. 1995; Douady et al. 2003; Winchell et al. 2004).

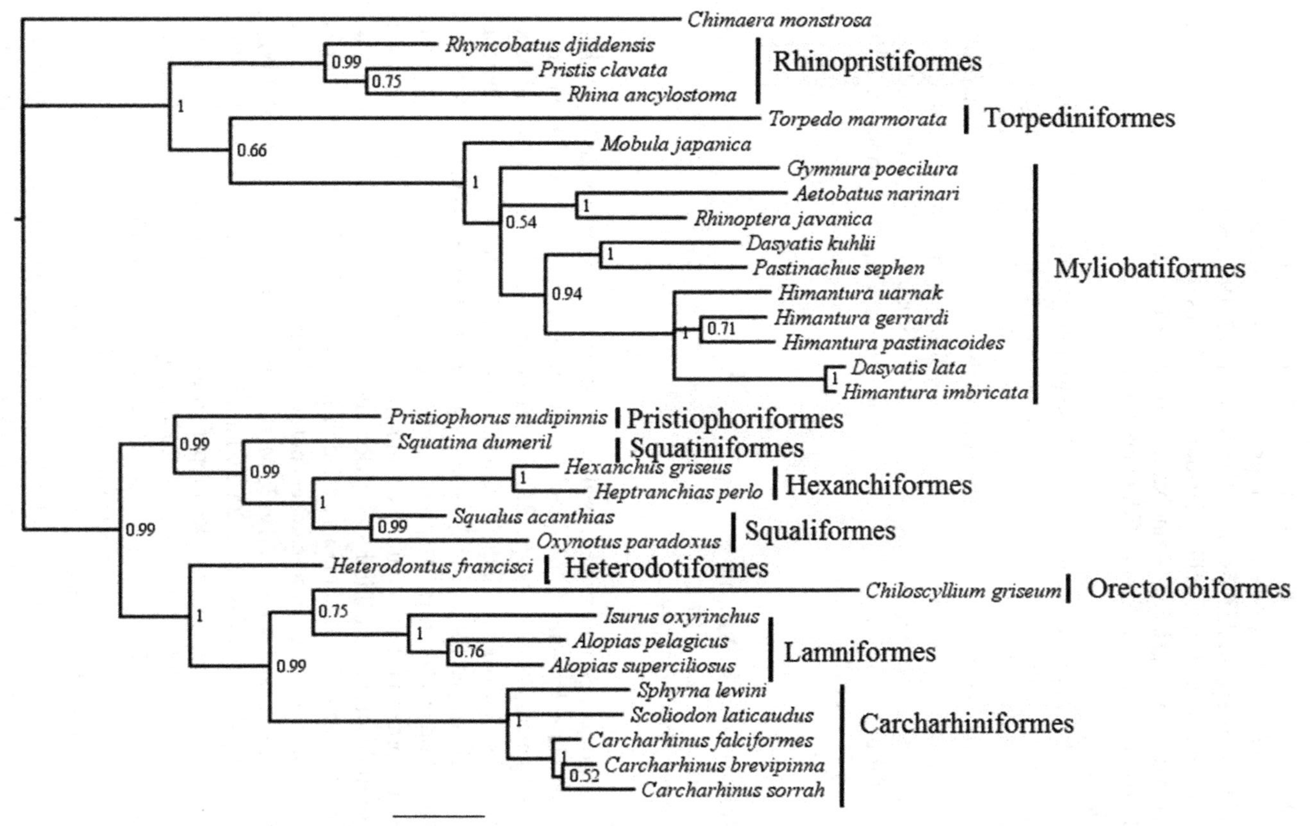

Fig. 4 Bayesian inference-based phylogeny depicting the relationship among major orders of elasmobranch inferred from mitochondrial markers (Adapted from Pavan-Kumar et al. 2014)

Molecular phylogenetic analyses revealed the monophyletic nature of all the orders except guitar fishes (Naylor et al. 2012). Skates (Rajiformes) were recovered as sister to all other batoids and sawfish (*Pristis clavata*) was nested within the guitarfish clades. Whereas morphological studies suggested that sawfishes are sister to all other batoids (Nishida 1990; Shirai 1992, 1996; Kriwet 2004). Based on phylogenetic analysis, families Pristidae (sawfishes), Rhinidae (shark rays), Rhynchobatidae (wedgefish), Rhinobatidae (guitarfish) and Zanobatidae were grouped together and an order "Rhinopristiformes" was proposed by Naylor et al. (2012).

Evolutionary relationships among the four major lineages of batoids (Rajiformes, Torpediniformes, Rhinopristiformes and Myliobatiformes) have shown slight variations as per the number of taxa and genes used for reconstructing phylogeny (Naylor et al. 2012). In a few studies, Torpediniformes formed as a basal to a group consisting of the Myliobatiformes and the Rhinopristiformes + Rajiformes (McEachran et al. 1996, Rocco et al. 2007). Shirai (1996) proposed the Rhinopristiformes to be basal to a group comprising Torpediniformes and the Rajiformes + Myliobatiformes.

4 Intraordinal Phylogeny

4.1 Carcharhiniformes

Carcharhiniformes (ground sharks) is the largest order of sharks with more than 290 species classified into 9 families: Carcharhinidae (12 genera, 59 species), Triakidae (9 genera, 49 species), Pentanchidae (8 genera, 89 species), Scyliorhinidae (17 genera, 69 species), Proscyllidae (3 genera, 6 species), Pseudotriakidae (3 genera, 5 species), Sphyrnidae (2 genera, 10 species), Hemigaleidae (4 genera, 8 species) and Leptocathridae (1 genera, 1 species) (Fricke et al. 2018). The monophyletic nature of this order has been confirmed and supported by morphological characters as well as molecular data (Shirai 1996; Douady et al. 2003, Winchell et al. 2004; Iglésias et al. 2005). The evolutionary relationship among families of this order has been investigated by different researchers but with limited species coverage (Martin et al. 1992; Martin and Palumbi 1993; Martin 1993, 1995; Douady et al. 2003; Winchell et al. 2004; Iglésias et al. 2005). Family Scyliorhinidae (catsharks) is the basal group of this order and has been reported to have paraphylies (Maisey et al. 2004; Iglésias et al. 2005; Human et al. 2006; Naylor et al. 2012). Based on the phylogenetic studies using 12S ribosomal RNA, valine tRNA, 16S rRNA and RAG gene, Iglésias et al. (2005) resurrected a new family "Pentanchidae" to include a distinct clade of Scyliorhinidae. A comprehensive study including species of all families was carried by Naylor et al. (2012). The phylogenetic tree of Naylor et al. (2012) showed a sister relationship between Carcharhinidae and Sphyrnidae families. Species of family Scyliorhinidae formed into three distinct clades and the clade consisted of species of *Cephaloscyllium, Poroderma* and *Scyliorhinus* formed as a basal group to Carcharhiniformes. Family Hemigaleidae formed as a sister group to the clade of Carcharhinidae and Sphyrnidae. Paraphylies were observed in family Triakidae (Iglésias et al. 2005; Lopez et al. 2006; Naylor et al. 2012). This phylogenetic tree formed soft polytomies and could not resolve the relationship among Triakidae, Pseudotriakidae and Scyliorhinidae. These polytomies can be resolved by implying nuclear markers along with the reported mitochondrial markers. Nevertheless, this study has provided robust and comprehensive phylogeny to date.

4.2 Squaliformes

Squaliform sharks are highly specialized sharks with more than 130 species classified into 24 genera (Ebert et al. 2013). Several studies have proved the monophyletic nature of families Centrophoridae, Dalatiidae, Etmopteridae and Squalidae (Shirai 1992, 1996; de Carvalho, 1996; Compagno, 1984; Garrick 1959; Maisey et al. 2004; Vélez-Zuazo and Agnarsson 2011; Naylor et al. 2012; Straube et al. 2010). The phylogenetic placement of family Echinorhinidae was resolved by Straube et al. (2015). His study showed this family is a sister group to Squatiniformes and Pristiophoriformes. In Straube et al. (2015) phylogenetic tree, Squalidae formed as a basal group whereas Etmopteridae was the derived group. Oxynotidae nested within Somniosidae indicating paraphyletic nature of the latter family. Family Centrophoridae formed as a sister group to the clade comprising families of Etmopteridae, Dalatiidae, Somniosidae and Oxynotidae. Dalatiidae was a sister group to the clade comprising Etmopteridae, Somniosidae and Oxynotidae.

4.3 Myliobatiformes

Myliobatiformes (stingrays) includes more than 200 species (42 genera) classified into 11 families (Fricke et al. 2018). They have a wide distribution in tropical and temperate waters throughout the world. Several morphological synapomorphies supported the monophyly of Myliobatiformes (Compagno 1977; Heemstra and Smith 1980; Lovejoy 1996; McEachran et al. 1996; Nishida 1990). The molecular phylogenetic approach has revised the placement of different families within this group and also assisted in resurrecting new families. McEachran and Aschliman (2004) placed families of Platyrhinidae and Zanobatidae under Myliobatiformes in contrast to Compagno's (1999) arrangement within Rhinobatiformes. Naylor et al. (2012) have provided molecular evidence to place these families under Rhinobatiformes. Based on molecular phylogeny and other morphological characters, the family of Aetobatidae was resurrected from this order and subfamily Mobulinae was elevated to family status as Mobulidae (White and Naylor 2016; Poortvliet et al. 2015; White et al. 2018). Naylor et al. (2012) have shown the monophyletic nature of families Potamotrygonidae, Urotrygonidae, Myliobatidae and Gymnuridae. The family Dasyatidae was reported to be paraphyletic (Naylor et al. 2012; Lim et al. 2015). Lim et al. (2015) proposed the resurrection of three new families "Himanturidae", "Neotrigonidae" and Pastinachidae. The sister relationship between Rhinopteridae and Mobulidae is reported by several molecular phylogenetic methods (Naylor et al. 2012; Lim et al. 2015, Dunn et al. 2003).

5 Family-level Phylogeny

5.1 *Orectolobidae*

Wobbegong sharks are dorso-ventrally flattened demersal sharks with unique dermal lobes on the lateral sides of the head (Compagno 2001). Orectolobidae comprises three genera, namely *Eucrossorhinus* Regan, 1908 (*Eucrossorhinus dasypogon*), *Sutorectus* Whitely, 1939 (*Sutorectus tentaculatus*) and *Orectolobus* Bonaparte, 1834 (*O. japoicus, O. wardi, O. maculates, O. ornatus, O. halei, O. hatchinsi, O. floridus, O. parvimaculatus*). Several morphology-based studies placed *Sutorectus* in an ancestral clade and showed that *Eucrossorhinus* was the most derived genus and sister to *Orectolobus* (Dingerkus 1986; Goto 2001). Corrigan and Beheregaray (2009) studied molecular phylogeny of this family and the phylogeny tree revealed *Eucrossorhinus* at the basal position and *Sutorectus* as a sister group to *O. floridus* in a recently derived clade. Based on the genetic divergence values and evolutionary relationship, Corrigan and Beheregaray (2009) recommended to revise the taxonomy of this family to include all species within a single monophyletic genus, *Orectolobus*.

5.2 *Scyliorhinidae*

Sharks of the family Scyliorhinidae (Order: Carcharhiniformes) are known as catsharks and consist of approximately 17 genera with 69 species (Fricke et al. 2018). Previously this family consisted of ~150 species which were classified into two subfamilies: Scyliorhininae (genera: *Atelomycterus, Aulohalaelurus, Cephaloscyllium, Poroderma, Schroederichthys* and *Scyliorhinus*) and Pentanchinae (genera: *Apristurus, Asymbolus, Bythalaelurus, Cephalurus, Figaro, Galeus, Halaelurus, Haploblepharus, Holohalaelurus, Parmatus* and *Pentanchus*). Initially, this family was known to be as monophyletic based on morphological characters (Maisey 1984; Compagno 1988; Sato 2000). Later, various studies using the molecular phylogenetic approach have proved the occurrence of paraphylies in this family (Iglésias et al. 2005; Human et al. 2006; Naylor et al. 2012). Based on the phylogenetic study, Iglésias et al. (2005) resurrected a new family "Pentanchidae" to name the species of subfamily Pentanchininae.

5.3 Sphyrnidae

Due to the laterally expanded, dorsal–ventrally compressed head (Cephalofoil) species of family Sphyrnidae (2 genera, 10 species) are known as Hammerhead sharks. Based on morphological similarities, Gilbert (1967) has classified these sharks into three genera, namely *Sphyrna* (large species: *S. zygaena*, *S. lewini*, *S. mokarran*), *Platysqualus* (Small species: *S. tiburo*, *S. tudes*, *S. media* and *S. corona*) and *Eusphyra* (*E. blochii*). The structure and shape of cephalofoil are different in different groups of sharks (Mara et al. 2015). Laterally minimum expanded head is observed in the *Sphyrna tiburo* (bonnethead shark) while the *Eusphyra blochii* (winghead shark) has the most laterally expanded head. Gilbert (1967) placed the species *E. blochii* as a basal to this group. Later, Compagno (1988) revised the classification by erecting a new subgenus "*Mesozygaena*" to accommodate small species (*S. tudes*, *S. corona* and *S. media*). He assigned *S. tiburo* to subgenus "*Platysqulaus*" and other large species (*S. lewini*, *S. zygaena* and *S. mokarran*) were grouped under subgenus *Sphyrna.* In contrast to the Gilbert hypothesis, Compagno (1988) study revealed a phylogenetic tree with *S. tiburo* (laterally least expanded head) as a basal group while *E. blocii* (laterally most expanded head) as a nested taxon. By using the molecular phylogenetic approach, Martin (1993) showed *E. blochii* as a basal group while *S. tiburo* as a derived group. Lim et al. (2010) have used both mitochondrial and nuclear markers to reconstruct the phylogeny between species of this family. This phylogenetic tree showed a sister relationship between *S.tiburo-S.corona* and *S.tudes-S.media. Sphyrna lewini* formed as a sister group to the clade of ((*S.tudes*: *S.media*) (*S.tiburo*:*S.corona*)). *E. blochii* formed as a basal group to the tree. Lim et al. (2010) concluded that *"proposed subgenera remain paraphyletic. Continued recognition of two distinct genera (Eusphyra and Sphyrna) makes sense given the monophyly of the genus Sphyrna and the degree of divergence between Eusphyra* and *Sphyrna"*. The major application/contribution of the phylogenetic tree to this group is that it provided a valid hypothesis for the evolution of cephalofoil as the most laterally expanded forms were the ancestors for extant sharks while the least laterally expanded forms are derived species. However, Naylor et al. (2012) showed *E. blochii* as a nested group using mitochondrial NADH2 gene for the phylogenetic tree.

5.4 Triakidae

Species of family Triakidae (9 genera, 47 species) are classified into two subfamilies viz., Triakinae (Genera: *Triakis, Mustelus, Scylliogaleus*) and Galeorhininae (*Galeorhinus, Hypogaleus, Iago, Furgaleus, Hemitriakis* and *Gogolia)*. The species of this family are considered as an intermediate group between the basal Scyliorhinidae family and derived Carcharhinidae and Sphyrnidae. Lopez et al. (2006) showed the paraphyletic nature of this family and the genera *Iago* formed a separate clade (Naylor et al. 2012). Different studies have proved the paraphyly of the genera of *Triakis* and *Mustelus* (Lopez et al. 2006; Naylor et al. 2012).

5.5 Mobulidae

Mobulidae family consists of 11 species under genera *Manta* and *Mobula*. However, in molecular phylogenetic trees, the species of *Manta* was nested within a clade of species of *Mobula* (Poortvliet et al. 2015; Naylor et al. 2012). In this family, primarily three clades were resolved in phylogenetic trees: Clade one contained *Mobula japonica*, *M. mobular*, *M. tarapacana*, *Manta birostris* and *M. alfredi*. *Mobula eregoodootenkee* and *M. kuhlii* formed as a second clade. Third clade was composed of *M. munkiana*, *M. hypostoma* and *M. rochebrunei* (Poortvliet et al. 2015). With a large sample and genome coverage, White et al. (2018) have constructed the phylogenetic tree of Mobulidae family and their phylogenetic tree provided evidence to synonymize the genera *Manta* with *Mobula*. Further, this study synonymized *Mobula eregoodootenkee*, *M. japanica* and *M. rochebrunei* with *M. kuhlii, M. mobular* and *M. hypostoma*, respectively.

5.6 Studies on Trait Evolution and Phylogenetic Diversity

The molecular phylogenetic approach was used to investigate the evolution of bioluminescence trait in Squaliform sharks. Straube et al. (2015) showed that the photophores might have originated during the transition of the lower to upper Cretaceous time while families Dalatiidae, Etmopteridae, Oxynotidae and Somniosidae were splitting from their common ancestor. Chen and Kishino (2015) investigated patterns of phylogenetic diversity (PD) of sharks across the world and advocated to consider PD for formulating conservation measures. Gruber et al. (2016) reported the evolution of biofluorescence in families Urotrygonidae, Orectolobidae and Scyliorhinidae using the phylogenetic approach.

References

Aschliman NC (2014) Interrelationships of the durophagous stingrays (Batoidea: Myliobatidae). Environ Biol Fish 97:967–979

Aschliman NC, Nishida M, Miya M, Inoue JG, Rosana KM, Naylor GJ (2012a) Body plan convergence in the evolution of skates and rays (Chondrichthyes:Batoidea). Mol Phylogenet Evol 63:28–42

Aschliman NC, Claeson KM, McEachran JD (2012b) Phylogeny of batoidea. In: Carrier JC, Musick JA, Heithaus MR (eds) Biology of sharks and their relatives, 2nd edn. CRC Press, Boca Raton, FL

Bernardi G, Sordino P, Powers DA (1992) Nucleotide sequence of the 18S ribosomal ribonucleic acid gene from two teleosts and two sharks and their molecular phylogeny. Mol Mar Biol Biotechnol 3:187–194
Chen H, Kishino H (2015) Global pattern of phylogenetic species composition of shark and its conservation priority. Ecol Evol 5:4455–4465
Compagno LJV (1973) Interrelationships of living elasmobranchs. Zool J Linnean Soc 53:15–61
Compagno LJV (1977) Phyletic relationships of living sharks and rays. Am Zool 17:303–322
Compagno LJV (1984) FAO species catalogue, vol. 4, part 1. Sharks of the world. FAO publication 125. FAO, Rome
Compagno LJV (1988) Sharks of the order Carcharhiniformes. Princeton University Press, Princeton, NJ, pp 1–542
Compagno LJV (1999) Checklist of living elasmobranchs. In: Hamlett WC (ed) Sharks, skates, and rays: The biology of elasmobranch fishes. Johns Hopkins University Press, Baltimore, pp 471–498
Compagno LJV (2001) Sharks of the world. An annotated and illustrated catalogue of shark species known to date, vol 2. Bullhead, mackerel and carpet sharks (Heterodontiformes, Lamniformes and Orectolobiformes). FAO, Rome
Corrigan S, Beheregaray LB (2009) A recent shark radiation: molecular phylogeny, biogeography and speciation of wobbegong sharks (family: Orectolobidae). Mol Phy Evol 52:205–216
de Carvalho MR (1996) Higher-level elasmobranch phylogeny, basal squaleans, and paraphyly. In: Stiassny MLJ, Parenti LR, Johnson GD (eds) Interrelationships of fishes. Academic, San Diego, pp 35–62
de Carvalho MR, Maisey JG (1996) Phylogenetic relationships of the late Jurassic shark Protospinax Woodward, 1919 (Chondrichthyes: Elasmobranchii). In: Arratia G, Viohl G (eds) Mesozoic fishes: systematics and paleoecology. Verlag Dr. Friedrich Pfiel, Munich, pp 9–46
Dingerkus G (1986) Interrelationships of orectolobiform sharks (Chondrichthyes: Selachii). In: Proceedings of the 2nd international conference on indo-pacific fishes. Ichthyological Society of Japan, Tokyo, pp 227–245
Dosay-Akbulut M (2008) The phylogenetic relationship within the genus Carcharhinus. C R Biol 331:500–509
Douady CJ, Dosay M, Shivji MS, Stanhope MJ (2003) Molecular phylogenetic evidence refuting the hypothesis of Batoidea (rays and skates) as derived sharks. Mol Phylogenet Evol 26:215–221
Dunn KA, McEachran JD, Honeycutt RL (2003) Molecular phylogenetics of myliobatiform fishes (Chondrichthyes: Myliobatiformes), with comments on the effects of missing data on parsimony and likelihood. Mol Phylogenet Evol 27:259–270
Ebert DA, Fowler S, Compagno LJV (2013) Sharks of the world – a fully illustrated guide, Wild Nature Press, Plymouth, p 528
Eldredge N, Cracraft J (1980) Phylogenetic patterns and the evolutionary process. Methods and theory in contemparative biology. Columbia University Press, New York, p 349
Fricke R, Eschmeyer WN, van der Laan R (2018) Catalog of fishes: genera, species. https://researcharchive.calacademy.org/research/ichthyology/catalog/fishcatmain.asp. Electronic version Accessed 10 Oct 2018
Garrick JAF (1959) Studies on New Zealand Elasmobranchii. Part VII. The identity of specimens of Centrophorus from New Zealand. Trans Roy Soc New Zealand 86:127–141
Gilbert CR (1967) A revision of the hammerhead sharks (family Sphyrnidae). Pro US Natl Mus 119:1–87
Goto T (2001) Comparative anatomy, phylogeny, and cladistic classification of the order Orectolobiformes (Chondrichthyes: Elasmobranchii). Mem Grad Sch Fish Sci Hokkaido Univ 48:1–100
Grogan E, Lund R (2004) The origin and relationships of early chondrichthyes. In: Carrier J, Musick J, Heithaus M (eds) Biology of sharks and their relatives. CRC, Washington, DC, pp 3–31

Gruber DF et al (2016) Biofluorescence in Catsharks (Scyliorhinidae): fundamental description and relevance for elasmobranch visual ecology. Sci Rep 6:24751. https://doi.org/10.1038/srep24751

Heemstra PC, Smith MM (1980) Hexatrygonidae, a new family of stingrays (Myliobatiformes: Batoidea) from South Africa, with comments on the classification of batoid fishes. Ichthyol bull J.L.B. Smith Inst Ichthyol 43:1–17

Heinicke MP, Naylor GJP, Hedges SB (2009) Cartilaginous fishes (Chondrichthyes). In: Hedges SB, Kumar S (eds) The time tree of life. Oxford University Press, New York, p 320

Human BA, Owen EP, Compagno LJV, Harley EH (2006) Testing morphologically based phylogenetic theories within the cartilaginous fishes with molecular data, with special reference to the catshark family (Chondrichthyes; Scyliorhinidae) and the interrelationships within them. Mol Phylogenet Evol 39:384–391

Iglésias SP, Lecointre G, Sellos DY (2005) Extensive paraphylies within sharks of the order Carcharhiniformes inferred from nuclear and mitochondrial genes. Mol Phylogenet Evol 34:569–583

IUCN (2016) The IUCN red list of threatened species. Version 2016-3. http://www.iucnredlist.org. Downloaded on 15 March 2017.

Kriwet J (2004) The systematic position of the cretaceous sclerorhynchid sawfishes (Elasmobranchii, Pristiorajea). In: Arriata G, Tintori A (eds) Mesozoic fishes 3. München, Verlag Dr Friedrich Pfiel, pp 57–73

Lawson R, Burch SJ, Oughterson SM, Heath S, Davies DH (1995) Evolutionary relationships of cartilaginous fishes: an immunological study. J Zool 237:101–106

Lim DD, Motta P, Mara K, Martin AP (2010) Phylogeny of hammerhead sharks (family Sphyrnidae) inferred from mitochondrial and nuclear genes. Mol Phylogenet Evol 55:572–579

Lim KC, Lim PE, Chong VC, Loh KH (2015) Molecular and morphological analyses reveal phylogenetic relationships of stingrays focusing on the family Dasyatidae (Myliobatiformes). PLoS One 10(4):e0120518

Lopez JA, Ryburn JA, Fedrigo O, Naylor GJ (2006) Phylogeny of sharks of the family Triakidae (Carcharhiniformes) and its implications for the evolution of carcharhiniform placental viviparity. Mol Phylogenet Evol 40:50–60

Lovejoy NR (1996) Systematics of myliobatoid elasmobranchs: with emphasis on the phylogeny and historical biogeography of neotropical freshwater stingrays (Potamotrygonidae: Rajiformes). Zool J Linnean Soc 117:207–257

Maisey JG (1980) An evaluation of jaw suspension in sharks. Am Mus Novit 2706:1–17

Maisey JG (1984) Higher elasmobranch phylogeny and biostratigraphy. Zool J Linnean Soc 82:33–54

Maisey JG, Naylor GJP, Ward DJ (2004) Mesozoic elasmobranchs, neoselachian phylogeny, and the rise of modern neoselachian diversity. In: Arratia G, Tintori A (eds) Mesozic fishes III. Systematics, Paleoenvironments and biodiversity. Verlag Pfeil, Munich

Mara KR, Motto PJ, Martin AP, Hueter RE (2015) Constructional morphology within the head of hammerhead sharks (sphyrnidae). J Morphol 276:526–539

Marra JN, Richards VP, Early A, Bogdanowicz SM, Pavinski Bitar PD, Stanhope MJ, Shivji SM (2017) Comparative transcriptomics of elasmobranchs and teleosts highlight important processes in adaptive immunity and regional endothermy. BMC Genomics 18:87

Martin A (1993) Hammerhead shark origins. Nature 364:494

Martin AP (1995) Mitochondrial DNA sequence evolution in sharks: rates, patterns, and phylogenetic inferences. Mol Biol Evol 12:1114–1123

Martin AP, Palumbi SR (1993) Protein evolution in divergent cellular environments: cytochrome b in sharks and mammals. Mol Biol Evol 10:873–891

Martin AP, Naylor GJP, Palumbi SR (1992) Rates of mitochondrial DNA evolution in sharks are slow compared with mammals. Nature 357:153–155

McEachran JD, Aschliman N (2004) Phylogeny of Batoidea. In: Carrier JC, Musick JA, Heithaus MR (eds) Biology of sharks and their relatives. CRC, Boca Raton, FL, pp 79–113

McEachran JD, Dunn KA (1998) Phylogenetic analysis of skates, a morphologically conservative clade of elasmobranchs (Chondrichthyes: Rajidae). Copeia 2:271–290

McEachran JD, Dunn KA, Miyake T (1996) Interrelationships of the batoid fishes (Chondrichthyes: Batoidea). In: Stiassny MJ, Parenti LR, Johnson GD (eds) Interrelationships of fishes. Academic Press, London, pp 63–82

Naylor GJP, Ryburn JA, Fedrigo O, López JA (2005) Phylogenetic relationships among the major lineages of modern elasmobranchs. In: Hamlett WC, Jamieson BGM (eds) Reproductive biology and phylogeny, vol 3. Science, Enfield, NH, pp 1–25

Naylor GJP, Caira JN, Jensen K, Rosana KAM, Straube N, Lakner C (2012) Elasmobranch phylogeny: a mitochondrial estimate based on 595 species. In: Carrier JC, Musick JA, Heithaus MR (eds) Biology of sharks and their relatives. CRC, Boca Raton, pp 31–56

Nelson JS (2006) Fishes of the world, 4th edn. Wiley, Hoboken, NJ, p 601

Nelson G, Platnick NI (1981) Systematics and biogeography: cladistics and vicariance. Columbia University, New York, p 567

Nishida K (1990) Phylogeny of the suborder Myliobatoidei. Mem Fac Fish Hokkaido Uni 37:1–108

Pavan-Kumar A, Gireesh-Babu P, Suresh Babu PP, Jaiswar AK et al (2014) Molecular phylogeny of elasmobranchs inferred from mitochondrial and nuclear markers. Mol Bio Rep 41:447–457

Peach MB, Rouse GW (2004) Phylogenetic trends in the abundance and distribution of pit organs of elasmobranchs. Acta Zool 85:233–244

Poortvliet M, Olsen JL, Croll DA, Bernardi G, Newton K, Kollias S, O'Sullivan J, Fernando D, Stevens G et al (2015) A dated molecular phylogeny of manta and devil rays (Mobulidae) based on mitogenome and nuclear sequences. Mol Phylogenet Evol 83:72–85

Regan CT (1906) A classification of the selachian fishes. Proc Zool Soc London 1906:722–758

Rocco L, Liguori I, Costagliola D, Morescalchi MA, Tinti F, Stingo V (2007) Molecular and karyological aspects of Baoidea (Chondricthyes, Elasmobranchi) phylogeny. Gene 389:80–86

Sato K (2000) Phylogenetic systematics of the deep-water catshark genus Apristurus (Chondrichthyes, Carcharhiniformes, Scyliorhinidae). PhD. Thesis, Faculty of Fisheries, Hokkaido University

Shirai S (1992) Squalean phylogeny: a new framework of squaloid sharks and related taxa. Hokkaido University Press, Sapporo

Shirai S (1996) Phylogenetic interrelationships of neoselachians (Chondrichthyes: uselachii). In: Stiassny MLJ, Parenti LR, Johnson GD (eds) Interrelationships of fishes. Academic, San Diego, pp 9–34

Sorenson L, Sanini F, Alfaro ME (2014) The effect of habitat on modern shark diversification. J Evol Biol 27:1536–1548

Straube N, Iglésias SP, Sellos DY, Kriwet J, Schliewen UK (2010) Molecular phylogeny and node time estimation of bioluminescent lantern sharks (Elasmobranchii: Etmopteridae). Mol Phylogenet Evol 56:905–917

Straube N, Li C, Claes JM, Corrigan S, Naylor GJP (2015) Molecular phylogeny of Squaliformes and first occurrence of bioluminescence in sharks. BMC Evol Biol 15:62

Vélez-Zuazo X, Agnarsson I (2011) Shark tales: a molecular species-level phylogeny of sharks (Selachimorpha, Chondrichthyes). Mol Phylogenet Evol 58:207–217

Venkatesh B, Kirkness EF, Loh YH, Halpern AL, Lee AP et al (2007) Survey sequencing and comparative analysis of the elephant shark (*Callorhinchus milii*) genome. PLoS Biol 5(4):e101. https://doi.org/10.1371/journal.pbio.0050101

White EG (1937) Interrelationships of the elasmobranchs with a key to the order Galea. Bull Am Mus Nat Hist 74:25–138

White WT, Naylor GJP (2016) Resurrection of the family Aetobatidae (Myliobatiformes) for the pelagic eagle rays, genus Aetobatus. Zootaxa 4139:435–438

White TW, Corrigan S, Yang L, Henderson CA, Bazinet AL, Swofford DL, Naylor GJP (2018) Phylogeny of the manta and devilrays (Chondricthyes: Mobulidae), with an updated taxonomic arrangement for the family. Zool J Linnean Soc 182:50–75

Winchell CJ, Martin AP, Mallatt J (2004) Phylogeny of elasmobranchs based on LSU and SSU ribosomal RNA genes. Mol Phylogenet Evol 31:214–224

A Review on DNA Barcoding on Fish Taxonomy in India

V. Sachithanandam and P. M. Mohan

Abstract DNA barcoding has been promoted as an efficient tool in the identification and discovery of species through the use of a short gene, standardised mitochondrial cytochrome c oxidase I (COI) gene region. Fishes are a highly diverse group of vertebrates; the identification of fish species through a DNA barcoding tool will provide new perspectives in ecology and systematics of fish taxonomy sciences. The identification of fishes can be a problematic and time-consuming process through morphological taxonomy, even for experts due to numerous reasons. DNA barcoding is proving to be a useful and effective tool for species identification at the gene level. The Fish Barcode of Life campaign (FISH-BOL), an international research collaboration centre, was established as a DNA barcode library for reference sequence repository and to monitor the DNA barcode project progress at the regional level. The DNA barcode sequence from any fish specimen's tissue, fin, egg or larva can be matched with the online platform of NCBI and BOLD systems for species discrimination/identification. This chapter aims to investigate the current status of fish barcoding, approaches and future direction of DNA barcoding in fishery sciences. The current status of barcoding studies with reference to fish taxonomy in India has been evaluated, and a detailed review of the existing literature has been carried out at the regional, national and global levels. The study results elucidated that marine invertebrates' DNA barcoding study is still in its infancy in India.

Keywords Biodiversity · COI · Cytochrome c oxidase · DNA barcoding · Species identification

---

V. Sachithanandam (✉)
National Centre for Sustainable Coastal Management, Ministry of Environment, Forests and Climate Change, Anna University Campus, Chennai, India

Department of Ocean Studies and Marine Biology, Pondicherry University, Brookshabad Campus, Port Blair, India

P. M. Mohan
Department of Ocean Studies and Marine Biology, Pondicherry University, Brookshabad Campus, Port Blair, India

S. Trivedi et al. (eds.), *DNA Barcoding and Molecular Phylogeny*,
https://doi.org/10.1007/978-3-030-50075-7_10

1 Background

The biodiversity of the earth has not been explored extensively to date. It has been estimated that around 1.7 million species have been named, and yet another eight million species still have to be described ((Stoeckle and Hebert 2008). Lavina-Vincent et al. (2016) reported that during this decade, around a third of the inhabitants of the earth would be extinct. The technique used by a short standardised fragment of genomic DNA to identify the organism has been named DNA barcoding. This method will hasten the cataloguing of millions of species yet to be named. However, without proper traditional taxonomy, this is a very difficult task. Mitochondrial genes like 16S, COI, etc. are ideal gene targets to be used as DNA barcodes (Saccone et al. 1999) for the taxonomy of animals. Around 50,000 species (30% of all known species) have been barcoded so far, and it is available as open-source data (approximately 4,30,000 barcodes) in the public domain of NCBI, BOLD systems, etc. (Silva-Brandao et al. 2009; International Barcode of Life 2010; Lavina-Vincent et al. 2016). Fish have high diversity and morphological ambiguity, which leads to the misidentification of species (sibling level also). In these circumstances, the DNA barcoding tool emerged and can be used as an efficient and cost-effective method for the taxonomical identification/classification of fishes (Hubert et al. 2008; Ward et al. 2005, 2009; Sachithanandam et al. 2012, 2015). Almost 50% of the vertebrates are represented by fishes (six classes, sixty-two orders and 540 families), and approximately 32,100 existence fish species have been catalogued worldwide (Eschmeyer 2010; Sachithanandam et al. 2014). Among these catalogued fishes, around 6000 species of fishes were DNA barcoded, of which 400 species were from New Zealand, 207 species from Australia, 250 species from South Africa, 100 fish species from Pacific Canada water and 127 fishes species from Indian waters (Ward et al. 2009; Lakra et al. 2007; Sachithanandam et al. 2012).

The COI gene sequence is considered as a taxonomy tool, and the characterisation of fishes using this method is gaining momentum in taxonomy science, because of its accuracy in species identification and authentication without the help of morphological characters (Sachithanandam et al. 2012, 2014, 2015). Identification of different fauna, such as birds, fishes and bats, through the COI gene sequence, has been used as an inter alia among the confamily level, well documented throughout the world (Hebert et al. 2003a, 2004b; Ward et al. 2005; Hubert et al. 2008; Persis et al. 2009; Steinke et al. 2009a; Lakra et al. 2010; Sachithanandam et al. 2011, 2012, 2015). The effectiveness of this method has been demonstrated in the identification of diverse taxa, such as butterflies (Hebert et al. 2003a), birds (Hebert et al. 2004b), Australian fishes (Ward et al. 2005), invertebrates Mollusca (Bivalvia from Gulf of Mexico) (Jarnegren et al. 2007), mammals (Clare et al. 2007), Pacific Canada's fishes (Steinke et al. 2009a) and Indian fishes (Persis et al. 2009; Lakra et al. 2010; Sachithanandam et al. 2011, 2012, 2014, 2015). DNA barcoding is also being applied in the multidisciplinary fields such as fish conservation (Holmes et al. 2009), fishery managements and sustainable use (Rasmussen et al. 2009) and food safety analysis where mislabelling of commercially important species has been effectively detected by the mtDNA COI sequence (Wong and Hanner 2008).

The FISH-BOL was set up to establish a comprehensive repository of COI gene sequences of fishes from different regions, to enable global taxonomic identification of fishes (Ward et al. 2009; Eschmeyer 2010; Becker et al. 2011). The prime objective of the FISH-BOL is the collection of fish samples from different geographical regions by scientists specialising in fish both taxonomy (classical and barcoding approaches) and establishing the taxonomy first by traditional procedures and later by COI gene sequencing. This has provided a good amount of COI gene sequence results. The result shows that around 74 and 50% of the fishes in the Arctic and Antarctic regions respectively have been identified and barcoded by different studies. Twenty per cent of fishes in the Australian water have been barcoded by the COI sequence as a pilot study conducted by Ward et al. (2005, 2009). The tropical to the subtropical marine environment, which includes species-rich regions of Asia, South America and Africa display, does not have more data on DNA barcode studies reviewed by Becker et al., (2011). Around 11,023 fish species (fresh and marine waters) have been identified in India, which are morphologically identified and reported (Nelson 2006; Mecklenburg et al. 2011; Sachithanandam et al. 2014, 2015); however, the barcode of the COI gene sequence has been reported in only around 1918 species, i.e. 17.4%, these data have been gathered from information in public domains such as NCBI and BOLD databases (Becker et al. 2011; Sachithanandam et al. 2015). These data represent a great opportunity to study the global status, approaches and future direction of DNA barcoding in fishery sciences. The current study has been undertaken to present a research chapter on the status of barcoding of fishes with reference to fish taxonomy in the Indian coastal waters with a detailed review of existing literature at the regional and national levels and also worldwide.

2 DNA Barcoding in Fishes and Other Animals: Status

Taxonomy is a branch of science where species are classified based on the morphology characters, geography, etc.; currently, very limited expertise people are pursuing the science through convectional taxonomy, which has in turn led to a global shortfall in trained taxonomists, and this has led to a great shortfall in skilled manpower for the estimation of biodiversity and exploration of new species in the earth (Novacek and Cleland 2001; Bellwood et al. 2004). It is important to classify the species before the rapid loss of marine and terrestrial biodiversity due to different environmental issues including climate change factors also. The morphological identification of species has been declining due to rapid decline of taxonomic skills (Hopkins and Freckleton 2002; Rubinoff 2006; Hebert et al. 2003a; Packer et al. 2009) and also due to the decline in funding by the government and regional bodies for undertaking studies on taxonomy sciences. In these circumstances, Hebert et al. (2003b) estimated that around 15,000 taxonomists will be required for the identification of 10–15 million unknown species. It will be cumbersome to identify the new species, which have morphological ambiguities and environment identities. A valiant effort was undertaken by the Census of Marine Life (CoML) programme to catalogue the distribution and abundance of particular species in the marine realm by assessing and the development of the baseline database for conservation and management of marine resources. Currently, researchers are applying advanced molecular methods (genetic tools) for resource identification and extraction in the place of basic sciences, like taxonomy, ecology and biology.

This in turn has led to the use of molecular techniques, like mtDNA barcoding using a short DNA sequence, and these have been increasingly adopted in taxonomic identification studies (Hebert et al. 2003a; Ball et al. 2005; Hebert and Gregory 2005; Coyne and Orr 2004; Packer et al. 2009). Currently, these methods are being applied to a wide range of taxonomic groups including birds, bats, sponges, molluscs, polychaetes, fishes, mammals, etc. (Hebert et al. 2004b; Ward et al. 2005; Meyer et al. 2008). This initiative has helped in the discovery of new species and complementary to classical taxonomy, many of which have been shown to be morphologically ambiguous and cryptic, which has considerably improved the biodiversity assessment (Blaxter et al. 2004; Derycke et al. 2005; Leasi and Todaro 2009). A careful perusal of literature revealed that many works had been undertaken using DNA barcoding molecular tools, to identify many species that include sibling species without any ambiguities. Different molecular markers are available globally, these include 18S nuclear ribosome (Colgan et al. 2001; Bleidorn et al. 2006), two mitochondrial genes 12S rDNA and 16S rDNA (Envall et al. 2006) and one nuclear gene 18S rDNA, and these are used for species identification and phylogenetic analysis. However, due to the lack of specific gene details, the genetic database was not able to identify the same species from other regions using these methods. Many studies were undertaken for the phylogenetic analysis of species using18S rDNA and new 16S rDNA sequence data (Sjolin et al. 2005; Thomas et al. 2005; Satheeshkumar and Jagadeesan 2010; Erseus et al. 2010).

The DNA barcoding tool is a well-accepted taxonomic method, which uses a short gene sequence (COI) to facilitate identification of a particular species accurately even by a novice who is not trained in classical taxonomy through biotechnology approaches (Hebert et al. 2003a, b, 2004a; Ward et al. 2005). Barcoding is a compliment for current research undertaken in taxonomic studies, by providing detailed information helpful for the identification of taxa in the marine environment (Hajibabaei et al. 2006a; Hebert et al. 2004b; Ball et al. 2005; Saunders 2005; Ward et al. 2005; Hajibabaei et al. 2007a; Sachithanandam et al. 2012). Hubert et al. (2008) explained the efficiency of DNA barcoding hinges on the degree of sequence divergence among species and intraspecies level identifications from different ecosystems. Escalante et al. (2011) and Baird and Sweeney (2011) reported the use of the COI gene sequence for the bio-monitoring and diversity assessment in benthology ecosystem. Sweeney et al. (2011) elucidated the importance of re-examination of morphological data in certain cases where ambiguity existed for species clarification/identification, and this can be done using the additional molecular data. Min and Hickey (2007a) proved that barcoding was sufficient for species identification among the fungal diversity. DNA barcoding achieved a new milestone when mini-barcoding studies were carried out by Meusnier et al. (2008) for specimens that were stored in formalin for prolonged periods. The COI gene sequence has been adopted for the identification of characters for some groups of organisms where identification carried solely by morphology failed due to morphological ambiguity. This clearly indicates that barcoding is an ideal tool for species identification (Mitchell 2008; Pages et al. 2009; Zettler et al. 2002; Abriouel et al. 2008; Hebert et al. 2003a). A detailed gene and gene classification used for different species taxonomy have been reviewed and are provided in Table 1.

mtDNA profiling studies are carried out in the cytoplasmic mtDNA, which is inherited from the female parent, and hence, each copy is identical. It offered valuable insights into the population structure and greatly contributed to the establishment of phylogeographic information (Avise et al. 1987). The development of "Universal" primers (Kocher et al. 1989) made DNA amplification of COI gene using PCR and the direct sequencing for many numbers of phyla possible.

Also, the sequences of mtDNA COI diverged quickly by its information and the gene order whose compositions are relatively uniform (Simon et al. 1994, 2006). Mitochondrial genome lacks introns when compared to nuclear genome, which restricted the exposure for recombination and hence has a haploid mode of inheritance (Saccone et al. 1999); this has been an added advantage for the DNA barcoding gene of COI and its sequences. Mitochondrial genes are useful to study the species that diverged recently because they have a high rate of substitution. However, if the divergence event was not recent, nuclear genes are ideal for phylogenetic analysis (Lin and Danforth 2004). mtDNA COI sequence divergences had been successfully used to discriminate phylogenetic differences within the same species and between different species of North American birds (Hebert et al. 2004b), spiders (Hebert and Barrett 2005), cryptic species of butterflies (Hebert et al. 2004a), mosquitoes (Besansky et al. 2003), leeches (Siddall and Budinoff 2005), springtails (Stevens and Hogg 2003; Hogg and Hebert 2004), beetles (Monaghan et al. 2005), oligochaetes (Nylander et al. 1999), naidid worms (Bely and Wray 2004), extinct moas (Lambert et al. 2005) and various other species of vertebrates and invertebrates (Saccone et al. 1999; Hebert et al. 2003b).

The barcode system was based on COI sequence diversity in a single short gene (Schander and Willassen 2005) region, i.e. a section of the mitochondrial DNA cytochrome C oxidase I gene (COI) about 650 base pairs. These sequences demonstrated higher-order genetic relationship and shallower divergence observed at intra- and interspecies levels (Seifert et al. 2007; Clare et al. 2008). This technique provides higher flexibility for the identification of species in large taxonomic assemblages with specificity (Caterino and Tishechkin 2006; Pegg et al. 2006; Hajibabaei et al. 2006c, 2007b). The ability to use COI gene to identify species would enable the identification of cryptic and polymorphic taxa and also identify and associate individuals of life stages other than the adult with their correct species (Schander and Willassen 2005; Kartavtsev et al. 2009; Kochzius et al. 2010; Duo et al. 2010). Further, COI partial sequence analysis has proven to be an economically

Table 1 Synoptic view of DNA barcoding, which employed COI gene to identify animal taxa and its K2P values

Sl. No	References	Taxa name and No.	% of the identified species	K2P (%) intra vs. interspecific	Remark
1	Hebert et al. (2003a)	2238 Annelida, Arthopoda, Chordata, Cnidaria, Echinodermata, Mollusca, Nematoda, Platyhelminthes	>98%	Overall <2 vs. 11.3	Efficacy of the COI sequence in identifying species from eight major groups and several minor phyla plus a variety of arthropod classes assessed. Cnidarians showed less genetic variation between species and then all other taxonomic groups, 94.1% vs. 1.9% showing <2 K2P between spp. ($p < 0.0001$)
2	Remigio et al. (2003)	70 gastropod spp.	98%		To identify gastropod species genetic divergence observed
3	Hebert et al. (2004b)	260 avian spp.	100%	0.43 vs. 7.93	Cryptic species genetic diversity was identified
4	Penton et al. (2004)	2 Daphnia spp.	100%		Identification of cryptic species with overlapping distribution of crustacean
5	Hebert and Barrett (2005)	203 arachnid spp.	100%	1.4 vs. 16.4	Mean intra- and interspecific nucleotide divergences did not overlap
6	Ward et al. (2005)	207 marine fishes from Australia	100%	0.39 vs. 9.93	Effectiveness of COI at fish species identifying and taxonomic relationships assessed
7	Dooh et al. (2006)	4 crustacean spp.	96–98%	1.5 vs. 27	Using barcodes to examine the phycology of two glacial relict crustacean taxa in North America

(continued)

Table 1 (continued)

Sl. No	References	Taxa name and No.	% of the identified species	K2P (%) intra vs. interspecific	Remark
8	Hajibabaei et al. (2006a)	Lepidopteran	100%	0.1–0.4 vs. 5.4–8.9	Using mini-barcodes to identify specimens at the species level
9	Hajibabaei et al. (2006b)	521 lepidopteron spp.	97.90%	0.17–0.46 vs. 4–6	Morphologically distinct sympatric species from three families identified
10	Costa et al. (2007)				DNA barcoding of crustacean species
11	Elias-Gutierrez et al. (2008)	2 Cladoceran spp.	96–98%	14.3%	A new cryptic species (Crustacea, Chydoridae) from the desert region of Mexican
12	Moura et al. (2008)	2 genera of elasmobranchs	97%		DNA barcoding used to resolve within genera identification problems in deep water sharks
13	Holmes et al. (2009)	20 shark spp. 7 ray spp.	91.5%		Identifying shark species from dried fins for conservation purposes
14	Radulovici et al. (2010)	Marine crustaceans from the Gulf of St. Lawrence	95–97%		COI gene sequence used to identify eighty Mollusca species
15	Steinke et al. (2009a)	391 fish spp.	100%	0.42 vs. 10.81	Producing a DNA database of ornamental fish from Pacific Canada
16	Vargas et al. (2009)	5 sea turtle spp.	100%		DNA barcoding of Brazilian sea turtles

feasible option in the taxonomic study carried out by Ball et al. (2005). The use of the COI sequence is a powerful tool for species identification and complementary of phylogenetic information of individually isolated fish eggs, larvae and fillets and fins from different coastal regions as reported by Ward et al. (2005), Hubert et al. (2008) and Steinke et al. (2009b). Nguyen and Seifert (2008) discovered three new species of *Leohumicola* using COI sequences (barcodes) analysed. According to Min and Hickey (2007b), the reducing sequence (mini barcode concept developed) length had a profound effect on the accuracy, resulting in NJ tree species assemblage and relationship, but surprisingly short sequences still yielded accurate species.

Radulovici et al. (2010) suggested that the DNA barcoding approach is the emerging tool to study the marine biodiversity assessment at the species level in the stipulated time line. Further, polychaete identification and phylogenetic classification are attempted using the COI gene (Meyer et al. 2008). Steinke et al. (2009a) described a software tool for DNA barcoding sequences used to study genetic distance methods or threshold of species genetic divergence (intra- and interspecies levels) and calculated K2P values. Siddall and Budinoff (2005) made an attempt on molecular identification and phylogenetic analysis of leech samples, which were carried out in South America using DNA barcoding. This study provided widespread information on particular species, which would be helpful for observing the diversity index.

It would be beneficial to have a standard segment (such as the barcoding region), which can be used for routine identifications. DNA barcodes have the potential to be used for identification of species without the aid of a taxonomist in certain situations. Barcoding had been proposed as a quality control measure to confirm the identity of specimens (Mitchell 2008). The commercial use of barcoding, in agricultural aspects of pest species identification, invasive species detection and fishery management (without morphology character), is also worth studying (Mitchell 2008; Rock et al. 2008). The fish taxonomy mainly consists of environmental factors and its dynamics of larval dispersal based upon the water current movement. Many unanswered issues in the marine ecology and evolution of fish population centre on how far planktonic larvae disperse away from their parents (Levin et al. 2006). Regardless of the importance of the ecological processes affected by larval fish dynamics, the inability of unambiguous taxonomic identification of early life stages of many taxa is still a major burden that impairs the proficient management of these populations. Early larval studies attempted by several scientific groups faced great difficulties in distinguishing larvae identification to the genus and species levels (Chow and Walsh 1992; Victor et al. 2009). To overcome the above issues, the COI gene sequence-based identification tool giving a solution of larvae was identified without morphology characters.

DNA barcoding has been proven to be controversial in some scientific circles (Will and Rubinoff 2004) despite its promising potential. Recent results illustrated potential benefits by the use of a standardised molecular approach for identification (Hebert et al. 2003a; Hebert and Gregory 2005). Intraspecific phenotypic variation often overlapped that of sister taxa in nature, which leads to incorrect identifications, when based on the phenotype only (Pfenninger et al. 2006). The recently introduced next-generation sequencing (NGS) approaches in biodiversity science have the potential to further extend the application of DNA information (Hajibabaei et al. 2011). Ratnasingham and Hebert (2007) reported that DNA technology can be used to gather barcode sequences in minutes and used an onboard barcode reference library to generate identifications through the development of portable devices.

The taxonomic ambiguity existing for several fish genera/species and proper identification were imperative for the management and trade of commercially important resources. Eschmeyer et al. (1998), Wiens and Servedio (2000), Hebert et al. (2003b), Hebert and Gregory (2005), Nelson (2006), Ward et al. (2009) and Sachithanandam et al. (2012) have shown that the DNA barcoding methodology resolved this problem and also helped in the discovery of new/cryptic species information. Morphology-based studies required a number of taxonomists who are dwindling due to the nature of hard work (Steinke et al. 2009a). So, the barcode methods may provide new taxonomist—gene (mtDNA COI)-based identification. The barcode studies also provide an understanding of the eco-diversity, especially for the marine fishes, which cannot be monitored continuously by traditional taxonomy (Dasmahapatra and Mallet 2006). Moreover, the fishery managers and scientists are struggling with the lack of basic information for many shark and ray species. These can be identified through barcode and lead to the confiscation of materials like shark meat, fin, bones, etc. This methodology provides details that can be used for scientific conservation and management (Holmes et al. 2009).

Larval fishes were frequently not identified through classical taxonomy method up to the species level due to their small size and limited morphological developments (Webb et al. 2006; Richardson et al. 2006). This problem leads to difficulties in the understanding of the life histories of fishes at the juvenile stage, specifically marine fishes. Very few genera have been explored (in terms of commercially important fish species) for the taxonomic identification, early life history and phylogenetic relationships (Lutjanidae, Serranidae); these studies were far from complete and have been continually reviewed (Rivas 1949; Chow and Walsh 1992; Leis 1986, 2005; Miller and Cribb 2007). Shirak et al. (2009) studied the barcoding and taxonomic analysis of the five *Tilapiine* species, and its results depicted COI gene sequences discriminating species identification without any ambiguity. The larvae and newly settled juveniles of the Cubera snapper were identified using DNA barcoding approached by Victor et al. (2009). The stomatopod (marine crustaceans) larvae were studied by barcode technique (Barber and Boyce 2006) for us to understand the accurate biodiversity assessment of gonodactylid (mantis shrimp), etc. Zemlak et al. (2009) studied the Indo-pacific fishes, for a substantial number, which were overlooked under the category of commercially important fishes. Moreover, the COI gene sequence for marine fishes was successfully carried out in Australia (Ward et al. 2005), Pacific Canada (Steinke et al. 2009a), North Atlantic (Ward et al. 2008), South Africa (Zemlak et al. 2009), Finland (Salokannel et al. 2010) and the Great Barrier Reef fish (Pegg et al. 2006). The faunal studies, which used the COI sequence for their taxonomy identification, are listed in Table 2.

Fishes are among the most studied marine groups when compared to other groups such as invertebrate and are currently being barcoded by two global campaigns such as FISH-BOL and SHARK-BOL (Ward et al. 2009). Existing fish barcoding study results were reviewed, and lacunae were presented worldwide (Ward et al. 2009; Radulovici et al. 2010). Barcoding studies in Indian marine fishery are very few and far between, and there are no comprehensive data available from the Indian coast, specifically for grouper species. According to Johnson and Keener (1984), the earlier taxonomy works based on the grouper fishes had been intended to increase the

Table 2 Details of K2P values reported for the COI gene used for species identification carried out in different regions

Sl. No.	Authors and Year	Intraspecific K2P values (%)	Intergeneric K2P values (%)	Within family and order K2P values (%)	Animal included in DNA barcoding paper tile
1	Hebert et al. (2004b)	Within species mean K2P −0.27	Within genus 7.93	Within family 12.71	Identification of birds through DNA barcode
2	Clare et al. (2007)	Intraspecific genetic variation mean values −0.60	Congenera mean values −7.80	Within family genetic value −21.26	DNA barcoding of neo tropical bats
3	Ward et al. (2008)	Intraspecific K2P values – 0.028 ($n = 22$) of *Z. faber*; Avg.0.20–0.23			DNA barcoding of shared fish species from the North Atlantic and Australasia
4	Steinke et al. (2009b)	Mean intra-specific genetic values −0.21	Congeneric K2P values – 10.81		DNA identifications for the ornamental fish trade
5	Khedkar et al. (2009)	Molecular variance analysis revealed that 96%/6 haplotypes per species			DNA barcoding of fish in the Godavari river, India
6	Odeny et al. (2009)	Within species was 0.11 of genetic variation			Para taxonomy for fishery surveys using the DNA barcoding tool
7	Oliveira et al. (2009)	Average = 0.6 within species	Avg. values of K2P −8.7 within genera	Avg K2P value −17.61	DNA barcode of freshwater fish of Brazil
8	Rasmussen et al. (2009)	Mean intra-species k2P value −0.26 (range 0.04–1.09)	Congeneric K2P values −8.22 (range 3.42–12.67) −32 times fold greater		Commercially important Salmon fish using the DNA barcoding tool from North America
9	Zemlak et al. (2009)	K2P distances between African & Australian (mean values −5.10)			DNA barcoding reveals overlooked marine fishes from different continental regions

(continued)

Table 2 (continued)

Sl. No.	Authors and Year	Intraspecific K2P values (%)	Intergeneric K2P values (%)	Within family and order K2P values (%)	Animal included in DNA barcoding paper tile
10	Aliabadian et al. (2009)	**COI:** Intraspecific K2P distances averaged 0.24 (SD = 0.59)	**COI:** Intrageneric K2P 24-fold higher than the mean intraspecific K2P distances	**COI:** Mean K2P within families = 11.46 and order level −15.80	Molecular identification of birds: Performance of distance-based DNA barcoding tool
11	Steinke et al. (2009a)	Intraspecific pacific Canada fish K2P value −0.25	Mean genetic level of genera 3.75		Pacific Canada's fishes discrimination using COI gene sequences
12	Kartavtsev et al. (2009)	Average intra-species, 0.11 ± 0.04	Intra-genus 1.87 ± 0.68	Intra-family 12.67 ± 0.28; intra-order 16.52 ± 0.10	Molecular phylogenetics of prickle backs and other percoid fishes from the sea of Japan
13	Persis et al. (2009)	Mean intraspecific K2P values is 0.24	Avg. congeneric K2P values −17.2	Mean K2P family = 0.875%	DNA barcoding of carangid fishes from Andhra coast, India
14	Lakra et al. (2010)	Mean K2P value of the intraspecies level −0.30	Within genus mean K2P value −6.66	Family level mean K2P −9.91, order mean K2P −16.00	Indian marine fishes barcoding 115 species, 79 genera and 37 families
15	Zhang (2011)	Avg. K2P −0.319% for intraspecific individuals	Avg. K2P −15.742 among congeners		DNA barcoding of marine fishes in South China Sea
16	Zhang and Hanner (2011)	Among conspecifics average K2P of 0.3%	Genetic distances averaged 17.6% among congeners		DNA barcoding of marine fishes from Japan

(continued)

Table 2 (continued)

Sl. No.	Authors and Year	Intraspecific K2P values (%)	Intergeneric K2P values (%)	Within family and order K2P values (%)	Animal included in DNA barcoding paper tile
17	Hubert et al. (2011) Unpublished data	Within species 0.001–0.004; *S. caudimaculatum* & *S. spiniferum* that diverged only by 0.007 on average; this result reinforces the view that no canonical threshold applies to the frontier separating populations and species in fishes (Hubert et al. 2008)	Among genera from an average of 0.063 in *Myripristis*. 0–0.11 on average for *Acanthurus*		Hubert et al.
18	Pramual et al. (2011)	Intraspecific K2P −0–9.28, with a mean of 2.75	Interspecific K2P 0.34–16.05		Cryptic biodiversity & phylogenetic relationships revealed by DNA barcoding of oriental black
19	Carolan et al. (2012)	Within species *B. cryptarum*, *B. lucorum* and *B. magnus* for 0.004, 0.001 and 0.001	Interspecific K2P: 0.033–0.044		DNA barcoded of bumblebee species complex and colour patterns do not diagnose species
20.	Lorz et al. (2012)	Within the species of *R. chathamensis* & *R.abyssalis* (K2P = 0.000); 24 specimen *R. aculeate* (0.0089); 9 specimen *R. helleri* (K2P = 0.0003); 13individual *R. inflata* (K2P = 0.037)	Inter-clade 0.143–0.370; an overall average divergence: 0.284		Description of a new species Crustacea, Amphipoda *Rhachotropis* using COI gene sequences

(continued)

Table 2 (continued)

Sl. No.	Authors and Year	Intraspecific K2P values (%)	Intergeneric K2P values (%)	Within family and order K2P values (%)	Animal included in DNA barcoding paper tile
21	Turanov et al. (2012)	0.06% among the species K2P value	Mean K2P within genus 0.37	Within family K2P −11.83; within order −15.22	Molecular phylogenetic study of eel-pout fishes from eastern seas

understanding of the relationship of phylogenetic analysis between species studies with related families attempted. However, there are large gaps on grouper species taxonomy validation and phylogenetic analysis at the regional level because of morphological ambiguity. Few studies attempted in grouper fish identification are as follows: Maggio et al. (2005) carried out mitochondrial DNA cytochrome b (cyto b) and 16S rDNA sequence analysis for species phylogenetic assemblage between the close genus (*Epinephelus* and *Mycteroperca*) from the Atlantic groupers. These study results suggested that level genetic diversity is observed in close related genus (*Epinephelus* and *Mycteroperca*). Craig and Hastings (2007) carried out molecular approaches to understand the phylogenetic relationships among the fishes in the perciform tribe Epinephelinae (Serranidae), which had been poorly understood from Indian waters. Further, Smith and Craig (2007) studied the limits and relationships of Serranid and Percid fishes using the nucleotide character analysis. Koedprang et al. (2007) studied the genetic diversity among the commercially important grouper species (eight species) using microsatellite marker for species identification.

3 Fish Diversity and DNA Barcode Status in India

India has a wide coastal region and continental shelf with 8000 km long coastline, which is a treasure trove of huge fishery resources. Coastal waters of India are known for their rich and diverse fish species (Venkataraman and Wafar 2005). Major contributions to Indian ichthyology have been made by the pioneering work of Day (1878) since the eighteenth century. Latter and Talwar (1990) reported about 2546 fish species belonging to 969 genera, 254 families and 40 orders reported from the Indian coast. Fifty-seven per cent of marine fish genera are common in the Indo-Pacific and Atlantic and Mediterranean regions (Venkataraman and Wafar 2005). Fish diversity in the seas of the Andaman and Nicobar Islands is also of special interest in terms of marine zoogeography and coral reef ecosystem because they lie at the confluence of the Andaman Sea with the western Pacific and the Indian Ocean (Rajan 2010). A total of about 1485 species of fish under 603 genera belonging to

177 families are represented from these islands, of which about 400 species have commercial significance. Among the fish species, 1089 (about 73.38%) are recorded from the coral reef environment, 277 from mangroves ecosystem, 152 from seagrass meadows, 23 from freshwater streams and 101 from the offshore (deep sea fish) environment, while 158 species are commonly observed in mangrove, seagrass, coral reefs and offshore ecosystems (Rao 2003; Rajan and Sreeraj et al. 2012). The catalogued endemic biodiversity of the islands comprises 16 species of fish, 31 species of reptiles and 8 species of amphibians (Ramakrishna and Sivaperuman 2010). Tropical fishery in Indian coast has high diversity that there are several misnomers among the identified fishes and certain similar species (misidentification) are recorded as similar/same for even scientific work (Sachithanandam et al. 2012, 2014, 2015).

An exploratory survey along the southwest coastal waters of India revealed the presence of certain new species recorded akin to their nearest generic counterparts but appearing distinct for a classical taxonomist. Eastern part of India includes the Bay of Bengal, and the Andaman group of Islands is rich in the diversity of fishes. It is an isolated virgin area in terms of exploration as well as the exploitation of fishery resources. An attempt has been carried out to barcode the fishes from Andaman waters for understanding the genetic divergences of species and the evolution of fishes and also for developing a faster and reliable tool for commercialisation. DNA barcoding was performed on the marine fishes in the east coast and west coast of India. These barcode results covered very minimal and reported that the average K2P levels within species, genera, families and orders were 0.30%, 6.60%, 9.91% and 16.00%, respectively (Lakra et al. 2010). Persis et al. (2009) studied the Carangid fishes for genetic diversity and identification effectiveness based on the COI sequence approach from the Kakinada coastal region, east coast of India. Jhon et al. (2010) studied *Stolephorus* spp. genetic divergence using mtDNA COI method for species identification from Parangipettai coastal areas, southeast coast of India. DNA barcode studies on coral reef fishes especially grouper from the Andaman Islands were very limited, which were carried out by Sachithanandam et al. (2011, 2012). Fish species identification studies using RAPD methods were applied to discriminate the diversity of grouper species from the Indian coast by Govindaraju and Jayasankar (2004). They reported that phenotypic identification of grouper genus *Epinephelus* was from samples drawn from southeast and southwest coasts of India, which has low genetic divergence. Jayasankar et al. (2004) carried out a comprehensive study of the Indian mackerel in the east and west coasts of India using the RAPD techniques, and the study revealed significant genetic differences between species. Lakra et al. (2007) studied five Indian Sciaenids using the RAPD markers for species identification.

DNA barcoding of *Lates calcarifer* was studied in the Porto Nova coastal region using COI gene for the phylogenetic analysis, and genetic distances were compared with worldwide species (Jhon et al. 2010). Kumar et al. (2011) worked in cryptic species of Mugilidae fishes using DNA barcoding tools for phylogenetic and haplotype diversity as revealed by intra-species genetic distances observed. Sachithanandam et al. (2011, 2012) brought out barcode work details for the grouper subfamily Epinephelinae species of Andaman waters. Based on the above review of literature, Sachithanandam et al. (2012) suggested that the study on monophyly of fish barcode work was not carried out systematically in Indian fishery sciences. So, this chapter's aim was to investigate global status, approaches and future direction of DNA barcoding in fishery sciences. The works from existing literature, which have been carried out at the regional, national and global levels, and the studies carried out in the Indian coastal regions have been reviewed in detail to identify the gaps in specific groups (Tables 1 and 2). Based on the review of literature survey data, it clearly indicates that further DNA barcoding studies need to be carried out to estimate the Indian marine biodiversity in a systematic manner, with special attention to invertebrates for diversity assessment.

4 DNA Barcoding Emphases

Application of DNA barcoding tools is emerging in the fields of fish conservation and management, which help in the estimation of marine living resource quota, by-catch monitoring and sustainable fishery monitoring science (Holmes et al. 2009; Steinke et al. 2009a, b; Rasmussen et al. 2009). DNA barcoding has demonstrated that 25% of fish samples from markets and restaurants in USA and Canada were mislabelled or substituted as observed from COI gene sequences (Wong and Hanner 2008), proving that it can be used as an efficient tool in the food safety industry. DNA barcoding can also be applied successfully to cooked or processed fish meat (Smith et al. 2008), grilled or deep-fried fillets (Wong and Hanner 2008), boiled samples (Shirak et al. 2009) and canning samples by using shorter fragments called minibarcodes (Hajibabaei et al. 2006a; Meusnier et al. 2008; Rasmussen et al. 2009; Ward et al. 2009).

5 DNA Barcoding Progress

The primary goal of FISH-BOL is to generate and gather the DNA barcode records for all the fishes in the world, which include around 32,500 species (Ward et al. 2009; Eschmeyer 2010). The aim of the FISH-BOL is to develop COI gene sequences for all the fishes worldwide. 74 and 50% of fish were barcoded in the Arctic and Antarctic regions, respectively. Other regions such as Australia, America and Oceania indicated a coverage of 20%. However, extremely species-rich regions such as Asia and Africa showed lower coverage. This clearly denotes that there has been a bias towards the processing of data of marine species. This is proven by the following information: of the 7800 species recorded as barcoded in FISH-BOL, about 5700 (73.1%) are marine (Eschmeyer 2010). 11,023 species morphologically of fish fauna have been reported in Indian waters, but only 1918 species have been barcoded so far which work to around 17.4%, and this has been reported by

Mecklenburg et al. (2011). Therefore, sampling studies in the future should focus on the collection of marine and freshwater species in Indian and Southeast Asian regions. Becker et al. (2011) clearly stated that more barcoding sampling campaigns towards neglected orders and families of Indian fish fauna have to be performed in future.

6 DNA Barcoding Success Rate

The barcoding success rate was found to be 98% in the fish species identification carried out in marine species (Ward et al. 2005, 2009). It is possible to identify species using the DNA barcoding methodology developed by the fishbowl database. DNA barcoding COI gene sequences indicated that regional genetic differentiation and shared haplotype genetic differences are due to the different habitats and local environment changes (Hubert et al. 2008; Ward et al. 2009; Sachithanandam et al. 2012).

7 Some Limitation and Cautions

Ward et al. (2009) analysed mtDNA COI sequences in fishes and indicated that the success for barcoding depended upon recent speciation, incorrect morphological taxonomy and species hybridisation, where barcoding could not differentiate inter-species. There are many drawbacks in the use of a barcoding tool for species identification, and so, the scientific community must be cautious in accepting the above factors and use additional nuclear gene for further clarification. Generally, biological phenomena, such as hybridisation process of close species, natural introgression process, error in specimen identification using classical taxonomy and recent speciation process, are known to strongly interfere with DNA barcoding process, and these phenomena are known to occur at different degrees depending on the animal groups and datasets (Hebert et al. 2003b, 2004a; Mitchell 2008; Ward et al. 2009; Rubinoff 2006; Rock et al. 2008; Langhoff et al. 2009). So, it is authoritative to have more databases on individual species COI gene sequences from different geographical locations and correct identification of species through traditional taxonomy, as well as uploading of error-free barcoding sequences for correct species, which will give the best results from barcoding approaches for accurate species identification without any ambiguity.

Acknowledgements This work was supported by the Ministry of Earth Sciences (MoES), New Delhi, India, under Benthic programme and the higher authorities of Pondicherry University, Puducherry, India, for which the authors are grateful.

References

Abriouel HPAM, Maqueda M, Valdivia E, Bueno MM (2008) Biodiversity of the microbial community in a Spanish farmhouse cheese as revealed by culture-dependent and culture-independent methods. Int J Food Microbiol 127(3):200–208

Aliabadian M, Kaboli M, Nijman V, Vences M (2009) Molecular identification of birds: performance of distance-based DNA barcoding in three genes to delimit parapatric species. PLoS One 4(1):e4119. https://doi.org/10.1371/journal.pone.0004119

Avise JC, Arnold J, Ball RM, Bermingham E, Lamb T, Neigel JE, Reeb CA, Saunders NC (1987) Intraspecific phylogeography: the mitochondrial DNA bridge between population genetics and systematics. Annu Rev Ecol Syst 18:489–522

Baird DJ, Sweeney WB (2011) Applying DNA barcoding in benthology: the state of the science. J North Am Benthol Soc 30(1):122–124

Ball SL, Hebert PDN, Burian SK, Webb JM (2005) Biological identification of mayflies (Ephemeroptera) using DNA barcodes. J N Am Benthol Soc 24:508–524

Barber P, Boyce SL (2006) Estimating diversity of indo-Pacific coral reef stomatopods through DNA barcoding of stomatopod larvae. Proc Royal Soc Biol Sci 273:2053–2064. https://doi.org/10.1098/rspb

Becker S, Hanner R, Steinke D (2011) Five years of FISH-BOL: brief status report. Mitochondrial DNA 22(S.1):3–9. https://doi.org/10.3109/19401736.2010.535528

Bellwood DR, Hughes TP, Folke C, Nystrom M (2004) Confronting the coral reef crisis. Nature 429:827–833

Bely AE, Wray GA (2004) Molecular phylogeny of naidid worms (Annelida: Citellata) based of cytochrome oxidase I. Mol Phylogenet Evol 30:50–63

Besansky NJ, Severson DW, Ferdig MT (2003) DNA barcoding of parasites and invertebrate disease vectors: what you don't know can hurt you. Trends Parasitol 19(12):545–546

Blaxter M, Elsworth B, Daub J (2004) DNA taxonomy of a neglected animal phylum: An unexpected diversity of tardigrades. Proc R Soc Lond Biol Sci 271:S189–S192

Bleidorn C, Podsiadlowski L, Bartolomaeus T (2006) The complete mitochondrial genome of the orbiniid polychaete Orbinialatreillii (Annelida Orbiniidae) – A novel gene order for Annelida and implications for annelid phylogeny. Gene 370:96–103

Carolan JC, Murray TE, Fitzpatrick U, Crossley J, Schmidt H (2012) Colour patterns do not diagnose species: quantitative evaluation of a DNA barcoded cryptic bumblebee complex. PLoS One 7(1):e29251. https://doi.org/10.1371/journal.pone.0029251

Caterino MS, Tishechkin AK (2006) DNA identification and morphological description of the first confirmed larvae of Hetaeriinae (Coleoptera: Histeridae). Syst Entomol 31:405–418

Chow S, Walsh PJ (1992) Biochemical and morphometric analyses for phylogenetic relationships between seven snapper species (subfamily Lutjanidae) of the Western Atlantic. Bull Mar Sci 50:508–519

Clare EL, Lim BK, Engstrom MD, Eger JL, Hebert PDN (2007) DNA barcoding of neo tropical bats: species identification and discovery within Guyana. Mol Ecol Notes 7:184–190. https://doi.org/10.1111/j.1471-8286.2006.01657.x

Clare EL, Kerr KC, Von-Konigslow TE, Wilson JJ, Hebert PDN (2008) Diagnosing mitochondrial DNA diversity: applications of a sentinel gene approach. J Mol Evol 66:362–367. https://doi.org/10.1007/s00239-008-9088-2

Colgan DJ, Hutchings P, Brown S (2001) Phylogenetic relationships within the Terebellomorpha. J Mar Biol Ass UK 81:765–773

Costa F, DeWaard JR, Boutilier J, Ratnasingham S, Dooh RT, Hajibabaei M, Hebert PDN (2007) Biological identifications through DNA barcodes: the case of the Crustacea. Can J Fish Aquat Sci 64(2):272–295

Coyne JA, Orr HA (2004) Speciation. Sinauer Associates, Sunderland, MA

Craig MT, Hastings PA (2007) A molecular phylogeny of the groupers of the subfamily Epinephelinae (Serranidae) with a revised classification of the Epinephelinae. Ichthyol Res 54:1–17. https://doi.org/10.1007/s10228-006-0367-x

Dasmahapatra KK, Mallet J (2006) DNA barcodes: recent successes and future prospects. Heredity 97:1–2. https://doi.org/10.1038/sj.hdy.6800858

Day F (1878) The fishes of India; Being a natural history of the fishes known to inhabit the seasand fresh waters of India, Burma, and Ceylon. Parts 1–4. William Dawson and Sons, London, p 778

Derycke S, Remerie T, Vierstraete A, Backeljau T, Vanfleteren J, Vincx M, Moens T (2005) Mitochondrial DNA variation and cryptic speciation within the freeliving marine nematode Pellioditis marina. Mar Ecol Prog Ser 300:91–103

Dooh RT, Adamowicz SJ, Hebert PDN (2006) Comparative phylogeography of two North American 'glacial relict' crustaceans. Mol Ecol 15(14):4459–4475

Duo WZ, Yusong G, Wei T, Lu L, EnPu T, ChuWu L, Yun L (2010) DNA barcoding, phylogenetic relationships and speciation of snappers (genus *Lutjanus*). Sci China Life Sci 53(8):1025–1030. https://doi.org/10.1007/s11427-010-4034-0

Elias-Gutierrez M, Jeronimo FM, Iavanova NV, Valdes-Moreno M, Hebert PDN (2008) DNA barcodes for *Cladocera* and *Copepoda* from Mexico and Guantemala, highlights and new discoveries. Zootaxa 1839:1–42

Envall I, Källersj M, Erseus C (2006) Molecular evidence for the non-monophyletic status of Naidinae (Annelida:Clitellata:Tubificidae). Mol Phylogenet Evol 40:570–584

Erseus C, Rota E, Matamoros L, De Wit P (2010) Molecular phylogeny of Enchytraeidae (Annelida: Clitellata). Mol Phylogenet Evol 57:849–858

Escalante P, Vazquez AI, Escobar PR (2011) Tropical montane nymphalids in Mexico: DNA barcodes reveal greater diversity. Mitochondrial DNA 21(S1):30–37

Eschmeyer WN (2010) Catalog of fishes. http://research.calacademy.org/research/ichthyology/catalog/speciesby family.asp

Eschmeyer WN, Ferraris JCJ, Hoang MD, Long DJ (1998) Part 1. Species of fishes. In: Eschmeyer WN (ed) Catalog of fishes. California Academy of Sciences, San Francisco

Govindaraju GS, Jayasankar P (2004) Taxonomic relationship among seven species of groupers (genus Epinephelus; family serranidae) as revealed by RAPD Fingerprinting. Mar Biotechnol 6:229–237. https://doi.org/10.1007/s10126-003-0021-9

Hajibabaei M, Janzen DH, Burns JM, Hallwachs W, Hebert PDN (2006a) DNA barcodes distinguish species of tropical Lepidoptera. Proc Nat Acad Sci U S A 103:968–971

Hajibabaei M, Smith MA, Janzen DH, Rodriguez JJ, Whitefield JB, Hebart PDN (2006b) A minimalist barcode can identify a specimen whose DNA is degraded. Mol Ecol 6:959–964. https://doi.org/10.1111/j.1471-8286.2006.01470.x

Hajibabaei M, Singer GAC, Hickey DA (2006c) Benchmarking DNA barcodes: an assessment using available primate sequences. Genome 49:851–854

Hajibabaei M, Singer GAC, Clare EL, Hebert PDN (2007a) Design and applicability of DNA arrays and DNA barcodes in biodiversity monitoring. BMC Biol 5:24. https://doi.org/10.1186/1741-7007-5-24

Hajibabaei M, Singer GAC, Hebert PDN, Hickey DA (2007b) DNA barcoding: how it complements taxonomy, molecular phylogenetic and population genetics. Trends Genet 23 (4):167–172. https://doi.org/10.1016/j.tig.2007.02.001

Hajibabaei M, Shokralla S, Zhou X, Singer GAC, Baird DJ (2011) Environmental barcoding: A next-generation sequencing approach for biomonitoring applications using river benthos. PLoS One 6(4):e17497. https://doi.org/10.1371/journal.pone.0017497

Hebert PDN, Barrett RDH (2005) Identifying spiders through DNA barcodes. Can J Zool 83:481–491

Hebert PDN, Gregory TR (2005) The promise of DNA barcoding for taxonomy. Syst Biol 54:852–859. https://doi.org/10.1080/10635150500354886

Hebert PDN, Cywinska A, Ball SL, DeWaard JR (2003a) Biological identifications through DNA barcodes. Proc R Soc Lond Ser B 270:313–321

Hebert PDN, Ratnasingham S, Ward JR (2003b) Barcoding animal life: cytochrome c oxidase subunit1 divergences among closely related species. Proc Royal Soc B 270:96–99

Hebert PDN, Penton EH, Burns JDH, Hallwachs W (2004a) Ten species in one: DNA barcoding reveals cryptic species in the neotropical skipper butterfly *Astraptes fulgerator*. Proc Nat Acad Sci U S A 101(41):14812–14817

Hebert PDN, Stoeckle MY, Zemlak TS, Francis CM (2004b) Identification of Birds through DNA Barcodes. PLoS Biol 2(10): e312. https://doi.org/10.1371/journal.pbio.0020312

Hogg ID, Hebert PDN (2004) Biological identification of springtails (Collembola: Hexapoda) from the Canadian Arctic, using mitochondrial DNA barcodes Canadian. J Zool 82:1–6

Holmes BH, Steinke D, Ward RD (2009) Identification of shark and ray fins using DNA barcoding. Fish Res 95:280–288

Hopkins GW, Freckleton RP (2002) Declines in the numbers of amateur and professional taxonomists: implications for conservation. Anim Conserv 5:245–249

Hubert N, Hanner R, Holm E, Mandrak NE, Taylor E (2008) Identifying Canadian freshwater fishes through DNA barcodes. PLoS One 3(6):e2490. https://doi.org/10.1371/journal.pone.0002490

Hubert N, Trottin ED, Irisson JO, Meyer C, Planes S (2011, Unpublished) Identification coral reef fish larvae through DNA barcoding: a test case with the families Acanthuridae and Holocentridae, Mol Phylogenet Evol 55(3):1195–1203

International Barcode of Life (2010) Lepidoptera barcode of life. [Cited 15 October 2010] Available from http://www.lepbarcoding.org/

Jarnegren J, Schander C, Sneli JA, Ronningen V, Young CM (2007) Four genes, morphology and ecology: distinguishing a new species of Acesta (Mollusca; Bivalvia) from the Gulf of Mexico. Mar Biol 152:43–55

Jayasankar P, Thomas PC, Paulton MP, Mathew J (2004) Morphometric and genetic analyzes of Indian mackerel (*Rastrelliger kanagurta*) from peninsular India. Asian Fish Sci 17:201–215

Jhon A, Prasanaakumar C, Lyla PS, Khan A, Jalal KCA (2010) DNA barcoding of *Lates calcarifer* (Bloch, 1970). Res J Biol Sci 5(6):414–419

Johnson GD, Keener P (1984) Aid to identification of American grouper larvae. Bull Mar Sci 34:106–134

Kartavtsev YP, Sharina SN, Goto T, Rutenko OA, Zemnukhov VV, Semenchenko AA, Pitruk DL, Hanzawa N (2009) Molecular phylogenetics of pricklebacks and other percoid fishes from the sea of Japan. Aquat Biol 8:95–103. https://doi.org/10.3354/ab00205

Khedkar GD, Lutzky S, Kalyankar VB, Reddy ACS, David L (2009) DNA barcoding reveals a discontinuous genetic diversity pattern of fish in the Godavari River, India. In: 3rd International barcode of life conference, pp 80–81

Kocher TD, Thomas WK, Meyer A, Edwards SV, Pabo S, Villablanca FX, Wilson AC (1989) Dynamics of mitochondrial DNA evolution in animals: amplification and sequencing with conserved primers. Proc Nat Acad Sci U S A 86:6196–6200

Kochzius M, Seidel C, Antoniou A, Botla SK, Campo D (2010) Identifying fishes through DNA barcodes and microarrays. PLoS One 5(9):e12620. https://doi.org/10.1371/journal.pone.0012620

Koedprang W, Nakorn UN, Nakajima M, Taniguchi H (2007) Evaluation of genetic diversity of eight grouper species *Epinephelus* Spp. based on microsatellite variations fisheries. Science 73:227–236

Kumar CP, John BA, Khan SA, Lyla PS, Murugan S, Rozihan M, Jalal KCA (2011) Efficiency of universal barcode gene (Coxi) on morphologically cryptic Mugilidae fishes delineation. Trends Appl Sci Res 6(9):1028–1036. https://doi.org/10.3923/tasr.2011.1028.1036

Lakra WS, Goswami M, Mohindra V, Lal KK, Punia P (2007) Molecular identification of five Indian Sciaenids (Pisces: perciformes, Sciaenidae) using RAPD markers. Hydrobiologia 583:359–363. https://doi.org/10.1007/s10750-006-0480-x

Lakra WS, Verma MS, Goswami M, Lal KK, Mohindra V, Punia P, Gopalakrishnan A, Singh KV, Ward RD, Hebert PDN (2010) DNA barcoding Indian marine fishes. Mol Ecol Res 11:60–71

Lambert DM, Baker A, Huynen L, Haddrath O, Hebert PDN, Millar CD (2005) Is a large-scale DNA-based inventory of ancient life possible. J Hered 96(3):279–284

Langhoff P, Authier A, Buckley TR, Dugdale JS, Rodrigo A, Newcomb RD (2009) DNA barcoding of the endemic New Zealand leafroller moth genera, *Ctenopseustis* and *Planotortrix*. Mol Ecol Res 9:691–698

Lavina-Vincent C, Dutta S, Banerjee G, Saravanan R, Sachithanandam V, Krishnan P (2016) A molecular analysis of selected marine fishes from the southwest coast of India for species delineation. Mar Biodivers 47(2):413–419. https://doi.org/10.1007/s12526-016-0495-0

Leasi F, Todaro MA (2009) Meiofaunal cryptic species revealed by confocal microscopy: the case of *Xenotrichula intermedia* (Gastrotricha). Mar Biol 156:1335–1346

Leis JM (1986) Larval development in four species of indo-Pacific coral trout *Plectropomus*(Pisces: Serranidae: Epinephelinae) with an analysis of the relationships of the genus. Bull Mar Sci 38:525–552

Leis JM (2005) A larva of the elelinelutjanid, *Randallichythysfilamentosus* (Pisece: Perciformes), with comments on phylogenetic implications of larval morphology of basal lutjanids. Zootaxa 1008:57–64

Levin DM, Quinones ML, Povona MM, Linton Y (2006) DNA barcoding reveals both known and novel taxa in the Albitarsis group (*Anopheles: Nyssorhynchus*) of Neotropical malaria vectors. Parasit Vector 5:44. https://doi.org/10.1186/1756-3305-5-44

Lin CP, Danforth BN (2004) How do insect nuclear and mitochondrial gene substitution patterns differ? Insights from Bayesian analyses of combined data sets. Mol Phylogenet Evol 30 (3):686–702

Lorz AN, Linse K, Smith PJ, Steinke D (2012) First molecular evidence for underestimated biodiversity of Rhachotropis (Crustacea, Amphipoda), with description of a new species. PLoS One 7(3):e32365. https://doi.org/10.1371/journal.pone.0032365

Maggio T, Andaloro F, Hemida F, Arculeo M (2005) A molecular analysis of some eastern Atlantic grouper from the *Epinephelus* and *Mycteroperca* genus. J Exp Mar Biol Ecol 321:83–92

Mecklenburg CW, Moller PR, Steinke D (2011) Biodiversity of arctic marine fishes: taxonomy and zoogeography. Mar Biodivers 41(1):109–140

Meusnier I, Singer ACG, Landry JF, Hickey AD, Hebert PDN, Hajibabaei M (2008) A universal DNA mini-barcoding for biodiversity analysis. BMC Genomics 9:214–218. https://doi.org/10.1186/147-2164-9-214

Meyer M, Briggs AW, Maricic T, Hober B, Hoffner B, Krause J, Weihmann A, Paabo S, Hofreiter M (2008) From micrograms to picograms: quantitative PCR reduces the material demands of high-throughput sequencing. Nucleic Acids Res 36:e5

Miller TL, Cribb TH (2007) Phylogenetic relationship of some common indo-Pacific snapper (Perciformes: Lutjanidae) based on mitochondrial DNA sequences, with comments on the taxonomic position of the Caesioninae. Mol Phylogenet Evol 44:450–460

Min XJ, Hickey DA (2007a) Assessing the effect of varying sequence length on DNA barcoding of fungi. Mol Ecol Notes 7:365. https://doi.org/10.1111/j.1471-8286.2007.01698.x

Min XJ, Hickey DA (2007b) DNA barcodes provide a quick preview of mitochondrial genome composition. PLoS One 2(3):e325. https://doi.org/10.1371/journal.pone.0000325

Mitchell A (2008) DNA barcoding demystified. Aust J Entomol 47:169–173

Monaghan MT, Balke M, Gregory TR, Volger AP (2005) DNA-based species delineation in tropical beetles using mitochondrial and nuclear markers. Philos Trans Royal Soc Lond B Biol Sci 360(1462):1925–1933. https://doi.org/10.1098/rstb.2005.1724

Moura T, Silva MC, Flgueiredo I, Neves A, Munoz PD, Coelho MM, Gordo LS (2008) Molecular bar-coding of north-East Atlantic deep-water sharks: species identification and application to fisheries management and conservation. Mar Freshw Res 59(3):214–223

Nelson JS (2006) Fishes of the world, 4th edn. Wiley, Hoboken, NJ

Nguyen HDT, Seifert KA (2008) Description and DNA barcoding of three new species of *Leohumicola* from South Africa and the United States. Persoonia 21:57–69. https://doi.org/10.3767/003158508X361334

Novacek M, Cleland EE (2001) The current biodiversity extinction event: scenarios for mitigation and recovery. Proc Natl Acad Sci U S A 98:5466–5470

Nylander JAA, Erséus C, Källersjo M (1999) A test of monophyly of the gutless Phallodrilinae (Oligochaeta, Tubificidae) and the use of a 573-bp region of the mitochondrial cytochrome oxidase I gene in analysis of annelid phylogeny. Zoologica Script 28(3–4):305–313

Odeny DO, Oyieke HA, Ojwang WO (2009) DNA barcoding: refining para taxonomy for fishery surveys. In: 3rd International barcode of life conference, pp 81–82

Oliveira C, Pereira LHG, Henriques JM, Foresti F (2009) DNA barcode of freshwater fishes from upper Parana Basin, Brazil. In: 3rd international barcode of life conference, Mexico

Packer L, Gibbs J, Sheffield C, Hanner R (2009) DNA barcoding and the mediocrity of morphology. Mol Ecol Res 9:42–50

Pages N, Muñoz-Muñoz F, Talavera S, Sarto V, Lorca C, Núñez JI (2009) Identification of cryptic species of *Culicoides* (Diptera: Ceratopogonidae) in the subgenus *Culicoides* and development of species-specific PCR assays based on barcode regions. Vet Parasitol 165:298–310

Pegg CG, Sinclair BA, Briskey L, Aspden WJ (2006) MtDNA barcode identification of fish larvae in the southern great barrier reef. Aust Sci Mar 70:7–12

Penton EH, Hebert PDN, Crease TJ (2004) Mitochondria, DNA variation in North American populations of *Daphnia obtusa*: continentalism or cryptic endemism? Mol Ecol 13(1):97–107

Persis M, Reddy ACS, Rao LM, Khedkar GD, Ravinder K, Nasruddin K (2009) COI (cytochrome oxidase – I) sequence based studies of carangid fishes from Kakinada coast. India Mol Biol Rep 36:1733–1740. https://doi.org/10.1007/s11033-008-9375-4

Pfenninger M, Cordellier M, Streit B (2006) Comparing the efficacy of morphologic and DNA-based taxonomy in the freshwater gastropod genus Radix (Basommatophora, Pulmonata). BMC Evol Biol 6:100

Pramual P, Wongpakam K, Adler PH (2011) Cryptic biodiversity and phylogenetic relationships revealed by DNA barcoding of oriental black flies in the subgenus *Gomphostilbia* (Diptera: Simuliidae). Genome 54:1–9. https://doi.org/10.1139/G10-100

Radulovici AE, Archambault P, Dufresne F (2010) Review DNA barcodes for marine biodiversity: moving fast forward. Diversity 2:450–472. https://doi.org/10.3390/d2040450

Rajan PT (2010) Diversity of butterfly fishes (Chaetodontidae) of Andaman and Nicobar Islands: indicators in coral reef habitat monitoring and managements. In: Ramakrishna CR, Sivaperuman C (eds) Recent trends in biodiversity of Andaman, Nicobar Islands. Zoological Survey of India, Kolkata, pp 337–342

Rajan PT, Sreeraj CR (2012) Structure of reef fish communities of seven islands of Andaman and Nicobar Islands, India. In: Venkataraman K, Raghunathan C, Sivaperuman C (eds) Ecology of faunal communities on the Andaman, Nicobar Islands. Springer, Berlin, pp 146–127

Ramakrishna CR, Sivaperuman C (2010) Biodiversity of Andaman and Nicobar Islands – an overview. In: Ramakrishna CR, Sivaperuman C (eds) Recent trends in biodiversity of Andaman, Nicobar Islands. Zoological Survey of India, Kolkata, pp 1–42

Rao DV (2003) Guide to reef fishes of Andaman and Nicobar Islands. Zoological Survey of India, Kolkata, p 555

Rasmussen RS, Morrissey MT, Hebert PDN (2009) DNA barcoding of commercially important Salmon and Trout species in North America. In: 3rd International barcode of life conference. Mexico City, pp 84–85

Ratnasingham S, Hebert PDN (2007) BOLD: the barcode of life data system (www.barcodinglife.org). Mol Ecol 7:355–364. https://doi.org/10.1111/j.1471-8286.2006.01678.x

Remigio EA, Hebert PDN (2003) Testing the utility of partial COI sequences for phylogenetic estimates of gastropod relationships. Mol Phylogenet Evol 29(3):641–647

Richardson DE, Vanwye JD, Exum AM, Cowen RT, Crawford DL (2006) High-throughput species identification: from DNA isolation to bioinformatics. DOI 7:199. https://doi.org/10.1111/j.1471-8286.2006.01620.x

Rivas LR (1949) A record of the Lutjanid fish (*Lutjanus cyanopterus)* for the Atlantic coast of the United States, with notes on related species of the genus. Copeia 2:150–152

Rock J, Costa FO, Walker DI, North AW, Hutchinson WF, Carvalho GR (2008) DNA barcodes of fish of the Scotia Sea, Antarctica indicate priority groups for taxonomic and systematic focus. Antarct Sci 20(3):253–262

Rubinoff PD (2006) Utility of mitochondrial DNA barcodes in species conservation. Conserv Biol 20(4):1026–1033

Saccone C, Giorgi CD, Gissi C, Pesole G, Reyes A (1999) Evolutionary genomics in Metazoa: the mitochondrial DNA as a model system. Gene 238:195–209

Sachithanandam V, Mohan PM, Dhivya P, Muruganandam N, Baskaran R, Chaaithanya IK, Vijayachari P (2011) DNA barcoding, phylogenetic relationships and speciation of Genus: *Plectropomus* in Andaman coast. J Res Biol 3:179–183

Sachithanandam V, Mohan PM, Muruganandam N, Chaaithanya IK, Dhivya P, Baskaran R (2012) DNA barcoding, Phylogenetic Study of *Epinephelus*spp from andaman coastal region. Indian J Mar Sci 41(2):203–211

Sachithanandam V, Mohan PM, Muruganandam N (2014) DNA barcoding of marine venomous and poisonous fish of family Scorpaenidae and Tetraodontidae from Andaman water. In: Venkataraman K, Sivaperuman C (eds) Marine faunal diversity in India taxonomy, ecology and conservation. Academic Press Elsevier, London

Sachithanandam V, Mohan PM, Muruganandam N (2015) DNA barcoding of marine venomous and poisonous fish of families Scorpaenidae and Tetraodontidae from Andaman waters. In: Venkataraman K, Sivaperuman C (eds) Marine faunal diversity in India - taxonomy, ecology and conservation. Academic Press, London, pp 351–372. https://doi.org/10.1016/B978-0-12-801948-1.00020-3

Salokannel J, Rantala M, Wahlberg N (2010) DNA-barcoding clarifies species definitions of Finnish *Apatania* (Trichoptera: Apataniidae). Entomol Fenn 2:1–11

Satheeshkumar P, Jagadeesan L (2010) Phylogenetic position and genetic diversity of Neridae – Polychaeta based on molecular data from 16S rRNA sequences. Middle East J Sci Res 6 (6):550–555

Saunders G (2005) Applying DNA barcoding to red macroalgae, a preliminary appraisal holds promise for future applications. Philos Trans R Soc Biol 360:1879–1888

Schander C, Willassen E (2005) What can biological barcoding do for marine biology. Mar Biol Res 1:79–83

Seifert KA, Samson RA, Dewaard JR, Houbraken J, Levesque CA, Moncalvo JM, Seize GL, Hebert PDN (2007) Prospects for fungus identification using CO1 DNA barcodes, with Penicillium as a test case. Proc Nat Acad Sci U S A 104(10):3901–3906

Shirak A, Zinder MC, Barroso RM, Seroussi E, Ron M, Hulata G (2009) DNA barcoding of Israeli indigenous and introduced cichlids. Isr J Aquacult Bamid 61(2):83–88

Siddall ME, Budinoff RB (2005) DNA-barcoding evidence for widespread introductions of a leech from south American *Helobdellatriserialis*complex. Conserv Genet 6:467–472

Silva-Brandao KL, Lyra ML, Freitas AVL (2009) Barcoding Lepidoptera: current situation and perspectives on the usefulness of a contentious technique. Neotrop Entomol 38:441–451

Simon C et al (1994) Evolution, weighting, and phylogenetic utility of mitochondrial gene sequences and a compilation of conserved polymerase chain reaction primers. Ann Entomol Soc Am 87(6):651–701

Simon C et al (2006) Incorporating molecular evolution into phylogenetic analysis, and a new compilation of conserved polymerase chain reaction primers for animal mitochondrial DNA annual review of ecology. Evol Syst 37:545–579

Sjolin E, Erseus C, Kallersjo M (2005) Phylogeny of TubiWcidae (Annelida: Clitellata) based on mitochondrial and nuclear sequence data. Mol Phylogenet Evol 35:431–441

Smith WL, Craig MT (2007) Casting the Percomorph net widely: the importance of broad taxonomic sampling in the search for the placement of Serranidand percid fishes. Copeia 1:35–55

Smith PJ, Mcveagh SM, Steinke D (2008) DNA barcoding for the identification of smoked fish products. J Fish Biol 72(2):464–471

Steinke D, Zemlak TS, Boutillier JA, Hebert PDN (2009a) DNA barcoding of pacific Canada's fishes. Mar Biol 156:2641–2647

Steinke D, Zemlak TS, Hebert PDN (2009b) Barcoding nemo: DNA-based identifications for the ornamental fish trade. PLoS One 4:e6300

Stevens MI, Hogg ID (2003) Blackwell Publishing Ltd. Long-term isolation and recent range expansion from glacial refugia revealed for the endemic springtail Gomphiocephalus hodgsoni from Victoria Land, Antarctica. Mol Ecol 12:2357–2369. https://doi.org/10.1046/j.1365-294X.2003.01907.x

Stoeckle MY, Hebert PDN (2008) Bar code of life: DNA tags help classify animals. Sci Am 299 (4):66–71

Sweeney BW, Battle JM, Jackson JK, Dapkey T (2011) Can DNA barcodes of stream macro invertebrates improve descriptions of community structure and water quality. J N Am Benthol Soc 30:195–216

Talwar PK (1990) Fishes of the Andaman and Nicobar Islands: A synoptic analysis. J Andaman Sci Assoc 6(2):71–102

Thomas M, Raharivololoniaina L, Glaw F, Vences M, Vieites DR (2005) Montane tadpoles in Madagascar: molecular identification and description of the larval stages of *Mantidactylus elegans*, *Mantidactylus madecassus*, and *Boophislaurenti* from the Andringitra massif. Copeia pp 2005:174–183

Turanov SV, Kartavtseva IF, Zemnukhov VV (2012) Molecular phylogenetic study of several eelpout fishes (Perciformes, Zoarcoidei) from far eastern seas on the basis of the nucleotide sequence of the mitochondrial cytochrome oxidase 1 gene (co-1). Genetika 48(2):235–252

Vargas SM, Araujo FCF, Santos FR (2009) DNA barcoding of Brazilian Sea turtles (Testudines). Genet Mol Biol 32(3):608–612

Venkataraman K, Wafar M (2005) Coastal and marine biodiversity of India. Ind J Mar Sci 34 (1):57–75

Victor BC, Hanner R, Shivji M, Hyde J, Caldow C (2009) Identification of the larval and juvenile stages of the Cubera snapper, Lutjanus cyanopterus, using DNA barcoding. Zootaxa 2215:24–36

Ward RD, Zemlak TS, Innes BH, Last PR, Hebert PDN (2005) DNA barcoding Australia's fish species. Philos Trans R Soc Biol 360:1847–1857

Ward RD, Costa FO, Holmes BH, Steinke D (2008) DNA barcoding of shared fish species from the North Atlantic and Australasia: minimal divergence for most taxa, but Zeus faberand *Lepidopuscaudatus* each probably constitute two species. Aquat Biol 3:71–78

Ward RD, Hanner R, Hebert PDN (2009) The campaign to DNA barcode all fishes. Fish-Bol J Fish Biol 74:329–356

Webb KE, Barnes DKA, Clark MS, Bowden DA (2006) DNA barcoding: A molecular tool to identify Antarctic marine larvae. Deep Sea Res 53(II):1053–1060

Wiens JJ, Servedio MR (2000) Species delimitation in systematics: inferring diagnostic differences between species. Proc R Soc Lond B Biol 267:631–636

Will KW, Rubinoff D (2004) Myth of the molecule: DNA barcodes for species cannot replace morphology for identification and classification. Cladistics 20:47–55

Wong EHK, Hanner R (2008) DNA barcoding detects market substitution in North American seafood. Food Res Int 41:828–837

Zemlak TS, Ward RD, Connell AD, Holmes BH, Hebert PDN (2009) Barcoding vertebrates DNA barcoding reveals overlooked marine fishes. Mol Ecol Res 9(1):237–242. https://doi.org/10.1111/j.1755-0998.2009.02649.x

Zettler L, Amaral A, Gómez F, Zettler E, Brendan GK, Amils R, Mitchell LS (2002) Eukaryotic diversity in Spain's river of fire. Nature 417:137

Zhang J (2011) Species identification of marine fishes in China with DNA barcoding. Evid Based Compl Alt Mede 2011:10. https://doi.org/10.1155/2011/978253

Zhang JB, Hanner R (2011) DNA barcoding is a useful tool for the identification of marine fishes from Japan. Biochem Syst Ecol 39:31–42

Applications of DNA Barcoding in Fisheries

A. Pavan-Kumar, A. K. Jaiswar, P. Gireesh-Babu, A. Chaudhari, and G. Krishna

Abstract Sustainable management of fish resources requires accurate identification of species for precise assessment of the stock size and recruitment. Molecular markers would complement morphological tools to differentiate species more accurately. Mitochondrial cytochrome *c* oxidase subunit I has been standardized as a barcode gene for discriminating fishes. DNA barcoding has been applied in fisheries to document fish diversity, to identify ichthyoplankton, prey items, invasive species, parasites, and to authenticate processed fish products. Furthermore, with the advent of next generation sequencing technology, it is possible to identify the presence of invasive species in environmental DNA collected from water and soil. In culture fisheries, some of the fish larvae survival is low due to the lack of knowledge on their prey items. The DNA barcoding approach with NGS technology could be useful to identify the species from samples including thermally-processed fish products, gut content, and environmental samples using DNA mini barcodes.

Keywords Cytochrome *c* oxidase subunit I · Invasive species · Seafood mislabelling · Conservation · Aquaculture

1 Introduction

Fisheries including capture and culture is one of the important sources of quality food, animal protein, and provides livelihood to millions of people around the world. Globally, fish and fishery products represent 9% of the total agricultural exports with a production of 167.2 million tonnes (FAO 2016). Sustainability in fisheries depends on optimum exploitation of resources with effective management measures. Formulation of management measures relies on accurate identification of species for proper stock assessment, recruitment, and delineation. Molecular markers would complement morphological tools to delineate species more accurately.

A. Pavan-Kumar (✉) · A. K. Jaiswar · P. Gireesh-Babu · A. Chaudhari · G. Krishna
ICAR-Central Institute of Fisheries Education, Mumbai, Maharashtra, India

S. Trivedi et al. (eds.), *DNA Barcoding and Molecular Phylogeny*,
https://doi.org/10.1007/978-3-030-50075-7_11

Mitochondrial cytochrome *c* oxidase subunit I (COI) gene has been standardized as a barcode gene for discriminating metazoans (Hebert et al. 2003). After successful demonstration of the efficiency of DNA barcoding in delineating fishes (Ward et al. 2005), it has been applied in different fields of fisheries. Some of the major applications are discussed in this chapter.

2 DNA Barcoding Applications in Capture Fisheries

2.1 Species Identification

Fishes, one of the largest vertebrate groups with more than 32,000 species, display a range of morphological characters and most of the fishes are reported to show phenotypic plasticity (Muschick et al. 2011; Barry et al. 2016). Accurate identification and cataloguing of fishes are necessary for sustainable fishery management. DNA barcoding has been primarily used to discriminate the fish species based on intra- and interspecific genetic divergence values. Reference barcodes have been developed for 17,108 fish species (50%) and cryptic species have been reported in different fish groups (Barcode of Life Database 2017; Pavan-Kumar et al. 2016). DNA barcoding has been used as one of the discernible characters for describing new species and 32 new species have been reported over the last 2 years using this approach (Table 1). DNA barcodes can also be used to study the evolutionary relationship among the species to resolve the taxonomic ambiguity.

2.2 Ecosystem Management

The biodiversity of sensitive ecosystems such as coral reefs, mangroves and wetlands is under threat due to anthropogenic factors, invasive alien species, and climate change (Masters and Norgrove 2010). These factors lead to habitat destruction, loss of biodiversity, and subsequently reduction of resilience of the ecosystem. Regular monitoring of biodiversity from these ecosystems would provide vital information about the pattern of changes in biodiversity and the impact of the environment on biodiversity. To achieve this, baseline information on native biodiversity is essential and DNA barcoding/metabarcoding would be the best approach for developing ecosystem specific databases. Leray and Knowlton (2015) used the DNA barcoding and metabarcoding approaches to characterize benthic diversity on oyster reefs.

Table 1 List of new species described along with DNA barcodes

S. No	Order	Family	Species	References
1	Perciformes	Labridae	*Paracheilinus Paineorum*	Allen et al. (2016)
2			*Paracheilinus xanthocirritus*	
3			*Paracheilinus alfiani*	
4		Pempheridae	*Pempheris gasparinii*	Pinheiro et al. (2016)
5		Serranidae	*Tosanoides Obama*	Pyle et al. (2016)
6		Serranidae	*Chelidoperca santosi*	Williams and Carpenter (2015)
7		Cichlidae	*Ptychochromis mainty*	Martinez et al. (2015)
8		Badidae	*Badis britzi*	Dahanukar et al. (2015)
9	Anguilliformes	Nettastomatidae	*Saurenchelys gigas*	Lin et al. (2015)
10		Muraenidae	*Gymnothorax pseudomelanosomatus*	Loh et al. (2015)
11	Aulopiformes	Synodontidae	*Trachinocephalus gauguini*	Guimarães-Costa et al. (2016)
12	Clupeiformes	Engraulidae	*Stolephorus tamilensis*	Gangan et al. (2017)
13	Cypriniformes	Cyprinidae	*Squalius namak*	Khaefi et al. (2016)
14			*Danio htamanthinus*	Kullander et al. (2015)
15			*Garra mondica*	Sayyadzadeh et al. (2015)
16			*Coreoleuciscus aeruginos*	Song and Bang (2015)
17			*Metzia parva*	Luo et al. (2015)
18			*Danio annulosus*	Kullander et al. (2015)
19			*Osteobrama serrata*	Singh et al. (2016)
20			*Tor dongnaiensis*	Hoàng et al. (2015)
21			*Alburnoides Damghan*	Roudbar et al. (2015)
22			*Capoeta coadi*	Alwan et al. (2016)
23			*Hypselobarbus bicolor*	Knight et al. (2016)
24		Cobitidae	*Cobitis avicennae*	Mousavi-Sabet et al. (2015)
25		Nemacheilidae	*Eidinemacheilus proudlovei*	Freyhof et al. (2016)
26	Rajiformes	Rajiidae	*Raja parva*	Last and Séret (2016)
27	Myliobatiformes	Dasyatidae	*Maculabatis ambigua*	Last et al. (2016)

(continued)

Table 1 (continued)

S. No	Order	Family	Species	References
28	Decapoda	Palaemonidae	*Hamodactylus paraqabai*	Horka et al. (2016)
29			*H. Pseudaqabai*	Horka et al. (2016)
30		Alvinocarididae	*Alvinocaris kexueae*	Wang and Sha (2016)
31		Macrophthalmidae	*Macrophthalmus purpureocheir*	Teng and Shih (2015)
32	Pleurotomarioidea	Pleurotomariidae	*Bayerotrochus delicatu*	Zhang et al. (2016)

2.3 Ichthyoplankton Identification

To understand the processes that influence spatial distribution of species, population dynamics, and migration strategies, it is essential to study larval ecology (Bakun 1996; Cowen et al. 2006). Identification of ichthyoplankton up to species level would assist in mapping breeding habitats of respective fishes (Serafy et al. 2003; Govoni et al. 2003), estimation of population size (Ralston et al. 2003), and understanding the distribution of cryptic species (Richardson and Cowen 2004). Fish egg/larvae can be identified by comparing the COI sequences of eggs/larvae with reference sequence libraries such as BOLD (Barcode of Life Database). The sequences will be assigned to a taxon based on their similarity/genetic distance values with database sequences.

The DNA barcoding approach has been successfully applied for identification of ichthyoplankton from Antarctica (Webb et al. 2006), Great Barrier Reef (Pegg et al. 2006), Yucaton Peninsula, Mexico (Valdez-Moreno et al. 2010), and coral reefs (Hubert et al. 2015). Several studies implied this approach for describing early life-history traits of new species (Victor 2007; Victor et al. 2009) and for identifying fish spawning areas (Neira et al. 2014). Some of the studies have compared DNA barcoding approach with morphological identification techniques and reported higher efficiency of DNA barcoding in ichthyoplankton identification (Ko et al. 2013; Becker et al. 2015; Puncher et al. 2015, Overdyk et al. 2016).

2.4 Conservation

DNA barcoding is a cost-effective technique and can identify the species rapidly and accurately. DNA barcodes are useful for characterizing biodiversity of ecosystems that are species-rich, difficult to assess, and poorly catalogued. DNA barcodes can be used to estimate phylogenetic diversity values. Phylogenetic diversity (PD) values are the minimum total length of all the phylogenetic branches required to span a given set of taxa on the phylogenetic tree (Faith 1992). Faith and Baker (2007) advocated consideration of PD values for prioritization of geographical locations for conservation purpose. The areas that contribute much to the PD values could be given preference for conservation purpose.

As per the IUCN guidelines, criteria for assessing the conservation status of a species are its population size, extent of occurrence, and area of occupancy, etc. DNA barcoding would help in resolving taxonomy and assist in prioritizing species for conservation purpose. For example, consider two species "species A" and "species B" that are showing similar morphological characters and have a restricted distribution. If they are assessed based on morphological characters only, these two species could be considered as a single species and the conservation status may be assigned as "least concern." In this case, implementation of DNA barcoding can resolve the taxonomic ambiguity and species would be properly evaluated for conservation status.

2.5 *Mislabelling of the Seafood Products*

The nutrient profile of each fish species is different and the price could be at par with nutrient values and the processed products should be labelled properly (Mohanty et al. 2014; Bogard et al. 2015). However, mislabelling, adulteration, and replacement of high-value fish with low-value fish have been reported in the seafood processed products (Armani et al. 2011; Marko et al. 2004). Since processed fish products lack morphological characters, it is impossible to identify the species. DNA barcoding has been extensively used to authenticate seafood-processed products and reported various levels of mislabelling (Nagalaksmi et al. 2016; Cutarelli et al. 2014; Harris et al. 2016). The level of DNA degradation would be relatively high in the thermal-processed products and amplification of the entire barcode region (650 b) is relatively difficult. In this scenario, mini barcodes (250–300 bp) have been developed and successfully used for the identification of thermal processed products (Shokralla et al. 2015). In the case of mixed samples, DNA metabarcoding approach (integration of next generation sequencing with DNA barcoding) can be employed to identify the species composition of the product.

2.6 *Biosecurity and Invasive Species*

Biosecurity measures have to be implemented not only for disease-causing agents but also for invasive alien species, which cause huge economic and ecosystem loss (Williamson 1996). The risk of invasion by non-native species is increased by international trade and invasive species colonization is influenced by land use and climate change (Masters and Norgrove 2010). Several studies have reported the detrimental effects of invasive alien species on native biodiversity (Chapin et al. 2000; Pimentel et al. 2005). Rapid and accurate identification of traded biological materials up to the species level at the port of entry is one of the important measure for biosecurity. DNA barcodes would assist in precise identification of species and reference barcodes have been developed for several commercially important fish and shellfish (Bamaniya et al. 2016; Collins et al. 2012, BOLD 2017). Marescaux and Van Doninck (2013) discriminated two invasive species *Dreissena polymorpha* and *D. rostriformis bugensis* using DNA barcodes. Metabarcoding of environmental DNA (eDNA) has been successfully used to detect invasive species from different water bodies (Frédéric et al. 2015; Furlan et al. 2015; Salisbury et al. 2015).

2.7 Prey–predator Relationships

Understanding the predator–prey interactions helps in fisheries management. As per the Pope's study (1979), the overall maximum sustainable yield (MSY) of the system will be higher if the fish species are linked mainly through predator-prey relationships than species that are competing with each other. Predation during early life stages has been reported to be one of the main limiting factors for stock recruitment (Saitoh et al. 2003). Furthermore, predator–prey studies provide information on habitat use and help to identify critical foraging habitats (Peters et al. 2014). The information of prey items (size class-wise) would be useful to minimize impact of predator on the native fish populations. Jo et al. (2014) identified prey items of largemouth bass (*Micropterus salmoides*) in different size classes using the DNA barcoding approach and showed variation in prey preference as they mature (ontogenetic diet shift).

Inadvertent release of non-native species (invasive species) could cause restructuring of the ecosystem by changing food webs and altering the pattern of resource utilization (Stigall 2012). Most often, invasive species are generalists and are flexible with the prey items (Olden et al. 2004). Côté et al. (2013) characterized prey items of invasive Indo-pacific lionfish (*Pterois volitans)* from Bahamian coral reefs through the DNA barcoding approach and reported 37 species. Moran et al. (2016) characterized prey items of non-native Blue Catfish (*Ictalurus furcatus*) & Flathead Catfish (*Pylodictis olivaris)* and showed the presence of economically and ecologically important species in their gut. The metabarcoding approach, an extension of the DNA barcoding method wherein PCR amplicons from mixed/bulk samples sequenced using next generation sequencing platform has been successfully applied to characterize prey items from different fishes (Leray et al. 2013; Harms-Tuohy et al. 2016). Sousa et al. (2016) analyzed the gut content of ocean sunfish *Mola mola* (world's heaviest bony fish) through the metabarcoding approach. This study proves that *Mola mola* is a generalist predator and identified 41 prey items.

3 Potential Applications of DNA Barcoding in Aquaculture

3.1 *Seed Identification*

Accurate seed identification could be possible by metabarcoding of DNA collected from hatchery water.

3.2 *Fingerling Survival*

In Aquaculture, for some of the fishes, larvae or fingerling survival is low and one of the reasons for this is lack of proper nutrition/feed. Providing natural feed which the species used to get from the wild would be a promising approach for better survival. However, for most of the culture species, the information of prey items during different ontogenic stages is lacking. DNA metabarcoding of the gut content of fingerling/adult/brooder fishes would give insights about their natural prey items. The identified prey items can be cultured and provided as live feed to fishes.

3.3 *Improved Management/Production Management*

Biofloc technology (BFT) is going to revolutionize the aquaculture industry by promoting waste retention and its subsequent conversion to Biofloc as natural food for fish/shrimps (Crab et al. 2012). Biofloc is a consortium of microorganisms such as heterotrophic bacteria, algae (dinoflagellates and diatoms), fungi, ciliates, flagellates, rotifers, nematodes, and metazoans. They act synergistically to maintain the water quality and convert nitrogenous waste into protein. The species composition of biofloc varies with the carbon source and characterization of biofloc microscopic composition would be very useful for better management of culture ponds. Metabarcoding of biofloc could reveal the composition of biofloc and this information along with carbon source would act as a reference for implementing biofloc technology effectively. Different genes i.e. Internal Transcribed Spacer (ITS), COI, 16S rRNA, and LSU D1/D2 have been standardized for developing barcodes for fungi, protozoans/metazoans, and bacteria, respectively (Lebonah et al. 2014; Schoch et al. 2012).

3.4 *Health Management*

Pond ecosystem is dynamic and any alteration in biotic and abiotic factors could result in stress on fishes/shrimps and may increase the susceptibility of fish to infections. Metabarcoding of soil and water at regular intervals would give information about changes in bacteria/microbe composition over the time with response to uneaten, left over feed and other activities. This information would act as a reference for better management of ponds by keeping optimum water and soil quality parameters. Parasite (monogeneas, digeneans, and crustaceans) infection in fishes cause huge economic losses due to secondary bacterial infection and subsequent mortality. These parasites can be eradicated if early detection is possible. However, larvae and eggs of these parasites are very small in size and cannot be identified by the naked eye. Metabarcoding of DNA from pond water can be useful to identify early life stages of fish parasites. Furthermore, generated DNA barcodes are useful to study the evolution of Host–parasite interaction. Several authors have generated DNA barcodes for monogeneans (Hansen et al. 2007*;* Neeraja et al. 2016), digeneans (Locke et al. 2015), and Crustaceans (Ferozkhan et al. 2016).

4 Conclusion

Despite few limitations (lack of the DNA barcoding gap in recently evolved species, unable to identify Hybrid species), the DNA barcoding approach is the most promising approach to document the biodiversity and plays a major role in achieving Convention on Biodiversity Aichi targets. Implementation of the DNA barcoding approach in different fields of fisheries would assist in sustainable utilization of resources and increased culture production.

References

Allen GR, Erdmann MV, Yusmalinda NLA (2016) Review of the indo-Pacific Flasherwrasses of the genus Paracheilinus (Perciformes: Labridae), with descriptions of three new species. J Ocean Sci Found 19:18–90

Alwan NH, Zareian H, Esmaeili HR (2016) Capoeta coadi, a new species of cyprinid fish from the Karun River drainage, Iran based on morphological and molecular evidences (Teleostei, Cyprinidae). Zookeys 16:155–180

Armani A, Castigliego L, Tinacci L, Gianfaldoni D, Guidi A (2011) Molecular characterization of icefish (Salangidae family), using direct sequencing of mitochondrial cytochrome b gene. Food Control 22:888–895

Bakun A (1996) Patterns in the ocean: ocean processes and marine population dynamics. California Sea Grant College system report T-o37, La Jolla, California

Bamaniya DC, Pavan-Kumar A, Gireesh-Babu P, Sharma N, Reang D, Krishna G, Lakra WS (2016) DNA barcoding of marine ornamental fishes from India. Mitochondrial DNA A DNA Mapp Seq Anal 27:3093–3097

Barcode of Life Database (2017). http://boldsystems.org/ Accessed 10 Mar 2017

Barry J, Newton M, Dodd JA, Hooker OE, Boylan P, Lucas MC, Adams CE (2016) Foraging specialisms influence space use and movement patterns of the European eel *Anguilla anguilla*. Hydrobiologia 766:333–348

Becker RA, Sales NG, Santos GM, Santos GB, Carvalho DC (2015) DNA barcoding and morphological identification of neotropical ichthyoplankton from the upper Paraná and São Francisco. J Fish Biol 87:159–168

Bogard JR, Thilsted SH, Marks GC, Wahab MA, Hossain MAR, Jakobsen J, Stangoulis J (2015) Nutrient composition of important fish species in Bangladesh and potential contribution to recommended nutrient intakes. J Food Comp Anal 42:120–133

Chapin FS III, Zavaleta ES, Eviner VT, Naylor RL, Vitousek PM et al (2000) Consequences of changing biodiversity. Nature 405:234–242

Collins RA, Armstrong KF, Meier R, Yi Y, Brown SDJ et al (2012) Barcoding and border biosecurity: identifying cyprinid fishes in the aquarium trade. PLoS One 7(1):e28381. https://doi.org/10.1371/journal.pone.0028381

Côté IM, Green SJ, Morris JA Jr, Akins JL, Steinke D (2013) Diet richness of invasive indo-Pacific lionfish revealed by DNA barcoding. Mar Ecol Prog Ser 472:249–256

Cowen RK, Paris CB, Srinivasan A (2006) Scaling connectivity in marine populations. Science 311:522–527

Crab R, Defoirdt T, Bossier P, Verstraete W (2012) Biofloc technology in aquaculture: beneficial effects and future challenges. Aquaculture 356–357:351–356

Cutarelli A, Amoroso MG, De Roma A, Girardi S, Galiero G, Guarino A et al (2014) Italian market fish species identification and commercial frauds revealing by DNA sequencing. Food Control 37:46–50

Dahanukar N, Kumkar P, Katwate U, Raghavan R (2015) Badis britzi, a new percomorph fish (Teleostei: Badidae) from the Western Ghats of India. Zootaxa 3941:429–436

Faith DP (1992) Conservation evaluation and phylogenetic diversity. Biol Conserv 61:1–10

Faith DP, Baker AM (2007) Phylogenetic diversity (PD) and biodiversity conservation: some bioinformatics challenges. Evol Bioinformatics Online 2:121–128

FAO (2016) The state of world fisheries and aquaculture 2016. Contributing to food security and nutrition for all. FAO, Rome, p 200

Ferozkhan K, Sanker G, Prasanna Kumar C (2016) Linking eggs and adults of Argulus spp. using mitochondrial DNA barcodes. Mitochondrial DNA A DNA Mapp Seq Anal 27:3927–3931

Frédéric JJC, Brown E, MacIsaac H, Melania E (2015) Monitoring biodiversity for the early detection of aquatic invasive species using metabarcoding applied across Canadian ports in the Pacific, Arctic, Atlantic, and Great Lakes. In Scientific abstracts from the 6th International Barcode of Life Conference. Genome 58(5):245

Freyhof J, Abdullah YS, Ararat K, Ibrahim H, Geige MF (2016) Eidinemacheilus Proudlovei, a New Subterranean Loach From Iraqi Kurdistan (Teleostei; Nemacheilidae). Zootaxa 4173:225–236

Furlan E, Gleeson D, Hardy C, Duncan R (2015) A framework for estimating eDNA sensitivity. In Scientific abstracts from the 6th international barcode of life conference. Genome 58(5):218

Gangan S, Pavan-Kumar A, Jahageerdar S, Jaiswar AK (2017) A new fish species of the genus Stolephorus (Clupeiformes: Engraulidae) from the bay of Bengal, India. Submitted to Marine Biodiveristy Journal

Govoni JJ, Laban EH, Hare JA (2003) The early life history of swordfish (*Xiphias gladius*) in the western North Atlantic. Fish Bull 101:778–789

Guimarães-Costa A, Vallinoto M, Giarrizzo T, Pezold F, Schneider H, Sampaio I (2016) Molecular evidence of two new species of Eleotris (Gobiiformes: Eleotridae) in the western Atlantic. Mol Phylogenet Evol 98:52–60

Hansen H, Bakke TA, Bachmann L (2007) Mitochondrial haplotype diversity of *Gyrodactylus thymalli* (Platyhelminthes; Monogenea): extended geographic sampling in United Kingdom, Poland, and Norway reveals further lineages. Parasitol Res 100:1389–1394

Harms-Tuohy AC, Schizas VN, Appeldoorn SR (2016) Use of DNA metabarcoding for stomach content analysis in the invasive lionfish *Pterois volitans* in Puerto Rico. Mar Ecol Prog Ser 558:181–191

Harris JD, Rosado D, Xavier R (2016) DNA barcoding reveals extensive mislabeling in seafood sold in Portuguese supermarkets. J Aquat Food Prod T 25:1375–1380

Hebert PDN, Cywinska A, Ball SL (2003) Biological identifications through DNA barcodes. Proc Biol Sci R Soc 270(1512):313–321. https://doi.org/10.1098/rspb.2002.2218

Hoàng HĐ, Phạm HM, Durand JD, Trần NT, Phan PĐ (2015) Mahseers genera Tor and Neolissochilus (Teleostei: Cyprinidae) from southern Vietnam. Zootaxa 4006:551–568

Horka I, Fransen MJH, Duris Z (2016) Two new species of shrimp of the indo-West Pacific genus *Hamodactylus Holthuis*, 1952 (Crustacea: Decapoda: Palaemonidae). Eur J Taxon 188:1–26

Hubert N, Espiau B, Meyer C, Planes S (2015) Identifying the ichthyoplankton of a coral reef using DNA barcodes. Mol Ecol Resour 15:57–67

Jo H, Gim JA, Jeong KS, Kim HS, Joo GJ (2014) Application of DNA barcoding for identification of freshwater carnivorous fish diets: is number of prey items dependent on size class for *Micropterus salmoides*? Ecol Evol 4:219–229

Khaefi R, Esmaeili HR, Sayyadzadeh G, Geiger MF, Freyhof J (2016) Squalius namak, a new chub from Lake Namak basin in Iran (Teleostei: Cyprinidae). Zootaxa 4169:145–159

Knight JD, Rai A, D'Souza RK, Philip S, Dahanukar N (2016) Hypselobarbus bicolor, a new species of large barb (Teleostei: Cyprinidae) from the Western Ghats of India. Zootaxa 4184:316–328

Ko HL, Wang YT, Chiu TS, Lee MA, Leu MY, Chang KZ, Chen WY, Shao KT (2013) Evaluating the accuracy of morphological identification of larval fishes by applying DNA barcoding. PLoS One 8:e53451

Kullander SO, Rahman MM, Norén M, Mollah AR (2015) Danio annulosus, a new species of chain Danio from the Shuvolong falls in Bangladesh (Teleostei: Cyprinidae: Danioninae). Zootaxa 3994:053–068

Last PR, Séret B (2016) A new eastern Central Atlantic skate *Raja parva* sp. nov. (Rajoidei: Rajidae) belonging to the *Raja miraletus* species complex. Zootaxa 4147:477–489

Last PR, Bogorodsky SV, Alpermann TJ (2016) Maculabatis ambigua sp. nov., a new whipray (Myliobatiformes: Dasyatidae) from the Western Indian Ocean. Zootaxa 4154:66–78

Lebonah DE, Dileep A, Chandrasekhar K, Sreevani S, Sreedevi B, Kumari JP (2014) DNA barcoding on bacteria: a review. Adv Biol 2014:1–9

Leray M, Knowlton N (2015) DNA barcoding and metabarcoding of standardized samples reveal patterns of marine benthic diversity. Proc Natl Acad Sci U S A 112:2076–2081

Leray M, Yang JY, Meyer CP, Mills SC, Agudelo N et al (2013) A new versatile primer set targeting a short fragment of the mitochondrial COI region for metabarcoding metazoan diversity: application for characterizing coral reef fish gut contents. Front Zool 10:34

Lin J, Smith DG, Shao KT, Chen HM (2015) *Saurenchelys gigas* sp. nov., a new nettastomatid eel (Teleostei, Anguilliformes, Nettastomatidae) from the western Central Pacific. Zootaxa 4060:121–130

Locke SA, Al-Nasiri FS, Caffara M, Drago F, Kalbe M, Lapierre AR, McLaughlin JD (2015) Diversity, specificity and speciation in larval Diplostomidae (Platyhelminthes: Digenea) in the eyes of freshwater fish, as revealed by DNA barcodes. Int J Parasitol 45:841–855

Loh KH, Shao KT, Ho HC, Lim PE, Chen HM (2015) A new species of moray eel (Anguilliformes: Muraenidae) from Taiwan, with comments on related elongate unpatterned species. Zootaxa 4060:30–40

Luo W, Sullivan JP, Zhao HT, Peng ZG (2015) Metzia parva, a new cyprinid species (Teleostei: Cypriniformes) from South China. Zootaxa 3962:226–234

Marescaux J, Van Doninck K (2013) Using DNA barcoding to differentiate invasive Dreissena species (Mollusca, Bivalvia). In: Nagy ZT, Backeljau T, De Meyer M, Jordaens K (eds) DNA barcoding: a practical tool for fundamental and applied biodiversity research, ZooKeys, vol 365, pp 235–244

Marko PB, Lee SC, Rice AM, Gramling JM, Fitzhenry TM, McAlister JS, Harper GR, Moran AL (2004) Fisheries: mislabelling of a depleted reef fish. Nature 430:309–310

Martinez CM, Arroyave J, Sparks JS (2015) A new species of Ptychochromis from southeastern Madagascar (Teleostei: Cichlidae). Zootaxa 4044:79–92

Masters G, Norgrove L (2010) Climate change and invasive alien species. CABI working paper 1, p 30

Mohanty B, Mahanty A, Ganguly S et al (2014) Amino acid compositions of 27 food fishes and their importance in clinical nutrition. J Amino Acids 2014:1. https://doi.org/10.1155/2014/269797

Moran Z, Orth DJ, Schmitt JD, Hallerman EM, Aguilar R (2016) Effectiveness of DNA barcoding for identifying piscine prey items in stomach contents of piscivorous catfishes. Environ Biol Fish 99:161–167. https://doi.org/10.1007/s10641-015-0448-7

Mousavi-Sabet H, Vatandoust S, Esmaeili HR, Geiger MF, Freyhof J (2015) Cobitis avicennae, a new species of spined loach from the Tigris River drainage (Teleostei: Cobitidae). Zootaxa 3914:558–569

Muschick M, Barluenga M, Salzburger W, Meyer A (2011) Adaptive phenotypic plasticity in the Midas cichlid fish pharyngeal jaw and its relevance in adaptive radiation. BMC Evol Biol 11:116

Nagalaksmi K, Pavan-Kumar A, Venketashwarlu G, Gireesh-Babu P, Lakra WS (2016) Mislabeling in Indian seafood: an investigation using DNA barcoding. Food Control 59:196–200

Neeraja T, Tripathi G, Gireesh-Babu P, Pavan-Kumar A (2016) Molecular characterization and phylogeny of some mazocraeidean monogeneans from carangid fish. Acta Parasit 61:360–368

Neira FJ, Perry RA, Burridge CP, Lyle JM, Keane JP (2014) Molecular discrimination of shelf-spawned eggs of two co-occurring *Trachurus* spp.(Carangidae) in southeastern Australia: a key step to future egg-based biomass estimates. ICES J Mar Sci 72:614–624

Olden JD, Poff NL, Douglas MR, Douglas ME, Fausch KD (2004) Ecological and evolutionary consequences of biotic homogenization. Trends Ecol Evol 19:18–24

Overdyk LM, Holm E, Crawford SS, Hanner RH (2016) Increased taxonomic resolution of Laurentian Great Lakes ichthyoplankton through DNA barcoding: a case study comparison against visual identification of larval fishes from Stokes Bay, Lake Huron. J Great Lakes Res 42:812–818

Pavan-Kumar A, Gireesh-Babu P, Jaiswar AK, Chaudhari A, Krishna G, Lakra WS (2016) DNA barcoding of marine fishes: prospects and challenges. In: Trivedi S et al (eds) DNA barcoding in marine perspectives: assessment and conservation of biodiversity. Springer, Cham, pp 285–299

Pegg GG, Sinclair B, Briskey L, Aspden WJ (2006) MtDNA barcode identification of fish larvae in the southern great barrier reef–Australia. Sci Mar 70:7–12

Peters KJ, Ophelkeller K, Bott NJ, Deagle BE, Jarman SN, Goldsworthy SD (2014) Fine-scale diet of the Australian sea lion (*Neophoca cinerea*) using DNA-based analysis of faeces. Mar Ecol 36:347–367. https://doi.org/10.1111/maec.12145

Pimentel D, Zuniga R, Morrison D (2005) Update on the environmental and economic costs associated with alien-invasive species in the United States. Ecol Econ 52:273–288

Pinheiro HT, Bernardi G, Rocha LA (2016) Pempheris gasparinii, a new species of sweeper fish from Trindade Island, southwestern Atlantic (Teleostei, Pempheridae). Zookeys 8:105–115

Pope JG (1979) Stock assessment in multispecies fisheries, with special reference to the trawl fishery in the Gulf of Thailand. SCS/DEV.79/19, South China Sea Fish. Dev and Coord Prograrmne, p 106

Puncher G, Alemany F, Arrizabalaga H, Cariani A, Tinti F (2015) Misidentification of bluefin tuna larvae: a call for caution and taxonomic reform. Rev Fish Biol Fish 25:485–502

Pyle LR, Green DB, Kosaki KR (2016) Tosanoides Obama, a new basslet (Perciformes, Percoidei, Serranidae) from deep coral reefs in the northwestern Hawaiian islands. ZooKeys 641:165–181
Ralston S, Bence JR, Eldridge MB, Lenarz WH (2003) An approach to estimating rockfish biomass based on larval production, with application to *Sebastes jordani*. Fish Bull 101:129–146
Richardson DE, Cowen RK (2004) Diversity of leptocephalus larvae around the island of Barbados (West Indies): relevance to regional distributions. Mar Ecol Prog Ser 282:271–284
Roudbar AJ, Eagderi S, Esmaeili HR, Coad BW, Bogutskaya N (2015) A molecular approach to the genus Alburnoides using COI sequences data set and the description of a new species, A. *damghani*, from the Damghan River system (the Dasht-e Kavir Basin, Iran) (Actinopterygii, Cyprinidae). Zookeys 579:157–181
Saitoh K, Takagaki M, Yamashita Y (2003) Detection of Japanese flounder-specific DNA from gut contents of potential predators in the field. Fisheries Sci 69:473–477
Salisbury S, Shokralla S, Hajibabaei M (2015) The feasibility of detecting an Asian carp invasion using environmental DNA and next-generation sequencing. *In* Scientific abstracts from the 6th international barcode of life conference. Genome 58(5):275
Sayyadzadeh G, Esmaeili HR, Freyhof J (2015) Garra mondica, a new species from the Mond River drainage with remarks on the genus Garra from the Persian Gulf basin in Iran (Teleostei: Cyprinidae). Zootaxa 4048:75–89
Schoch CL, Seifert KA, Huhndorf S, Robert V, Spouge JL (2012) Nuclear ribosomal internal transcribed spacer (ITS) region as a universal DNA barcode marker for Fungi. Proc Natl Acad Sci U S A 109:6241–6246
Serafy JE, Cowen RK, Paris CB, Capo TR, Luthy SA (2003) Evidence of blue marlin, *Makaira nigricans*, spawning in the vicinity of Exuma sound, Bahamas. Mar Freshw Res 54:299–306
Shokralla S, Hellberg RS, Handy SM, King I, Hajibabaei MA (2015) DNA mini-barcoding system for authentication of processed fish products. Sci Rep 5:15894
Singh M, Verma R, Yumnam R, Vishwanath W (2016) Molecular phylogenetic analysis of genus Osteobrama Heckel, 1843 and discovery of Osteobrama serrata sp. nov. from north East India. Mitochondrial DNA A DNA Mapp Seq Anal 29(3):361–366. https://doi.org/10.1080/24701394.2017.1285289
Song HY, Bang IC (2015) Coreoleuciscus aeruginos (Teleostei: Cypriniformes: Cyprinidae), a new species from the Seomjin and Nakdong rivers, Korea. Zootaxa 3931:140–150
Sousa LL, Xavier R, Costa V, Humphries EN, Trueman C, Rosa R, Sims WD, Queiroz N (2016) DNA barcoding identifies a cosmopolitan diet in the ocean sunfish. Sci Rep 6:28762
Stigall LA (2012) Invasive species and evolution. Evol Educ Outreach 5:526–533
Teng SJ, Shih HT (2015) A new species of *Macrophthalmus* d*esmarest*, 1823 (Crustacea: Decapoda: Brachyura: Macrophthalmidae) from Taiwan, with notes on four new records. Zootaxa 4058:451–470
Valdez-Moreno M, Vásquez-Yeomans L, Elías-Gutiérrez M, Ivanova NV, Hebert PDN (2010) Using DNA barcodes to connect adults and early life stages of marine fishes from the Yucatan peninsula, Mexico: potential in fisheries management. Mar Freshw Res 61:655–671
Victor BC (2007) *Coryphopterus kuna*, a new goby (Perciformes: Gobiidae: Gobiinae) from the western Caribbean, with the identification of the late larval stage and an estimate of the pelagic larval duration. Zootaxa 1526:51–61
Victor BC, Hanne R, Shivji M, Hyde J, Caldow C (2009) Identification of the larval and juvenile stages of the cubera snapper, *Lutjanus cyanopterus,* using DNA barcoding. Zootaxa 2215:24–36
Wang YR, Sha ZL (2016) A new species of the genus Alvinocaris Williams and Chace, 1982 (Crustacea: Decapoda: Caridea: Alvinocarididae) from the Manus Basin hydrothermal vents, Southwest Pacific. Zootaxa 4226(1):zootaxa.4226.1.7. https://doi.org/10.11646/zootaxa.4226.1.7
Ward RD, Zemlak TS, Innes BH (2005) Barcoding Australia's fish species. Phil Trans R Soc Lond B 360:1847–1857

Webb KE, Barnes DK, Clark MS, Bowden DA (2006) DNA barcoding: a molecular tool to identify Antarctic marine larvae. Deep Sea Res Part 2 Top Stud Oceanogr 53:1053–1060

Williams TJ, Carpenter EK (2015) A new fish species of the subfamily Serraninae (Perciformes, Serranidae) from the Philippines. Zootaxa 3911:287–293

Williamson M (1996) Biological invasions. Chapman & Hall, London, p 244

Zhang S, Zhang S, Wei P (2016) Bayerotrochus delicatus, a new species of pleurotomariid from yap seamount, near Palau, Western Pacific (Gastropoda: Pleurotomariidae). Zootaxa 4161:252–260

DNA Barcoding in Avian Species with Special Reference to Taxonomically Wide Biogeographic Studies

Farhina Pasha

Abstract The establishment of DNA Barcode project in 2003, intending to construct a strong molecular identification tool via standardised genetic sequences, marked a new era of species identification and taxonomy. DNA barcoding so far proved to be simple and one of the excellent tools for identification of not only animals and plants but also the aves. The avian fauna represents an excellent contender for testing DNA barcode validation as aves or birds are amongst the most prominent groups in which a wide variety of morphological, genetic and behavioural studies have been conducted, thereby establishing a prime line of stable taxonomy. The idea of All Bird barcode initiative (ABBI) was conceived in 2005 with the intention to collect genetic data samples for deciphering a DNA barcode for over 10,000 known avian species. Regardless of hundreds of vigilant studies carried out during the past decade, there are still numerous avian species to be discovered and identified. ABBI is new hope for speedy identification of novel avian species and will also help in hundreds of new samples to be identified, thereby opening up new avenues for avian identification and its related scientific research. Adding on, with the advancement of the mt-DNA gene cytochrome *c* oxidase I (COI) library via DNA barcoding projects of avian species, there will be a better understanding of different avian realms and taxonomic territories. It will also serve as an unbiased taxonomic representation of different avian groups. The advantage is that the DNA barcode sequences deposited in these databases are of high quality and are standardised and therefore have fewer ambiguities, short sequence span, bidirectional sequencing and uniform sequence alignment. It has been observed that the rate of error on BOLD is much lower than on other databases, DNA barcoding in avian species in future will undoubtedly provide more specific species identification, recognition of cryptic species and tracing the line of avian evolution through different eras, deciphering their causes of divergence.

F. Pasha (✉)
Department of Biology, Faculty of Sciences, University of Tabuk, Tabuk, Saudi Arabia
e-mail: fbasha@ut.edu.sa

S. Trivedi et al. (eds.), *DNA Barcoding and Molecular Phylogeny*,
https://doi.org/10.1007/978-3-030-50075-7_12

Keywords Avian Species · DNA barcoding · COI gene library · Island birds · Sea birds · Migratory birds and continental birds

1 Introduction

The avian fauna represents an excellent contender for testing DNA barcode validation. Aves or birds are amongst the most prominent groups in which a wide variety of morphological, genetic and behavioural studies have been conducted to establish a prime line of stable taxonomy (Baker et al. 2009; Gill 2007). Carl Woese was the first research who applied the nucleotide sequence variation in a sole gene to trace the evolutionary relationships. On the basis of his research, he inferred that sequences differ in a conserved gene and the ribosomal RNA can be used to trace phylogenetic interactions and sequences as that revealed in *Archaea* furthermore leading to trace tree of life (Woese and Fox 1977). As a recent advance, the Polymerase chain reaction has emerged as an effective tool to examine sequence diversity in any gene. It is evident that the genes like r RNA-encoding genes evolve slowly. These genes are closely related and are crucial in tracing primeval relationships whereas the genes that evolve rapidly may overwrite but they do divulge divergence amongst closely related species (Woese 2000).

The sequence divergence in mitochondrial DNA (mtDNA) can be utilised to outline evolutionary history within the species, which was first revealed by John Avise. This can potentially be used for linking systematic and population genetics, thereby establishing phylogeography (Avise et al. 1987). Though some species show phylogeographic subdivisions, they are frequently in close proximity 'at distances much shorter than the internodal branch lengths of the species tree' as reported by Moore (Moore W.S. 1995), thereby concluding that the sequence divergences within the species are larger and therefore mtDNA confines the discontinuities in a gene identified by taxonomists as species and because of this reason taxonomic revisions use analysis of mtDNA divergences regularly. This has played a major role in the identification of native avian species on the basis of mtDNA divergences (Avise and Walker 1998; Gill and Slikas 1992; Murray et al. 1994; AOU 1998; Banks et al. 2000, 2002, 2003).

2 All Bird Barcoding Initiative (ABBI)

The idea of All Bird Barcode Initiative (ABBI) was conceived in 2005 with the intention to collect genetic data samples for deciphering DNA barcode for over 10,000 known avian species. Regardless of hundreds of vigilant studies carried out during the past decade, there are still numerous avian species to be discovered and identified. ABBI is new hope for speedy identification of novel avian species and will also help in hundreds of new samples to be identified thereby opening new avenues for avian identification and its related scientific research.

The establishment of DNA Barcode project in 2003, intending to construct a strong molecular identification tool via a standardised genetic fragment, marked a new era of species identification and taxonomy. The library of sequences so formed has reached almost five million sequences in 2015 (formally projected in 2009; Frézal and Leblois 2008). There are more than 47,000 sequences of 6000 avian species deposited here on the Barcode of Life Data (BOLD) (available on www.boldsystems.org; access 2016) from the entire world avian diversity of 10,473 species (Clements et al. 2015). ABBI stands as the first Global campaign for obtaining a specific taxonomic group of avian species, 34,000 avian DNA barcode records are available here with sequences of over 4280 species and 37 orders (Baker et al. 2005; Stoeckle 2005).

Avian species were amongst the first group selected for the testing of DNA barcoding because of their well-established taxonomy, which made it possible to recognise the genetic discrepancy and Linnaean species restrictions (Hebert et al. 2004). The growing ABBI open-access electronic library tends to link DNA Barcodes, different specimen and associated data collections by depositing records in BOLD/Gene Bank/EMBL/DDJB. This library will definitely be an important source for conservation management, biodiversity experts' ornithologists, ecologist's, public health officials and interested common public. Not only this, it will serve as a benchmark for identification of birds that strike the aeroplanes, helping airline safety and also identification and tracking divergence in migratory avian species (other animal and plant life). In future, these samples can be analysed regardless of their size, age and plumage/sex.

3 Modus Operandi

It has been well established now that the 648 bp region at the 5′ end of the mitochondrial DNA gene cytochrome *c* oxidase subunit 1 (COI or cox1) is ideal for most of the animal species including avian species (Hebert et al. 2003, 2004). Primarily, the samples available for DNA barcoding in avian species include fresh tissues of skin and skeleton, blood or feathers, most of which are conserved at low temperature or in ethanol. In the case of historical samples, vouchered museum specimens are the source of the sample, which usually includes skin and skeleton for various biochemical and DNA analyses as a recent practice (Arctander and Fjeldsa 1994; Seutin et al. 1991). Though the historical samples that usually are over two decades old are not so well preserved for DNA analysis, highly degraded DNA is obtained (Lindahl 1993). These samples are very difficult to amplify and to be used for further studies (Zimmermann et al. 2008). The solution to this has occurred by selecting a 100–200 bp fragment of the initially projected ~650 bp COI region known as the mini barcode (Hajibabaei et al. 2006). However, the success rate is very low as it works merely in the restricted taxonomic group. Other option lies in the use of DNA repair enzyme, which increases the success rate of amplification especially when the DNA is damaged by ageing or chemicals/preservatives (Evans 2007). The most common being oxidative and hydrolytic damage (Hoss et al. 1996). By targeting a small overlapping region from any section of DNA, an entire sequence can be gained from an ancient historical vouchered sample and hence DNA barcoding in avian species will be possible by a set of conserved primers, which allow proper amplification of the COI barcoding region (Millar et al. 2008).

To outline the pattern of COI sequence divergence in avian species, the initial sample includes single individual selected on the basis of convenience rather than a taxonomic concern to determine the COI divergence between species. At the next level, one to three more individuals are examined to provide information about intraspecific sequence divergence. Thereby, to obtain COI divergence at each level, the number of individuals is increased and a sample size of 20 years can be analysed using this method. A 2% sequence divergence was observed by Hebert et al. (2004) amongst individuals in "Birds of North America" using the above methodology.

The major difficulty encountered in identification of specie by Mitochondrial DNA studies is the interference of Nuclear Mitochondrial DNA Segments (NUMT) or pseudogenes, as they are non-coding sequences, which are accumulated via random mutations. Although the NUMT have a Stop codon and idles, which make it easy for their identification and deletion from the database, there are some cryptic peseudogenes that lack these characteristics and a slight genetic divergence is seen because of them instead of the original COI sequence. In Avian species, they are far more problematic as their common primers show the least resemblance to mitochondrial gene, for example, Tyrannidae (these are a large family of Passeriforme Order and North, Central and South Americas' birds are the prime representatives). Cryptic pseudogenes also occur as a result of amplification done via blood samples having a high number of nuclear DNA copies related to mitochondrial segments. As NUMT are placed in nuclear DNA, they tend to have a low evolutionary rate than mitochondrial COI (mtCOI) gene, which is 2–14 times greater as reported by Kerr 2010, where NUMT sequences were found to be 14.2%. Even though the COI pseudogenes show their presence occasionally, it may delude to the assumption of cryptic divergence and split conspecific in an additional Barcode Index Number (BIN). This problem can be solved by supplemented with the other barcodes especially the nuclear DNA barcode (Dasmahapatra and Mallet 2006).

4 DNA Barcoding in Island Birds

Avian species native to various islands have always attracted ornithologists all over the world. They attract the attention of researchers, interested in evolution and biogeographic studies globally. These aves are exceptionally unique in their behavioural patterns and are known for their reproductively isolated behaviour from the continental avian population. They also have a very specific colony pattern (Newton 2003). Though most of the avian species have a nuclear ancestral origin, many groups show a spectacular adaptive variation in food habitats, colony patterns and of some extent structural patterns too. As a consequence, a diverse group of closely related avian species have occurred, occupying wider habitats. There are many studies conducted on Island avian species and their population studies have been undertaken to understand their biogeographic orientation via DNA barcoding (Lohman et al. 2010; Campagna et al. 2012; Nishiumi and Kim 2015). A phylogenetic tree for identifying the cases of reverse colonisation amongst island aves was constructed by Nishiumi and Kim (2015) using the obtained DNA sequences of Holarctic avian species (i.e., from Japanese islands to mainland Asia) from BOLD and by perceiving the topology in relation to the geographic origin of the samples. The study identified 5 species with strong reverse colonisation pattern amongst the 118 samples, whereas 39 avian species had their genetic makeup as that of continental origin but were found breeding on Japanese islands. Overall, there is still a great need for studies to be undertaken for deciphering and obtaining more genetic markers especially of the remote island avian species, a wider sampling population and more ordain phylogenetic reconstruction, e.g. maximum likelihood and Bayesian analysis methods. The estimation of divergence time between the continental and island avian species also needs to be studied for obtaining accuracy in the reverse colonisation process.

5 DNA Barcoding in Continental Species

As compared to the island birds, the continental species exhibit more closely related genetic patterns exhibiting a familiar geological history amongst them (Newton 2003). Several studies had been undertaken to establish the various diversifying factors using the COI variation framework (Johnsen et al. 2010; Lijtmaer et al. 2011; Tavares et al. 2011; Milá et al. 2012; Chaves et al. 2015).

DNA barcoding can reliably identify the closely related sister species of birds by using the COI sequence. To test the efficacy of DNA barcodes for the identification of closely related species, Tavares and Baker (2008) compared *COI* in 60 (~650 bp) sister-species pairs from 10 orders of birds. Individuals of each species were of monophyly in a neighbour-joining (NJ) tree in all the pairs, each species were possessing fixed mutational differences differentiating them from their sister species (Tavares and Baker 2008).

In a study, Tavares (2011) joined the DNA barcode of Brazilian avian species with those of Argentina to study the genetic structure of Neotropical species. They observed 75% species had low intraspecific COI divergence and number of avian species in this territory and were having an outsized genetic split than in North American birds. It is more likely the result of a higher effect of glacial cycles isolating populations in North America and has generated speciation (Lijtmaer et al. 2011). In another study undertaken by Neotropics (Milá et al. 2012) genetic diversification amongst eastern and western Amazon population was observed and a large level of intraspecific divergence was reported. This genetic deviation amongst the population was not equivalent to phenotypic variations in plumage colouration (Milá et al. 2012). Another study conducted by Chaves et al. (2015) using Bayesian tree reconstructions of DNA barcode sequence for avian species of the Brazilian Cerrado and the Atlantic Forest showed 10.4% were non-monophyletic. It is well indicated by the fact that from 2004–2016 almost 116 articles have been published on pub med (search criteria: DNA barcoding in avian species), this definitely has marked a new era in identification of birds and their taxonomic coverage but credentials of cryptic species and diversifying factors still need to be deciphered for most of the geographical realms.

6 DNA Barcoding in Migratory Birds

Most of the avian species have a high metabolic rate thereby requiring plenteous food supply throughout. However, this supply may not occur throughout the year, in many geographic realms; therefore, birds have evolved and adapted very effective method of travelling interminability to food-rich grounds amid great energy-conserving efficiency. Though the characteristics of migratory birds do not differ totally from the non-migratory birds, many transitional groups exist amid them. These characteristics can be seen in a single local population exhibiting partial migration. The significance in studying these migratory birds lies in the fact that birds are very sensitive to environmental changes and therefore serve as very good biological indicators. Majority of them are at peak of their decline because of loss of habitat, immense hunting, introduction of foreign species and climatic changes resulting in disturbed rainfall pattern and high temperature. DNA Barcoding has played a significant role in species identification but in the case of Migratory birds it adds on to classify the distinct lineages and discovering new species. Not only this but through DNA barcoding, many cryptic species have also been identified which previously were classified as single species (Hajibabaei et al. 2005). DNA barcoding also helps in characterising the inter- and intraspecific genetic diversity, which is very important in governing the distinct lineage in migratory birds. Many studies have been conducted using DNA barcoding in migratory birds (Table 1). The researchers from many countries such as North America, Korea, Turkey, Argentina and Scandinavia have extensively worked using DNA barcoding as an effective tool for species identification and to visualise the patterns of distinct lineages (Hebert et al. 2004; Yoo et al. 2006; Kerr et al. 2009; Johnsen et al. 2010).

The basic protocol followed for DNA barcoding in birds/migratory birds begins with the collection of blood samples taken via a brachial vein. This is followed by extraction of DNA commonly using Genomic DNA extraction kits. Cytochrome oxidase I gene is usually used as a molecular marker as it is accepted by Consortium for Barcode of Life Database (CBOL) and ABBI as a unique DNA barcode marker for avian species. Standard primers are used for PCR amplification Bird F1 (TTCTCCAACC ACAAAGACATTGGCAC), Bird R1 (AC GT GGGA GATAATTCCAAATCCTG) and Bird R2 (ACTACATGTGA GATG ATTCC

Table 1 Recent publications on DNA Barcoding in Migratory Birds

S. No.	Title of publication	References	Search criteria
1	DNA Barcoding of Birds at a Migratory Hotspot in Eastern Turkey Highlights Continental Phylogeographic Relationships.	Bilgin et al. (2016)	DNA barcoding in migratory birds
2	Overseas seed dispersal by migratory birds.	Viana et al. (2016)	DNA barcoding in migratory birds
3	Avian haemosporidians from Neotropical highlands: Evidence from morphological and molecular data.	González et al. (2015)	DNA barcoding in migratory birds
4	Low resolution of mitochondrial COI barcodes for identifying species of the genus Larus (Charadriiformes: Laridae).	Kwon et al. (2012)	DNA barcoding in migratory birds
5	A DNA microarray for identification of selected Korean birds based on mitochondrial cytochrome c oxidase I gene sequences.	Chung et al. (2010)	DNA barcoding in migratory birds
6	DNA barcoding techniques for avian influenza virus surveillance in migratory bird habitats.	Lee et al. (2010a)	DNA barcoding in migratory birds
7	Application of DNA barcoding technique in avian influenza virus surveillance of wild bird habitats in Korea and Mongolia.	Lee et al. (2010b)	DNA barcoding in migratory birds

Search source: https://www.ncbi.nlm.nih.gov/pmc/?term=DNA+barcoding+in+migratory+birds

GAA TCCAG) (Johnsen et al. 2010). Whereas the protocol for COI amplification is in accordance with Hebert et al. (2004). The final PCR product is then subjected to commercial sequencing, and finally all obtained sequences are submitted to Consortium for Barcode of Life Database (BOLD). A neighbour-joining tree is then prepared by combining the sample and BOLD sequences and the intraspecific and interspecific distance is calculated (many software programs like MEGA 4.0, TSC 1.21 etc. are currently available for this).

7 DNA Barcoding in Sea Birds

Sea birds form a unique class in itself. They represent avian species, which are prominent sea dwellers but cannot be included in sea fauna. They spend most of their life in a marine environment, obtaining food from water and for reproduction and colonisation depending on the sandy beaches. The Netherlands is categorised as one of the best countries for ornithological studies where approximately 147 species of birds (including sea birds) have been DNA barcoded (Aliabadian et al. 2013). The birds are extremely sensitive to environmental changes especially the seabirds because global warming causing the continues rise of the sea level, sea salinity, which can play a crucial role in the extinction of many species of sea birds so DNA barcoding can help to keep the record of endangered species. Therefore, there is an urgent need for more studies on sea birds using DNA barcoding for their conservation.

8 DNA Barcoding for the Identification of Illegal Trades of Birds

DNA barcoding is not only used for the identification of species of a particular geographic area but also used to identify the illegal trades of birds.

Illegal wildlife trade imposes a negative impact on the survival of a species as it introduces the invasion of other species and the pathogens (Rosen and Smith 2010). Molecular markers play an important role in the identification of forensic samples. Parrots and cockatoos (Psittaciformes) are charming birds, their plumage makes them highly enviable pets. The illegal trade in parrots and cockatoos is a serious issue and poses a threat to the viability of local populations and their transport to non-endemic areas may genetically pollute the avifauna (Coghlan et al. 2012).

Priscila F. M and co-workers identified a case where a man was caught in a Brazilian airport who was illegally carrying 58 avian eggs. During the investigation, he claimed that the eggs were of quail but those were suspected to be parrot eggs. It is illegal to transport parrots or their eggs as 29% of parrot species are declared as endangered species, Priscila F. M and team conducted the barcoding of mtDNA (cytochrome c oxidase subunit I gene [COI] and 16S ribosomal DNA) of embryo samples. They compared the embryonic COI sequences with the BOLD (The Barcode of Life Data System) and the 16S sequences were compared with GenBank sequences. The results were surprising and on the basis of both the results, they identified that the 57 eggs were of parrots (Alipiopsitta xanthops, Ara ararauna, and the [Amazona aestiva/A. ochrocephala and 1 was of owl and 1 was an owl.

9 Conclusion

All Bird Barcode Initiative (ABBI) is a project to explore the biodiversity of birds all over the world. It was established in 2003 and from then thousands of ave species have been identified using the myDNA barcoding technique. ABBI seems to be new hope for the identification of novel avian species. There are many relevant applications of the COI sequence library submitted on BOLD. Although many studies have been conducted for avian species covering a wide geographical area, some areas are better covered than others. BOLD helps to identify those areas where DNA barcodes of avian species should be undertaken in future. With the advancement of the COI library via DNA barcoding projects of avian species, there will be a better understanding of different avian realms and taxonomic territories. It also serves as an unbiased taxonomic representation of different avian groups. The advantage is that the DNA barcode sequences deposited in these databases are of high quality and are standardised; therefore, have fewer ambiguities, short sequence span, bidirectional sequencing and uniform sequence alignment. It has been observed that the rate of error on BOLD is much lower than on other databases. DNA barcoding in avian species in future will undoubtedly provide more specific species identification, recognition of cryptic species and tracing the line of avian evolution through different eras and deciphering their causes of divergence. Many studies were already conducted to identify the new species of birds (migratory, seabirds or the birds of Iceland). DNA barcoding also plays a crucial role in the forensic investigation of illegal transport of birds or their eggs. Although this became a very important tool to identify species of aves, there are many species of aves to be identified in future.

Acknowledgement The authors would like to acknowledge the University of Tabuk, Tabuk, Saudi Arabia.

The author would also like to thank the Department of Biology, Faculty of Sciences, Saudi Digital Library and University Library for providing the facility for Literature survey and collection.

References

Aliabadian M, Beentjes KK, Roselaar CS (2013) DNA barcoding of Dutch birds. In: Nagy ZT, Backeljau T, De Meyer M, Jordaens K (eds) DNA barcoding: a practical tool for fundamental and applied biodiversity research, vol 365, pp 25–48. (ZooKeys)

American Ornithologists' Union (AOU) (1998) Check-list of North American birds, 7th edn. AOU Press, Lawrence, Kansas, p 829

Arctander P, Fjeldsa J (1994) Avian tissue collections for DNA analysis. Ibis 136:359360

Avise JC, Walker D (1998) Pleistocene phylogeographic effects on avian populations and the speciation process. Proc R Soc Lond B Bio Sci 265:457–463

Avise JC, Arnold J, Ball RM, Bermingham E, Lamb T (1987) Intraspecific phylogeography: the mitochondrial bridge between population genetics and systematics. Annu Rev Ecol Syst 18:489–522

Baker A, Edwards S, Hanken, J, Schindel D, Stoeckle M (2005) Barcoding life takes flight: all birds barcoding initiative (ABBI), Report of Inaugural Workshop

Baker AJ, Tavares ES, Elbourne RF (2009) Countering criticisms of single mitochondrial DNA gene barcoding in birds. Mol Ecol Resour 9:257268

Banks RC, Cicero C, Dunn JL, Kratter AW, Ouellet H (2000) Forty-second supplement to the Ornithologists' Union checklist of north American birds. Auk 117:847–858

Banks RC, Cicero C, Dunn JL, Kratter AW, Rasmussen PC (2002) Forty-third supplement to the Ornithologists' Union checklist of north American birds. Auk 119:897–906

Banks RC, Cicero C, Dunn JL, Kratter AW, Rasmussen PC (2003) Forty-fourth supplement to the Ornithologists' Union checklist of north American birds. Auk 120:923–931

Bilgin R, Ebeoğlu N, İnak S, Kırpık MA, Horns JJ, Şekercioğlu ÇH (2016) DNA barcoding of birds at a migratory hotspot in eastern Turkey highlights continental phylogeographic relationships. PLoS One 11(6):e0154454. https://doi.org/10.1371/journal.pone.0154454

Campagna L, St Clair JJH, Lougheed SC, Woods RW, Imberti S, Tubaro PL (2012) Divergence between passerine populations from the Malvinas – Falkland Islands and their continental counterparts: a comparative phylogeographical study. Biol J Linn Soc 106:865–879

Chaves BRN, Chaves AV, Nascimento ACA, Chevitarese J, Vasconcelos MF, Santos FR (2015) Barcoding Neotropical birds: assessing the impact of nonmonophyly in a highly diverse group. Mol Ecol Res 15(4):921–931

Chung IH, Yoo HS, Eah JY, Yoon HY, Jung JW, Hwang SY, Kim CB (2010) A DNA microarray for identification of selected Korean birds based on mitochondrial cytochrome C oxidase I gene sequences. Mol Cells 30(4):295–301. https://doi.org/10.1007/s10059-010-0118-8

Clements JF, Schulenberg TS, Iliff MJ, Roberson D, Fredericks TA, Sullivan BL, Wood CL (2015) The eBird/Clements checklist of birds of the world. Available from http://www.birds.cornell.edu/clementschecklist/download/

Coghlan ML, White NE, Parkinson L, Haile J, Spencer PB, Bunce M (2012) Egg forensics: an appraisal of DNA sequencing to assist in species identification of illegally smuggled eggs. Forensic Sci Int Genet 6(2):268–273

Dasmahapatra KK, Mallet J (2006) DNA barcodes: recent successes and future prospects. Heredity 97(4):254–255

Evans T (2007) DNA damage. NEB Exp 2(1):1–3

Frézal L, Leblois R (2008) Four years of DNA barcoding: current advances and prospects. Infect Genet Evol 8(5):727–736

Gill FB (2007) Ornithology, 3rd edn. W.H. Freeman, New York

Gill FB, Slikas B (1992) Patterns of mitochondrial DNA divergence in north American crested titmice. Condor 94:20–28

González AD, Lotta IA, García LF, Moncada LI, Matta NE (2015) Avian haemosporidians from neotropical highlands: evidence from morphological and molecular data. Parasitol Int 64(4):48–59. https://doi.org/10.1016/j.parint.2015.01.007

Hajibabaei M, deWaard JR, Ivanova NV, Ratnasingham S, Dooh RT, Kirk SL (2005) Critical factors for assembling a high volume of DNA barcodes. Philos Trans R Soc 360:1959–1967. https://doi.org/10.1098/rstb.1727

Hajibabaei M, Smith AM, Janzen DH (2006) A minimalist barcode can identify a specimen whose DNA is degraded. Mol Ecol Notes 6:959–964

Hebert PDN, Cywinska A, Ball SL, DeWaard JR (2003) Biological identifications through DNA barcodes. Proc R Soc Lond B Biol Sci 270:313–321

Hebert PDN, Stoeckle MY, Zemlak TS, Francis CM (2004) Identification of birds through DNA barcodes. PLoS Biol. https://doi.org/10.1371/journal.pbio.0020312

Hoss M, Jaruga P, Zastawny TH, Dizdaroglu M, Paabo S (1996) DNA damage and DNA sequence retrieval from ancient tissues. Nucleic Acids Res 24:1304–1307

Johnsen A, Rindal E, Ericson PGP, Zuccon D, Kerr KCR, Stoeckle MY (2010) DNA barcoding of Scandinavian birds reveals divergent lineages in trans-Atlantic species. J Ornithol 151:565–578. https://doi.org/10.1007/s10336-009-0490-3

Kerr K, Birks SM, Kalyakin MV, Red'kin YA, Koblik EA, Hebert PDN (2009) Filling the gap—COI barcode resolution in eastern Palearctic birds. Front Zool 6:10

Kwon YS, Kim JH, Choe JC, Park YC (2012) Low resolution of mitochondrial COI barcodes for identifying species of the genus larus (Charadriiformes: Laridae). Mitochondrial DNA 23 (2):157–166. https://doi.org/10.3109/19401736.2012.660921

Lee DH, Lee HJ, Lee YJ, Kang HM, Jeong OM, Kim MC, Kwon JS, Kwon JH, Kim CB, Lee JB, Park SY, Choi IS, Song CS (2010a) DNA barcoding techniques for avian influenza virus surveillance in migratory bird habitats. J Wildl Dis 46(2):649–654. https://doi.org/10.7589/0090-3558-46.2.649

Lee DH, Lee HJ, Lee YN, Lee YJ, Jeong OM, Kang HM, Kim MC, Kwon JS, Kwon JH, Lee JB, Park SK, Choi IS, Song CS (2010b) Application of DNA barcoding technique in avian influenza virus surveillance of wild bird habitats in Korea and Mongolia. Avian Dis 54(1 Suppl):677–681. https://doi.org/10.1637/8783-040109-ResNote.1

Lijtmaer DA, Kerr KC, Barreira AS, Hebert PD, Tubaro PL (2011) DNA barcode libraries provide insight into continental patterns of avian diversification. PLoS One 6(7):e20744

Lindahl T (1993) Instability and decay of the primary structure of DNA. Nature 362:709–715

Lohman DJ, Ingram KK, Prawiradilaga DM, Winker K, Sheldon FH, Moyle RG (2010) Cryptic genetic diversity in "widespread" southeast Asian bird species suggests that Philippine avian endemism is gravely underestimated. Biol Conserv 143:1885–1890

Milá B, Tavares ES, Muñoz Saldaña A, Karubian J, Smith TB, Baker AJ (2012) A trans-amazonian screening of mtDNA reveals deep intraspecific divergence in forest birds and suggests a vast underestimation of species diversity. PLoS One 7(7):e40541

Millar CD, Huynen L, Subramanian S, Mohandesan E, Lambert DM (2008) New developments in ancient genomics. Trends Ecol Evol 23:386–393

Murray BW, McGillivray WB, Barlow JC, Beech RN, Strobeck C (1994) The use of cytochrome *B* sequence variation in estimation of phylogeny in the *Vireonidae*. Condor 96:1037–1054

Newton I (2003) The speciation and biogeography of birds. Academic Press, London

Nishiumi I, Kim CH (2015) Assessing the potential for reverse colonization among Japanese birds by mining DNA barcode data. J Ornithol 156(S1):325–331

Rosen GE, Smith KF (2010) Summarizing the evidence on the international trade in illegal wildlife. EcoHealth 7:24–32

Seutin GW, Bradley N, Boag PT (1991) Preservation of avian blood and tissue samples for DNA analyses. Can J Zool 69:82–90

Stoeckle M (2005) Barcoding life takes flight: All Birds Barcoding Initiative (ABBI), Needs and Resources Statement

Tavares ES, Baker AJ (2008) Single mitochondrial gene barcodes reliably identify sister-species in diverse clades of birds. BMC Evol Biol 8(1):81

Tavares ES, Gonçalves P, Miyaki CY, Baker A (2011) DNA barcode detects high genetic structure within Neotropical bird species. PLoS One 6(12):e28543

Viana DS, Gangoso L, Bouten W (1822) Figuerola J (2016) Overseas seed dispersal by migratory birds. Proc Biol Sci 283:20152406. https://doi.org/10.1098/rspb.2015.2406

Woese CR, Fox GE (1977) Phylogenetic structure of the prokaryotic domain: the primary kingdoms. Proc Natl Acad Sci U S A 74:5088–5090

Woese CR (2000) Interpreting the universal phylogenetic tree. Proc Natl Acad Sci U S A 97:8392–8396

Yoo HY, Eah J, Kim JS, Young JK, Min M, Peak WK (2006) DNA barcoding Korean birds. Mol Cells 22:323–327. 17202861

Zimmermann J, Hajibabaei M, Blackburn D (2008) DNA damage in preserved specimens and tissue samples: a molecular assessment. Front Zool 5:18

Molecular Characterisation of Ruminant Mammals Using DNA Barcodes

Muniyandi Nagarajan, Koodali Nimisha, and Subhash Thomas

Abstract Ruminant mammals are widely distributed across the world and distinguished from other mammals by the presence of a four-chambered stomach. Most of the ruminants are wild while a few are domestic, which contribute significantly to the agricultural economy in the form of livestock resources. Characterisation of livestock breeds and the exact identification of wild ruminant species are imperative for developing improved breeds and wildlife conservation, respectively, though taxonomists determine breeds and species based on morphological traits, which is nugatory in the case of cryptic species or when unrelated species exhibit similar morphological traits. However, the emerging DNA-based techniques have overcome the challenges and limitations faced by conventional methods. DNA barcoding, specifically, the discovery of mitochondrial cytochrome *c* oxidase subunit I (COI) gene as a standard DNA barcode region for animals has transfigured the realm of molecular systematics by providing a platform to expeditiously find novel lineages and elucidate ruminant phylogeny. Despite this, DNA barcoding has huge applications including detection of adulteration and mislabelling of bushmeat, checking wildlife poaching and animal trafficking. This chapter provides an overview of ruminant mammals and the usefulness of COI in the identification of ruminant species.

Keywords Ruminant · Mammal · COI gene · Phylogeny

1 A Glimpse on Ruminant Mammals

Mammals can be basically defined as a group of vertebrate species possessing unique features such as the presence of mammary glands, a single bone in the lower jaw, and neocortex of the forebrain, which are found completely lacking in

M. Nagarajan (✉) · K. Nimisha · S. Thomas
Department of Genomic Science, School of Biological Sciences, Central University of Kerala, Kasaragod, Kerala, India
e-mail: nagarajan@cukerala.ac.in

S. Trivedi et al. (eds.), *DNA Barcoding and Molecular Phylogeny*,
https://doi.org/10.1007/978-3-030-50075-7_13

Fig. 1 The ruminant mammals. (**a**) Gaur (*Bos gaurus*). (**b**) African buffalo (*Syncerus caffer*). (**c**) Blackbuck (*Antilope cervicapra*). (**d**) Sheep (*Ovis aries*)

other vertebrates (Kemp 2004). Around the globe, a total of 5416 species of mammals have been reported so far and are placed under 154 families and 29 orders (Wilson and Reeder 2005). Of these, India possesses a total of 423 species, which makes 7.81% of the global mammalian species representing 48 families and 14 orders (Sharma et al. 2013). Amongst the various mammalian orders, the suborder Ruminantia (Order: Artiodactlya) holds a special position as most of them are livestock and thus are contributing indispensably to the world's economic growth. Ruminants are known for their complex digestive system consisting of 'rumen', a chamber specialised in microbial fermentation and regurgitation of food for its further breakdown and digestion (Fig. 1). These large terrestrial herbivores enjoy ubiquitous distribution and are native to almost all continents like Eurasia, Africa, North America and South America except Australia and Antarctica (Hassanin and Douzery 2003; Fernández and Vrba 2005; Hackmann and Spain 2010). As of now, there are around 200 extant ruminant species of which the widely domesticated species include cattle, buffalo, goat, sheep, mithun, reindeer, yak and camel. The suborder Ruminantia comprises of 6 extant families such as Trangulidae, Moschidae, Bovidae, Giraffidae, Antilocapridae and Cervidae. Except Trangulidae, latter five extant families, belonging to the infraorder Pecora, are popularly known as the true ruminants (Kulemzina et al. 2011). Amongst these families, the Bovidae family is the most diversified one (Kraus and Miyamoto 1991; Heywood 2010) and the majority of the livestock species are placed under this family. As per the 19th livestock census (2012) reports, total livestock population of India is 512.05 million, which includes Bovine (299.9 million), sheep (65.06 million) and goats (135.17 million).

First ruminants were believed to have evolved around 50 Million years BC in the middle of Eocene epoch at certain parts of Northern America and Eurasia (Gentry 1994; Hackmann and Spain 2010). Since pre-historic times, domesticated ruminants were closely associated with human population. Throughout the world, livestock are valued for their economic output such as meat (as a dietary protein source), dairy products, fibre (wool, hair, hooves and skin from sheep, goats), fertilisers (from animal bones), muscle power (for transportation and ploughing) and land management. Global demand for livestock and livestock products is increasing rapidly with expanding human population especially in developing nations (Thornton 2010).

To meet with the increasing demand for livestock and livestock products, extensive crossbreeding and selective breeding have been practised in many of the ruminant species. These practices often lead to a decline in the indigenous varieties of breeds mostly because they are not as productive as the crossbreeds or maybe their profitability under present market conditions is very low. Therefore, many of the local breeds are facing extinction. Also, the genetic variability in the available gene pool is getting eroded continuously. Under such circumstances, a proper assessment of existing breeds becomes essential for proper conservation and genetic improvement of the germplasm. Molecular approaches provide a wide opportunity for breed assessment, as there are numerous tools like DNA markers, which can greatly contribute to genetic variability evaluation within the breeds as well as between the breeds.

2 Trends in Molecular Phylogeny of Ruminant Mammals

Ruminants, being very diverse in their ecology, behaviour, physiology and bio-geographical distribution, are subjects of interest to many biologists. The diversity amongst them can be systematically studied using phylogeny. In general, phylogeny relies on variations at morphological or molecular levels to deduce the evolutionary relationship between taxonomic groups. Molecular phylogeny utilises inheritable structural or functional bio-molecular character sets for constructing a phylogenetic tree. A well-resolved phylogenetic tree can describe species relationship, population history and dynamics of evolution.

The construction of phylogenetic tree had always remained a challenge for taxonomists perhaps, due to morphological convergence of unrelated taxa or difficulty in discriminating cryptic species. The first molecular approach towards the identification of livestock diversity was based on protein polymorphisms (Avise 1994). A large number of studies have been carried out on livestock by characterising the allozyme system and blood groups. Since the level of polymorphism exhibited by blood groups and the allozyme system is low, the suitability of protein typing was found to be inadequate in identifying animals at the species level. In contrast, DNA polymorphisms have shown greater efficiency in detecting species precisely. The polymorphic DNA markers used in characterising livestock generally include RFLP (Restriction Fragment Length Polymorphism), AFLP (Amplified Fragment Length Polymorphism), RAPD (Random Amplified Polymorphic DNA), microsatellites, mitochondrial and Y chromosome markers. Numerous studies have been conducted on ruminants (goat, sheep, cattle, buffalo, mithun and camel) using RFLP, AFLP and RAPD markers, particularly, during the 1990s (Xiang-long et al. 1997; Ajmone-Marsan et al. 2001; de Mello Klocker Vasconcellos et al. 2003; Saifi et al. 2004; Sodhi et al. 2006; Khaldi et al. 2010; Mahrous and Ramadan 2011). Subsequent studies have characterised livestock that include small ruminants (goat and sheep) as well as large herbivores (buffalo, cattle, mithun, yak and camel) using microsatellite markers (Santos-Silva et al. 2008; Bozzi et al. 2009; Kumar et al. 2006; Nagarajan et al. 2009). As molecular tools for determining species developed, great effort was put forth to characterise variable regions of mitochondria such as the D-loop region, cytochrome *b* gene and COI gene (Lai et al. 2006; Kumar et al. 2007; Sanches et al. 2012; Nagarajan et al. 2015). However, amongst the three, COI gene stands out as the best molecular toolkit for carrying out genetic diversity studies since they are widely accepted as the universal DNA barcode region (exclusively for animals) due to a high evolutionary rate. The COI gene has opened up new avenues for the identification ruminant species and understanding the genetic relationship between various ruminants.

3 Emergence of DNA Barcoding as a Novel Tool

In 2003, it was proposed that mitochondrial DNA sequences of Cytochrome c oxidase subunit I gene can be used as 'barcodes' for the bio-identification of species globally (Hebert et al. 2003a, b). This novel technology uses the variations arising in the short DNA sequences as labels for species. Initially, the utility of this technique was established in Lepidopterans (Hebert et al. 2004a; Janzen et al. 2005; Hajibabaei et al. 2006), later it became widespread around the animal world, which included invertebrate species (Barrett and Hebert 2005; Wang et al. 2011), fishes (Ward et al. 2005), amphibians (Vences et al. 2005), reptiles (Vences et al. 2012; Nagy et al. 2012), birds (Hebert et al. 2004b) and mammals (Cai et al. 2011; Yan et al. 2013). The 651 bp fragment of mitochondrial DNA COI gene has been recognised as 'standard barcode' because of its peculiarities such as short size and availability of universal primers.

The mitochondrial DNA COI gene, which is responsible for carrying out oxidative phosphorylation, is highly conserved across the species (Waugh 2007). The nucleotide variation in the gene is efficient enough to distinguish animals at the interspecific level. However, the intraspecific variation has been reported to be less than 10%. In addition to this, COI gene rarely possesses any insertions or deletions (Waugh 2007). These properties make it COI gene ideal for DNA barcoding; a highly promising tool in disclosing the identity of cryptic species (Hebert et al. 2004a), distinguishing species at their juvenile stages (Valdez-Moreno et al. 2010) and in protecting endangered species (Elmeer et al. 2012). It is achieved by generating the barcode data of several species and depositing them on the public domain, further using these sequences for identification of unknown specimens. There are several international barcoding projects focused on developing Barcode sequence libraries for specific targeted species like 'Mammalia Barcode of Life' campaign, a project that aims to build a comprehensive reference library of DNA barcodes for the global mammal fauna.

The COI gene sequences have been widely used to identify ruminant species in different studies. Cai et al. (2011) analysed sequences of 223 individuals of Bovidae representing 18 species. The results showed that, except two species, all other species studied possessed unique COI sequences. Similarly, Arif et al. (2012) performed a comparative study to characterise 12 members of the family Bovidae using different mitochondrial markers including COI gene. In the phylogenetic tree, the genus *Bison* and *Bos* showed close relationship and supported the previous studies suggesting two lineages of the tribe Bovini-buffalo (*Bubalus*) and cattle (*Bos*). The study showed the main split of the 12 members of the Bovidae into bovine clade and non-bovine clade. In another study, COI barcodes helped in deciphering the identity of Tanzanian Bovidae species (Bitanyi et al. 2011). A 470 bp region of the COI gene was tested in 95 specimens representing 20 species of antelopes, wild buffalo and domesticated species of Bovidae (*Bos taurus, B. indicus, Ovis aries* and *Capra hircus*). In this particular study, even 50 bp COI sequence was proved to be effective in discriminating species as this region was found to be highly variable amongst the species, and also this variability was evenly distributed along with the gene fragment.

Yan et al. (2013) have reported that COI gene can be used as a potential tool for the identification of Bovidae (*Ovis aries, Capra hircus, Bos grunniens, B. taurus, Bubalus bubalis, Saiga tatarica and Procapra gutturosa*) and Cervidae (*Cervus elaphus, C. nippon* and *Elaphurus davidianus*) species even from traces of animal parts such as horns. The study was performed by analysing 223 mitochondrial COI sequences, which included sequences generated from 47 specimens of animal horn derived from 10 known species and 176 COI sequences retrieved from GenBank. The results showed less than 1.4% variation within species whereas above 2.0% variation between species, thus it was possible to identify and discriminate the species using samples of animal horn. Bondoc and Cerbito (2013) have shown the effectiveness of DNA barcoding in establishing the genetic relationship within sheep and goat breeds in the Philippines. Likewise, Ali et al. (2016) used COI gene sequences to distinguished the native goat breeds of Pakistan form the exotic goat breeds.

DNA barcoding (COI gene) has also served as a tool in forensic identification of specimens. A case study from South Africa utilised COI gene as genetic marker to determine the species identity of unknown samples of meat obtained from three different sources in different forms; dried sample, frozen sample and carcasses alleged to be from common reedbuck (*Redunca arundinum*). The analysis showed that the dried and frozen samples of meat belonged to that of domestic cattle (*Bos taurus*) and the carcasses were confirmed to be of common reedbuck as the COI gene sequences of the specimen matched with that of reference sequences in the database (Dalton and Kotze 2011). It was very difficult to identify the Chinese deer as they pronounce morphological similarities until Cai et al. (2015) reported unique barcodes for 21 species of the family Cervidae.

Use of COI makers has been extended to forensic identification of mislabelled or misbranded game meat or bushmeat products. Bushmeat or game meat is commonly referred to the meat of undomesticated animal's especially wild animals, which are hunted for food. Many of the ruminant species are exploited by hunting. Most common amongst them are antilopes like kudu, eland, deer and Cape buffalo. Though game meat hunting is legal in many nations like South Africa and Namibia, meat mislabelling is a common concern prevailing in these areas. Several studies have been reported with regard to meat mislabelling; however, identification of game meat becomes difficult due to conversion of poached game meat into unrecognisable form by removing body portions or by processing. A DNA barcoding study based on COI gene was conducted in 146 samples (14 beef and 132 game labels) of meat products to check the authenticity of commercial labels found in the local market. About 76.5% mislabelling was detected with the game meat samples, which showed substitution of the game meat with the meat from domestic cattle, pig, lamb and horse. Additionally, meats of less common species such as giraffe, waterbuck, bushbuck, duiker and mountain zebra were also found to be substituted with game meat (D'Amato et al. 2013). A similar study by Quinto et al. (2016) reported the game meat mislabelling in the US market. A 658 bp region of COI gene was used as a maker for analysing 54 game meat samples representing a variety of species. The results showed 10 out of 54 samples as mislabelled. Two products labelled as bison and yak were found to be of domestic cattle. The overall rate of mislabelling in this study was found to be 18.5%. Similarly, Kane and Hellberg (2016) analysed the meat purchased from three different sources; online sources, supermarkets and local butcher shops. The study showed a higher rate of mislabelling (38%) in meat products bought from online dealers compared to those of butcher shops (18%) and supermarkets (5.8%).

Syakalima et al. (2016) robustly identified 9 ruminant species (Puku, Eland, Impala, Kafue Lechwe, Bushbuck, Waterbuck, Buffalo, Wildebeest and Sable) from poached game meat of wild ruminant species by comparing the COI gene sequences with that of the reference database (BOLD). Similarly, illegal hunting of endangered Pampas deer has been reported from Brazil using COI (Sanches et al. 2012). Mbugua (2014) carried out DNA barcoding studies in 99 unknown meat samples collected from different parts of Kenya along with 5 wild species namely, impala (*Aepyceros melampus*), waterbuck (*Kobus ellipsiprimn*), African buffalo (*Syncerus caffer*), black rhino (*Diceros bicornis*) and elephant (*Loxodonta cyclotis*) in order to identify the prevalence of bushmeat and meat product substitution. COI gene successfully revealed the identity of the species from the tissue samples, which showed the meat substitutions in the samples analysed.

Understanding the relationship between hosts, parasites and their intermediate vectors is important in establishing the epidemiology of zoonotic diseases. Since the collection and identification of parasites is difficult using conventional methods, the DNA barcoding approach can be utilised for identifying the parasites as well as their diversity. Efficient protocols for DNA barcoding have been reported for identifying most of the parasites affecting ruminants including filarioid nematodes, ticks, mosquitoes and sarcocystis (Cywinska et al. 2006; Ferri et al. 2009; Gjerde 2013; Zhang and Zhang 2014). DNA barcoding through scat analysis is a convenient method for identifying parasites that bypass the euthanisation of animals. Being a non-invasive technique, it also extends the study to be conducted in live animals, unlike in conventional studies where the study could be performed only on diseased or dead animals.

4 A Composite Phylogenetic Analysis of Ruminants Using COI Gene

Even though previous studies have proven the efficiency of COI barcodes in delineating species, a comprehensive phylogenetic analysis has not been performed on ruminants using COI gene sequences, yet. Thus, to represent the inter-relationship between different ruminant species and to reaffirm the efficiency of COI markers in identification, COI gene sequences of 22 ruminant species were retrieved from the GenBank and analysed (Table 1). The sequences were aligned using ClustalW implemented in MEGA (Tamura et al. 2011) and subsequently, truncated to 523 bp to make it equal in size for further analysis. The maximum parsimony phylogenetic tree was constructed using 100 COI gene sequences with a bootstrap value of 1000 using the software MEGA. It formed three distinct major clades with a few internal subclades (Fig. 2). Amongst the three, the first clade consists of sheep (domestic sheep, Dall sheep and Bighorn sheep), goat (Siberian ibex, domestic goat), Himalayan blue sheep, African buffalo and domestic buffalo. The second clade forms a group exclusively of *Bos* and *Bison* species whereas clade 3 comprises members of the family Camelidae. Notably, there is no intermingling between species in the clades.

The MP tree topology clearly showed the close evolutionary relationship between sheep and goat with buffalo though they are classified under different subfamily-Caprinae and Bovinae, respectively, which were comparable to previously published reports (Fernández and Vrba (2005). The evolutionary relationship and systematic classification of Bovidae family remains as a dispute and different classification systems exists for bovids based on the phyletic relationship. Morphological and molecular examination has suggested considering the bovidae family as a monophyletic group, where all the descendants evolved from a common ancestor forms a clade. However, the establishment of monophyly has been weakly supported by previous studies. In contradiction with this statement, paraphyletic grouping (where

Table 1 Details of COI sequences used for phylogenetic tree construction

S. No	Species	No. of sequences	GenBank ID
1	*Bison bison*	6	JF443190-JF443195
2	*Bison bonasus*	2	EU623450, JF444283
3	*Bos frontalis*	1	HQ269429
4	*Bos gaurus*	1	KF808255
5	*Bos grunniens*	8	HQ269432, HQ269433, HQ269462- HQ269467
6	*Bos indicus*	10	KF952275 - KF952279, KF952281- KF952285
7	*Bos primigenius*	2	JQ735452, JQ735453
8	*Bos taurus*	10	HQ860420, JF700140, JF700141, JX567086, KF771228, KF799986, KU947025- KU947028
9	*Bubalus bubalis*	10	KU932060, KU932061, KU932067, KU932072 KU932073, KU932075, KU932096, KU932100, KU932106, KU932110
10	*Capra hircus*	10	HQ269454, JN850776, JN850777, JQ735456, KC679016-KC679018, KF317903, KF317905, KT750041
11	*Capra sibirica*	1	KU527896
12	*Ovis aries*	10	JN245995, JN850771- JN850773, JQ735465, JX567087, KF317902, KF317907, KT750038, KT750039
13	*Ovis canadensis*	3	JF443356- JF443358
14	*Ovis dalli*	2	JF443359, JF443360
15	*Pseudois nayaur*	4	HQ269457- HQ269459, KJ862174
16	*Syncerus caffer*	2	JN082178, KF482455
17	*Camelus dromedarius*	10	JN632608, AB753111- AB753115 AB753117- AB753120
18	*Camelus bactrianus*	5	AP003423, EF507798- EF507801
19	*Lama pacos*	2	AJ566364, DQ534055
20	*Lama glama*	1	DQ534054
21	*Lama guanicoe*	1	DQ534053
22	*Vicugna vicugna*	1	DQ534056

most of the descendants of a common evolutionary ancestor form a group excluding a few descendants which form a separate group) of family Bovidae has been described in the literature (Gatesy et al. 1992). In concordance with the study of Gatesy et al. (1992), our phylogenetic tree analysis quite apparently has represented two distinct lineages of the subfamily Bovinae, one which forms a distinct clade

Fig. 2 Maximum parsimony phylogenetic tree of selected ruminant mammals

(clade II) comprising bos and bison groups and the latter consisting of African and Indian buffalo that forms a different group with members of Caprinae family (Arif et al. 2012).

It is believed that cattle species have evolved from *Bos primigenius*. Studies have recorded that the ancestral species has already become extinct long back. The MP tree showed that *Bos indicus* is genetically closer to the extinct *Bos primigenius* than *B. taurus*. It also showed the close genetic relationship between *Bos* and *Bison* genera. The close relationship between these two genera is well supported by the evidence obtained from the paleontological and reproductive data (Arif et al. 2012). Fernández and Vrba (2005) have even suggested combining the genus *Bos* and *Bison* into a single genus. Several studies, based on microsatellites, mtDNA and nuclear DNA, including our analysis of CO I gene, have demonstrated that yaks (*Bos grunniens*) are genetically closer to American bison (*Bison bison*) and distanced from *Bison bonasus*, (Zhengchao et al. 1998; Tu et al. 2002; Hassanin and Ropiquet 2004; Lai et al. 2006; Xie et al. 2010; Qiu et al. 2012; Bai 2015). The third clade comprising members of Camelidae appears as a distinct group from the Bovids. Also, a clear distinction can be spotted between the Asian, African and New World camels. The Ilama (*Lama glama*) and Guanaco (*Lama guanicoe*) appear to be diverging from Alpaca (*Lama pacos*) and Vicugna (*Vicugna vicugna*) as reported in the previous studies using cytochrome *b* sequence (Stanley et al. 1994).

The phylogenetic analysis is necessary for understanding the evolution of diverse taxa and making the most probable prediction about cryptic or poorly known species. The COI gene has clearly distinguished different ruminant species and thus it can serve as an efficient tool for the identification of ruminants as well as systematic organisation of the species into various taxa.

5 Conclusion

DNA barcoding renders a potent identification system for ruminant species. Besides this, DNA barcodes have paved the way for labelling of livestock products especially meat and milk, which authenticate the composition of livestock products thereby protecting the consumer from being misguided and cheated by illegal substitutions. Amongst the various molecular markers used in molecular systematics, COI gene has received much attention amongst the molecular taxonomists as they are effective in discriminating congeneric taxa. Since most of the ruminants are wild animals, DNA barcoding serves as an effective method to develop strategies for conserving ruminant species facing extinction. On the other hand, it has also contributed in the identification of parasites that cause ailments in ruminant livestock. Parasites are identified mostly by analysing faeces of ruminants, which further provides a non-invasive alternative way of disease identification without disturbing the animal. A much more recent approach that integrates DNA barcoding with next generation sequencing technology has even made it possible to analyse the rumen contents of large ruminants with more precision generating sequences for multiple samples at once. We envision that in near future, ruminant research will be greatly benefitted by the advances in the DNA barcoding technique, which can serve as a key for the sustainable management of endangered wild species as well as indigenous livestock breeds.

Acknowledgements The senior author is thankful to the Science and Engineering Research Board (SERB), Department of Science and Technology, Government of India, New Delhi for the financial Assistance (EMR/2015/000937). ST is thankful to the Kerala State Council for Science Technology and Environment (KSCSTE) for the support in the form of research fellowship.

References

Ajmone-Marsan P, Negrini R, Crepaldi P et al (2001) Assessing genetic diversity in Italian goat populations using AFLP markers. Anim Genet 32:281–288

Ali A, Rehman A, William K (2016) Phylogenetic analysis of *Capra hircus* commonly found goat breeds of Pakistan using DNA barcode. J Bioresource Manage 3(1)

Arif IA, Bakir MA, Khan HA (2012) Inferring the phylogeny of bovidae using mitochondrial DNA sequences: resolving power of individual genes relative to complete genomes. Evol Bioinformatics Online 8:139–150. https://doi.org/10.4137/EBO.S8897

Avise JC (1994) History of molecular phylogenetics. In: Avise JC (ed) Molecular markers, natural history and evolution. Springer, New York, pp 16–43

Bai JL (2015) Phylotaxonomic position of Tianzhu white yak (*Poephagus grunniens*) based on nucleotide sequences of multiple subunits of cytochrome *c* oxidase. J Appl Anim Res 43:431–438. https://doi.org/10.1080/09712119.2014.980416

Barrett RDH, Hebert PDN (2005) Identifying spiders through DNA barcodes. Can J Zool 83:481–491. https://doi.org/10.1139/z05-024

Bitanyi S, Bjørnstad G, Ernest EM et al (2011) Species identification of Tanzanian antelopes using DNA barcoding. Mol Ecol Resour 11:442–449. https://doi.org/10.1111/j.1755-0998.2011.02980.x

Bondoc OL, Cerbito WA (2013) Genetic diversity and relationships of domestic goat and sheep breeds (Artiodactyla: Bovidae: Caprinae) in the Philippines based on DNA barcodes. J Vet Sci 50(2):64–74

Bozzi R, Degl'Innocenti P, Rivera Diaz P et al (2009) Genetic characterization and breed assignment in five Italian sheep breeds using microsatellite markers. Small Rumin Res 85:50–57. https://doi.org/10.1016/j.smallrumres.2009.07.005

Cai Y, Zhang L, Shen F et al (2011) DNA barcoding of 18 species of Bovidae. Chin Sci Bull 56:164–168. https://doi.org/10.1007/s11434-010-4302-1

Cai Y, Zhang L, Wang Y et al (2015) Identification of deer species (Cervidae, Cetartiodactyla) in China using mitochondrial cytochrome c oxidase subunit I (mtDNA COI). Mitochondrial DNA 1736:1–4. https://doi.org/10.3109/19401736.2014.1003919

Cywinska A, Hunter FF, Hebert PDN (2006) Identifying Canadian mosquito species through DNA barcodes. Med Vet Entomol 20:413–424. https://doi.org/10.1111/j.1365-2915.2006.00653.x

D'Amato ME, Alechine E, Cloete KW et al (2013) Where is the game? Wild meat products authentication in South Africa: a case study. Investig Genet 4:6. https://doi.org/10.1186/2041-2223-4-6

Dalton DL, Kotze A (2011) DNA barcoding as a tool for species identification in three forensic wildlife cases in South Africa. Forensic Sci Int 207:e51–e54. https://doi.org/10.1016/j.forsciint.2010.12.017

de Mello Klocker Vasconcellos LP, Tambasco-Talhari D, Pozzi Pereira A et al (2003) Genetic characterization of Aberdeen Angus cattle using molecular markers. Genet Mol Biol 26:133–137. https://doi.org/10.1590/S1415-47572003000200005

Elmeer K, Almalki A, Mohran KA et al (2012) DNA barcoding of *Oryx leucoryx* using the mitochondrial cytochrome C oxidase gene. Genet Mol Res 11(1):539–547

Fernández MH, Vrba ES (2005) A complete estimate of the phylogenetic relationships in Ruminantia: a dated species-level supertree of the extant ruminants. Biol Rev 80:269–302. https://doi.org/10.1017/S1464793104006670

Ferri E, Barbuto M, Bain O et al (2009) Integrated taxonomy: traditional approach and DNA barcoding for the identification of filarioid worms and related parasites (Nematoda). Front Zool 6:1. https://doi.org/10.1186/1742-9994-6-1

Gatesy J, Yelon D, Desall R, Vrba ES (1992) Phylogeny of the Bovidae (Artiodactyla, Mammalia), based on mitochondrial ribosomal DNA sequences. Mol Biol Evol 9(3):433–446

Gentry AW (1994) The Miocene differentiation of old world Pecora (Mammalia). Hist Biol 7:115–158. https://doi.org/10.1080/10292389409380449

Gjerde B (2013) Phylogenetic relationships among Sarcocystis species in cervids, cattle and sheep inferred from the mitochondrial cytochrome c oxidase subunit I gene. Int J Parasitol 43:579–591. https://doi.org/10.1016/j.ijpara.2013.02.004

Hackmann TJ, Spain JN (2010) Invited review: ruminant ecology and evolution: perspectives useful to ruminant livestock research and production. J Dairy Sci 93:1320–1334. https://doi.org/10.3168/jds.2009-2071

Hajibabaei M, Janzen DH, Burns JM et al (2006) DNA barcodes distinguish species of tropical Lepidoptera. Proc Natl Acad Sci U S A 103:968–971. https://doi.org/10.1073/pnas.0510466103

Hassanin A, Douzery EJP (2003) Molecular and morphological phylogenies of ruminantia and the alternative position of the moschidae. Syst Biol 52:206–228

Hassanin A, Ropiquet A (2004) Molecular phylogeny of the tribe Bovini (Bovidae, Bovinae) and the taxonomic status of the Kouprey, *Bos sauveli* Urbain 1937. Mol Phylogenet Evol 33:896–907. https://doi.org/10.1016/j.ympev.2004.08.009

Hebert PDN, Cywinska A, Ball SL, deWaard JR (2003a) Biological identifications through DNA barcodes. Proc Biol Sci 270:313–321. https://doi.org/10.1098/rspb.2002.2218

Hebert PDN, Ratnasingham S, Waard J (2003b) Barcoding animal life : cytochrome c oxidase subunit 1 divergences among closely related species barcoding animal life: cytochrome c oxidase subunit 1 divergences among closely related species. Proc R Soc Lond B 270:S96–S99. https://doi.org/10.1098/rsbl.2003.0025

Hebert PDN, Penton EH, Burns JM et al (2004a) Ten species in one: DNA barcoding reveals cryptic species in the neotropical skipper butterfly *Astraptes fulgerator*. Proc Natl Acad Sci United States Am 101:14812–14817

Hebert PDN, Stoeckle MY, Zemlak TS, Francis CM (2004b) Identification of birds through DNA barcodes. PLoS Biol 2(10):e312

Heywood JJN (2010) Explaining patterns in modern ruminant diversity: contingency or constraint? Biol J Linn Soc 99:657–672. https://doi.org/10.1111/j.1095-8312.2010.01436.x

Janzen DH, Hajibabaei M, Burns JM et al (2005) Wedding biodiversity inventory of a large and complex Lepidoptera fauna with DNA barcoding. Philos Trans R Soc Lond Ser B Biol Sci 360:1835–1845. https://doi.org/10.1098/rstb.2005.1715

Kane DE, Hellberg RS (2016) Identification of species in ground meat products sold on the U.S. commercial market using DNA-based methods. Food Control 59:158–163. https://doi.org/10.1016/j.foodcont.2015.05.020

Kemp TS (2004) The origin and evolution of mammals. Oxford University Press, New York

Khaldi Z, Rekik B, Haddad B, et al (2010) Genetic characterization of three ovine breeds in Tunisia using randomly amplified polymorphic DNA markers

Kraus F, Miyamoto MM (1991) Rapic Cladogenesis among the Pecoran ruminants: evidence from mitochondrial DNA sequences. Syst Zool 40:117–130

Kulemzina AI, Yang F, Trifonov VA et al (2011) Chromosome painting in Tragulidae facilitates the reconstruction of Ruminantia ancestral karyotype. Chromosom Res 19:531–539. https://doi.org/10.1007/s10577-011-9201-z

Kumar S, Gupta J, Kumar N, Dikshit K, Navani N, Jain P, Nagarajan M (2006) Genetic variation and relationships among eight Indian riverine buffalo breeds. Mol Ecol 15:593–600

Kumar S, Nagarajan M, Sandhu JS, Kumar N, Behl V, Nishanth G (2007) Mitochondrial DNA analyses of Indian water buffalo support a distinct genetic origin of river and swamp buffalo. Anim Genet 38:227–232

Lai SJ, Liu YP, Liu YX, Li XW, Yao YG (2006) Genetic diversity and origin of Chinese cattle revealed by mtDNA D-loop sequence variation. Mol Phylogenet Evol 38:146–154

Mahrous K, Ramadan H (2011) Genetic variations between camel breeds using microsatellite markers and RAPD techniques. J Appl Biosci 39:2626–2634

Mbugua D (2014) Use of cytochrome oxidase 1 gene region: a molecular tool for the domestic and wildlife industry in Kenya

Nagarajan M, Kumar N, Nishanth G et al (2009) Microsatellite markers of water buffalo, *Bubalus bubalis*- development, characterisation and linkage disequilibrium studies. BMC Genet 10:68. https://doi.org/10.1186/1471-2156-10-68

Nagarajan M, Nimisha K, Kumar S (2015) Mitochondrial DNA variability of Domestic River Buffalo (*Bubalus bubalis*) populations: genetic evidence for domestication of river Buffalo in Indian subcontinent. Genome Biol Evol 7:1252–1259. https://doi.org/10.1093/gbe/evv067

Nagy ZT, Sonet G, Glaw F, Vences M (2012) First large-scale DNA barcoding assessment of reptiles in the biodiversity hotspot of Madagascar, based on newly designed COI primers. PLoS One 7:e34506

Qiu Q, Zhang G, Ma T et al (2012) The yak genome and adaptation to life at high altitude. Nat Genet 44:946–949

Quinto CA, Tinoco R, Hellberg RS (2016) DNA barcoding reveals mislabeling of game meat species on the U.S. commercial market. Food Control 59:386–392. https://doi.org/10.1016/j.foodcont.2015.05.043

Saifi HW, Bhushan B, Kumar S et al (2004) Genetic identity between Bhadawari and Murrah breeds of Indian buffaloes (*Bubalus bubalis*) using RAPD-PCR. Asian-Australas J Anim Sci 17:603–607. https://doi.org/10.5713/ajas.2004.603

Sanches A, Tokumoto PM, Peres WAM et al (2012) Illegal hunting cases detected with molecular forensics in Brazil. Investig Genet 3:17. https://doi.org/10.1186/2041-2223-3-17

Santos-Silva F, Ivo RS, Sousa MCO et al (2008) Assessing genetic diversity and differentiation in Portuguese coarse-wool sheep breeds with microsatellite markers. Small Rumin Res 78:32–40. https://doi.org/10.1016/j.smallrumres.2008.04.006

Sharma G, Kamalakannan M, Venkataraman K (2013) A checklist of mammals of India with their distribution and conservation status. Director Zoological Survey of India

Sodhi M, Mukesh M, Anand A et al (2006) Assessment of genetic variability in two north Indian Buffalo breeds using random amplified polymorphic DNA (RAPD) markers. Asian-Australas J Anim Sci 19:1234–1239. https://doi.org/10.5713/ajas.2006.1234

Stanley HF, Kadwell M, Wheeler JC (1994) Molecular evolution of the family Camelidae: a mitochondrial DNA study. Proc Biol Sci 256:1–6. https://doi.org/10.1098/rspb.1994.0041

Syakalima M, Munyeme M, Yasuda J (2016) Cytochrome c oxidase sequences of Zambian wildlife helps to identify species of origin of meat. Int J Zool 2016:6. https://doi.org/10.1155/2016/1808912

Tamura K, Peterson D, Peterson N et al (2011) MEGA5: molecular evolutionary genetics analysis using maximum likelihood, evolutionary distance, and maximum parsimony methods. Mol Biol Evol 28:2731–2739. https://doi.org/10.1093/molbev/msr121

Thornton PK (2010) Livestock production: recent trends, future prospects. Philos trans R Soc London B Biol Sci 365:2853–2867. https://doi.org/10.1098/rstb.2010.0134

Tu Z-C, Qiu H, Zhang Y-P (2002) Polymorphism in mitochondrial DNA (mtDNA) of yak (*Bos grunniens*). Biochem Genet 40:187–193

Valdez-Moreno M, Vásquez-Yeomans L, Elías-Gutiérrez M et al (2010) Using DNA barcodes to connect adults and early life stages of marine fishes from the Yucatan peninsula, Mexico: potential in fisheries management. Mar Freshw Res 61:655–671

Vences M, Thomas M, Bonett RM, Vieites DR (2005) Deciphering amphibian diversity through DNA barcoding: chances and challenges. Philos Trans R Soc B Biol Sci 360:1859–1868. https://doi.org/10.1098/rstb.2005.1717

Vences M, Nagy ZT, Sonet G, Verheyen E (2012) DNA barcoding amphibians and reptiles. Methods Mol Biol 858:79–107. https://doi.org/10.1007/978-1-61779-591-6_5

Wang JJ-F, Jiang L-Y, Qiao G-X (2011) Use of a mitochondrial COI sequence to identify species of the subtribe Aphidina (Hemiptera, Aphididae). Zookeys 122:1–17. https://doi.org/10.3897/zookeys.122.1256

Ward RD, Zemlak TS, Innes BH et al (2005) DNA barcoding Australia's fish species. Philos Trans R Soc Lond Ser B Biol Sci 360:1847–1857. https://doi.org/10.1098/rstb.2005.1716

Waugh J (2007) DNA barcoding in animal species: Progress, potential and pitfalls. Bio Essays 29:188–197. https://doi.org/10.1002/bies.20529

Wilson DE, Reeder DAM (2005) Mammal species of the world: a taxonomic and geographic reference. Johns Hopkins University Press, Baltimore

Xiang-long L, Zhang Y, Shen-gou C et al (1997) Study on the mtDNA RFLP of goat breeds. Zool Res 18:421–428

Xie Y, Li Y, Zhao X, et al (2010) Origins of the Chinese yak: evidence from maternal and paternal inheritance. In: 2010 4th international conference on bioinformatics and biomedical engineering, Chengdu, China, pp 1–5

Yan D, Luo JY, Han YM et al (2013) Forensic DNA barcoding and bio-response studies of animal horn products used in traditional medicine. PLoS One 8:e55854

Zhang RL, Zhang B (2014) Prospects of using DNA barcoding for species identification and evaluation of the accuracy of sequence databases for ticks (Acari: Ixodida). Ticks Tick Borne Dis 5:352–358. https://doi.org/10.1016/j.ttbdis.2014.01.001

Zhengchao T, Yaping Z, Huai Q (1998) Mitochondrial DNA polymorphism and genetic diversity in Chinese yaks. Yi Chuan Xue Bao 25:205–212

Part V
Case Studies

DNA Barcoding and Molecular Phylogeny of Indigenous Bacteria in Fishes from a Tropical Tidal River in Malaysia

Mohammad Mustafizur Rahman, Mohd Haikal Izzuddin, Najmus Sakib Khan, Akbar John, and Mohd Azrul Naim

Abstract DNA barcoding along with molecular phylogeny can be used for taxonomic identification, characterization, discovery of species and understanding molecular relationships especially in terms of species divergence. Thus, they facilitate biodiversity studies. Some studies have addressed DNA barcoding of bacterial samples from various sources. Unfortunately, the DNA barcoding of fish bacterial diversity has not been studied especially in the tropical tidal river. Therefore, a study was conducted to (1) identify the observed bacterial isolates by comparing the partial sequence from an unknown sample to a collection of sequences from known reference samples, (2) know the taxonomic and phylogenetic identity of identified bacteria in fish and (3) know the abundance of fish bacteria in the Kuantan River. For this study, three commercially important fish namely *Pristipomoides filamentosus*, *Cyclocheilichthys apogon* and *Labiobarbus festivus* were captured with gill nets from the Kuantan River, Malaysia. Bacteria from skin, gill and gut in fish were cultured at 35 °C for 24 h in both nutrient and marine agar. Bacterial DNA was extracted using a Bacterial Genomic DNA Isolation Kit following manufacturer's specifications. Isolated DNA was quantified in NanoDrop 2000v and gel eluded in 1.5% agarose gel and visualized under a gel visualizer. PCR products were outsource sequenced at First BASE Laboratories Sdn Bhd using an ABI sequencer by the Sanger sequencing method. Sequences were trimmed using sequence scanner 2.0 V. The aligned sequences were inspected by the eye and edited to remove ambiguities based on PHRED scores and the chromatogram. Fully aligned sequences were subjected to BLAST for nucleotide similarity search against 16S rRNA database.

M. M. Rahman (✉)
Department of Marine Science, Faculty (Kulliyyah) of Science (KOS), International Islamic University Malaysia (IIUM), Kuantan, Pahang, Malaysia

INOCEM Research Station, KOS, IIUM, Kuantan, Malaysia
e-mail: mustafiz@iium.edu.my

M. H. Izzuddin · N. S. Khan · M. A. Naim
Department of Biotechnology, KOS, IIUM, Kuantan, Pahang, Malaysia

A. John
INOCEM Research Station, KOS, IIUM, Kuantan, Malaysia

S. Trivedi et al. (eds.), *DNA Barcoding and Molecular Phylogeny*,
https://doi.org/10.1007/978-3-030-50075-7_14

The best matched species were selected based on BLAST results and the lowest genetic distance between the known and unknown nucleotides. Genetic distances (sequence divergences) were calculated using the K2P (Kimura two parameter) distance model. Neighbour-joining (NJ) trees of K2P distances were created to provide a graphic representation of the patterning of divergence between species. This method identified a total of 11 fish bacteria, which are taxonomically classified into Enterobacteriales, Pseudomonadales, Actinomycetales and Bacillales. The range of pairwise genetic distances between species of Enterobacteriales was lower than Bacillales. Similarly, the within group mean genetic distance of Enterobacteriales (0.010) was lower than that of Bacillales (0.055). These results indicate that the identified bacterial species under Enterobacteriales are more closely related than the bacteria species under Bacillales. The mean genetic distances between groups were genetically almost equally close, which was confirmed by the overall mean diversity. Out of 11 species, 7 were identified as *Cyclocheilichthys apogon*, 8 as *Labiobarbus festivus* and 7 as *Pristipomoides filamentosus*. The overall mean bacterial abundance (CFU/g) was higher in *C. apogon* (6.68×10^3) compared to those in *L. festivus* (5.12×10^3) and *P. filamentosus* (5.20×10^3). Overall, the highest bacterial abundances were observed in fish gut (6.62×10^3), followed by fish gill (5.78×10^3) and fish skin (4.60×10^3).

Keywords *Cyclocheilichthys apogon* · *Labiobarbus festivus* · *Pristipomoides filamentosus* · Bacteria · Kuantan River · Malaysia

1 Introduction

Bacteria are the smallest and most active biological entities in all aquatic environments. They play the most important role in mineralization of organic matter and thus, they execute biogeochemical cycles (Rahman et al. 2008a, b). Subsequently, primary, secondary and tertiary production and aquatic food webs are executed (Rahman 2015a, b). Therefore, heterotrophic bacteria are a major constituent in all aquatic environments. Fish are continuously exposed to the bacteria present in water, which influences the bacterial flora on external surfaces of fish (El-Shafai et al. 2004). Similarly, the digestive tract receives water and food that are populated with different types of bacteria. However, colonization of bacteria starts at the egg and/or larval stage and continues with the development of fish (Olafsen 2001). Thus, the numbers and range of bacteria present in eggs, food and water influence the bacterial flora of the developing fish. However, determining the bacterial community composition in fish is one of the necessary steps in understanding fish bacterial ecology. Moreover, understanding the abundance of pathogenic bacteria in fish may help in preparing an appropriate management policy to protect fish and humans from the diseases caused by bacteria.

Bacterial floras of fish are grossly two categories. Numerous bacteria cause diseases in fish and shellfish. In general, marine fishes are susceptible to diseases caused by *Vibrio anguillarum* and freshwater species are susceptible to those caused by *Aeromonas hydrophila* and *A. salmonicida*. A majority of bacterial pathogens in fish are also capable of causing diseases in humans. They produce toxins that cause many lethal diseases such as paralytic shellfish poisoning, neurotoxic shellfish poisoning, diarrheic shellfish poisoning, amnesiac shellfish poisoning and ciguatera fish poisoning (Bienfang et al. 2011). Some bacteria play an important role with respect to the well-being and health of fish (Hasan et al. 2007). They help the digestion process and control harmful microflora of fish (Robertson et al. 2000). For example, the bacteria *Vibrio alginolyticus* reduces diseases in Atlantic salmon (*Salmo salar*) caused by infection of common pathogenic bacteria (*Aeromonas salmonicida*, *Vibrio anguillarum*) (Austin et al. 1995). However, interest is increasingly growing in the society to know the type and abundance of bacteria live in fish as it is an important part of the human diet and is the major source of animal protein in many countries (Smith and Prairie 2004; Van Horn et al. 2016). It is one of the preferred human foods in many countries in the world. Besides having a high nutritional value, fish and fish products play a very important role in preventing many diseases such as coronary heart disease, cancer, diabetes and inflammatory disease (Rahman and Verdegem 2007; Rahman and Meyer 2009; Rahman et al. 2009).

The type and abundance of bacteria in fish depend on many factors such as species, feeding pattern, age, size, season, geographical location and environmental condition (Novotny et al. 2004). There are some published information on the type and abundance of bacteria in river fishes. However, the study of bacteria in fish from a tropical tidal river is limited as the identification of bacteria is a very important issue in bacteriological studies. Most of the studies identified river fish bacteria through standard morphological, physiological and biochemical tests. For example, de Sousa and Silva-Souza (2001) observed a bacterial community associated with Congonhas river fish by morphological, physiological and biochemical tests. However, morphological, physiological and biochemical tests are unable to identify some groups of bacteria. According to Figueras et al. (2011), morphological and physiological characterization can provide imprecise results, and one cannot guarantee that a bacterial strain has been correctly identified at the species level, if it has not been verified using a reliable molecular method. Moreover, these methods do not provide any information about genetic closeness or divergence among various species of bacteria. This problem can be overcome by the DNA barcoding method, which identifies bacteria using a partial sequence of DNA (Hebert and Gregory 2005).

DNA barcoding can be used to diagnose taxa, increasing the speed, objectivity and efficiency of species identification (Hickerson et al. 2006; Stahlhut et al. 2012). DNA barcoding and molecular phylogenetic analysis also help in understanding molecular relationships especially in terms of species divergence among various fish bacteria (Hebert and Gregory 2005). According to Bhattacharya et al. (2016), DNA barcoding appears to be a promising approach for taxonomic identification, characterization and discovery of newer diverged species and thus facilitating biodiversity studies. It helps researchers to appreciate genetic and evolutionary associations by the collection of molecular, morphological and distributional data. According to

Begerow et al. (2010), DNA barcoding provides very important information to evaluate ecological sequences for resolving management priorities. Therefore, molecular phylogeny is necessary not only to identify species but also to understand the genetic closeness or relationship among various bacterial species. In this study, fish bacteria were identified through 16S rRNA sequencing, which is generally applied for the identification of bacteria (Purty and Chatterjee 2016).

All fishes were collected from a tropical tidal river. Tide and salinity are two important factors that greatly influence the dynamics of fish bacteria in the tidal river. In this study, to examine the type and abundance of fish bacteria in a tropical tidal river, the Kuantan river was taken as a model as this river plays a life-sustaining role for many inhabitants in Kuantan and is important in economy, ecology and recreation in the east coast of Peninsular Malaysia. The main objectives of the study were to (1) identify the observed bacterial isolates by comparing the partial sequence from an unknown sample to a collection of sequences from known reference samples, (2) know the taxonomic and phylogenetic identity of identified bacteria in the fish and (3) know the abundance of fish bacteria in the Kuantan River.

2 Materials and Methods

2.1 Collection and Preparation of Samples

For bacteriological analysis, three commercially important fish namely *Pristipomoides filamentosus*, *Cyclocheilichthys apogon* and *Labiobarbus festivus* were selected. *Pristipomoides filamentosus* is a saline water fish that was captured from the Kuantan river estuary. *Cyclocheilichthys apogon* and *Labiobarbus festivus* are freshwater fishes and were collected from 15 km upstream of the Kuantan River estuary, where water salinity fluctuated between 0 and 5 °C. All fishes were captured with gill nets and immediately transported to the laboratory, where they were dissected using scalpel, forceps, scissors and knives. Skin, gills and gut samples of fish were aseptically obtained by dissection. Then 1 g of each dissected organs were separately homogenized by a mortar and pestle. All apparatuses were sterilized prior to dissection.

2.2 Culture and Quantitative Estimation of Bacteria

The homogenized tissues from skin, gill and gut were serially diluted and cultured at 35 °C for 24 h in both nutrient and marine agar. The *aseptic technique was incorporated in all steps of the bacteriological work.* To avoid the problem of overcrowding of bacterial colonies on the agar plates, a few trials were conducted to find the best dilution method. All colonies were counted and expressed in CFU (colony forming unit) per g of sample. All bacterial colonies isolated from skin, gill and gut tissues were morphologically grouped based on their colony shape, size and color.

2.3 Isolation and Sequencing of DNA

A total of 14 morphologically different colonies (C1–14) were identified. Each bacterial colony was sub-cultured and subjected to DNA isolation using a Bacterial Genomic DNA Isolation Kit (Cat. No. 17900, Norgen Biotek Corp., Canada) following manufacturer's specifications. Isolated DNA was quantified in NanoDrop 2000v and gel eluded in 1.5% agarose gel and visualized under a gel visualizer. Samples with low DNA concentrations were further isolated to achieve ample DNA concentration for downstream applications. It can be noted that DNA isolation for C11 and C14 was unsuccessful even after multiple DNA isolation steps and hence excluded from the analysis. The extracted DNA was stored at −20 °C for further use. Isolated DNA was subjected to colony PCR under standard thermal cycling conditions. The PCR reaction mixture consisted of 1.5 mM $MgCl_2$, 0.2 mM of each DNTPs, 1 U of Taq DNA polymerase, 0.2 μM of forward primer (27b F – 5' AGAGTTTGATCCTGGCTCAG 3′) and 2 μl of DNA template in a volume of 25 μl. The following thermal cycling conditions were used in amplifying the target gene: 95 °C for 5 min for the initial hotstart, followed by 40 cycles at 95 °C for 30 sec, 56 °C for 30 sec for annihilation, 72 °C for 1 min and a final extension at 72 °C for 10 min. The PCR products were gel eluded in 1.5% agarose gel for further documentation. PCR products were outsource sequenced at First BASE Laboratories Sdn Bhd using an ABI sequencer by the Sanger sequencing method. Sequences were trimmed using sequence scanner 2.0 V. The sequence for sample C2 was not aligned as there were many mismatches and ambiguous base pairs observed throughout the sequence.

2.4 Data Analysis

Quantitative bacterial data were checked for normality using the *Shapiro*–Wilk *test* and homogeneity of variance was done using Levene's test. They were subjected to one-way ANOVA (analysis of variance) to understand their variation in skin, gills and gut of fishes. If an effect was significant, ANOVA was followed by Tukey's test for unplanned multiple comparisons of means. Statistically significant value was determined at $P \leq 0.05$. All statistical analyses were performed using *IBM SPSS* software for Windows (version 22.0).

To understand the nucleotide similarity, all sequences were aligned using sequence scanner 2 software. The aligned sequences were inspected by the eye and edited to remove ambiguities based on PHRED scores and the chromatogram. Fully aligned sequences were subjected to BLAST for nucleotide similarity search against the 16S rRNA database. For each sequence, a total of five closest nucleotides were selected to understand the closest species by genetic distances and phylogenetic trees using Mega 6 software (Tamura et al. 2013). The best matched species were selected based on BLAST results and the lowest genetic distance between the known and unknown nucleotides (Table 1). Genetic distances (sequence divergences) were calculated using the K2P (Kimura two parameter) distance model (Kimura 1980). Neighbour-joining (NJ) trees of K2P distances were created to provide a graphic representation of the patterning of divergence between species. Before calculating genetic distances and constructing phylogenetic trees, all sequenced DNAs were subjected to complete alignment using ClustalX software (version 2.1). By phylogenetic relatedness among known and unknown nucleotides based on the homology of 16S rRNA sequences, the closest affiliation of a new isolate was assigned. A phylogenetic relationship was observed within all known nucleotides to assign each species in a group. Neighbour-joining analysis was employed to calculate the numerical pairwise genetic distance.

3 Results and Discussion

This study provides qualitative and quantitative bacteriological information about fish bacteria in the Kuantan River. Qualitative and quantitative determination of bacterial communities are usually very challenging as only a small percentage of bacteria can be cultured under laboratory conditions. In this study, two types of agar media (nutrient and marine agars) were used to culture bacteria. The most important challenge of bacteriological studies is taxonomical identification up to the species level. Morphological, physiological and biochemical tests are traditionally used to identify bacteria but they are not precise methods to identify some groups of bacteria up to the species level. To overcome this challenge, the DNA barcoding technique using the genetic distance and a phylogenetic tree were applied to identify bacteria up to the species level. In the present study, to identify bacteria, a conserved region of 16S rRNA was amplified as these genes (1) are ubiquitous (present in all prokaryotes), (2) are structurally and functionally conserved and (3) contain variable and highly conserved regions (Hugenholtz 2002). According to Alperi et al. (2008), the 16S rRNA gene is essential to identify bacteria as this gene can be more discriminative especially in closely related strains. Therefore, this gene is commonly used as a molecular marker to identify bacteria (Han et al. 2011; Hidalgo and Figueras 2012; Table 2).

Figure 1 shows the phylogenetic tree that indicates genetic relationships and bootstrap values among known and unknown DNA sequences. Species ecology along with the lowest genetic distance and the highest bootstrap value between the known and unknown species led to the identification of a total of 11 species of fish bacteria. The mean nucleotide composition of the 11 identified species was estimated as $T = 19.4$, $C = 22.9$, $G = 31.7$ and $A = 26.0\%$. The phylogenetic relationships and pairwise genetic distances within identified bacteria are presented in Fig. 2 and Table 3, respectively. The phylogenetic tree divided all bacteria into four clades. Taxonomically, Clade 1, Clade 2, Clade 3 and Clade 4 belong to Enterobacteriales (*Enterobacter aerogenes*, *Enterobacter mori*, *Enterobacter cloacae* and *Edwardsiella tarda*), Pseudomonadales (*Psychrobacter sp.*),

Table 1 The best matched species and the second best matched species based on the 16S rRNA gene and its acquired Gene-Bank accession numbers

Sample number	Best matched species				2nd Best matched species			
	Species	Accession number	I (%)	Source	Species	Accession number	I (%)	source
C1	*Enterobacter aerogenes*	AM184247	99	River water	*Pantoea agglomerans*	KX684195	99	Wastewater
C3	*Leucobacter chromiireducens*	KT262983	97	River water	*Leucobacter salsicius*	NR117298	97	Fermented food
C4	*Edwardsiella tarda*	DQ855293	99	–	*Escherichia sp.*	KR063559	99	Wetland
C5	*Staphylococcus saprophyticus*	KT986100	100	Ocean water	*Staphylococcus saprophyticus*	KX343982	100	Acidic soil
C6	*Psychrobacter sp.*	AM422129	94	Drainage sample	*Psychrobacter faecalis*	KU364044	93	Seawater
C7	*Enterobacter cloacae*	KT328453	99	Intestine of fish	*Leclercia adecarboxylata*	NR104933	99	Drinking water
C8	*Enterobacter mori*	KF747680	99	Rhizosphere soil	*Enterobacter cloacae*	HM854373	99	Rhizosphere
C9	*Macrococcus caseolyticus*	KC969082	99	East China Sea mud	*Macrococcus caseolyticus*	KJ958217	99	Yellow croaker
C10	*Bacillus tequilensis*	KU356174	99	South China Sea	*Bacillus subtilis*	KT986107	99	Soil
C12	*Bacillus licheniformis*	KJ842640	99	Hot spring	*Bacillus licheniformis*	KT748523	99	Fecal matter
C13	*Bacillus pumilus*	KT986126	99	Ocean water	*Bacillus stratosphericus*	KM277362	99	Fish gut

"Ï" indicates identity

Table 2 Published information on DNA barcoding to identify various species of bacteria

Identified bacteria	Gene name	References
Bacteria within *Enterobacteriaceae*	16S rRNA	Hauben et al. (1998)
Enterobacteriaceae (some genera)	16 s rDNA	Sprber et al. (1998)
Cronobacter spp.	16S rRNA	Schmid et al. (2009)
Pantoea sp.	16S rRNA	Brady et al. (2008)
Pantoea agglomerans strains	16S rRNA	Deleetoile et al. (2009)
Erwinia and *Pantoea* species	Whole genome	Zhang and Qiu (2015)
Lactobacillus species	23S rRNA, 16S rRNA	Kwon et al. (2004)
Genera *Pectobacterium* and *Dickeya*	16S rRNA	Ma et al. (2007)
Enterobacteriaceae (some species)	16S rDNA	Drancourt et al. (2001)
Raoultella terrigena	16S rRNA	Wang et al. (2016)
Fish bacteria	16S rRNA	Alikunhi et al. (2017)
Fish spoilage bacteria	16S rDNA	Garcıa-Lopez et al. (2004)
River water bacteria	16S rRNA	Cottrell et al. (2005)
Psychrophilic bacteria	16S rDNA	Maruyama et al. (2000)
Fish bacteria	16S rRNA	Sebastiao et al. (2015)
Aeromonas (from drinking water)	16S rRNA	Alzahrani (2015)

Actinomycetales (*Leucobacter chromiireducens*) and Bacillales (*Bacillus tequilensis*, *Bacillus licheniformis*, *Bacillus pumilus*, *Macrococcus caseolyticus* and *Staphylococcus saprophyticus*), respectively. The range of pairwise genetic distances between species of Enterobacteriales (range: 0.001–0.015) was lower than Bacillales (range: 0.016–0.780) (Table 3). Similarly, the within group mean genetic distance of Enterobacteriales (0.010) was lower than that of Bacillales (0.055). These results indicate that the identified bacterial species under Enterobacteriales were more closely related than the bacterial species under Bacillales (Dunbar et al. 2002). The mean genetic distance between groups Enterobacteriales and Pseudomonadales was the lowest (0.229) compared to all other pair groups although the range of mean genetic distances between groups was 0.029–0.331 (Table 4). This indicates that all groups were genetically almost equally close and the identified bacteria were genetically less diversified (Dunbar et al. 2002; Rastogi and Sani 2011). This can be further confirmed by the overall mean diversity among all identified bacterial species. We observed the mean species diversity among all identified bacterial species as 0.209. According to Rastogi and Sani (2011), the typical clone libraries of 16S rRNA genes contain a lower number of sequences ($\leq$1000 sequences) and therefore reveal only a small portion of the microbial diversity present in a sample. Some studies have addressed DNA barcoding of bacterial samples from various sources (Table 2). Unfortunately, the DNA barcoding of fish bacterial diversity has not been studied especially in the tropical tidal river. Studies on bacterial identification is usually employing either one of the two gene sequencing methods such as 16S rRNA or chaperonin protein (cpn60). However, due to a larger barcode gap in cpn60, the 16 s rRNA gene is usually preferred by the framework established by International Barcode of Life

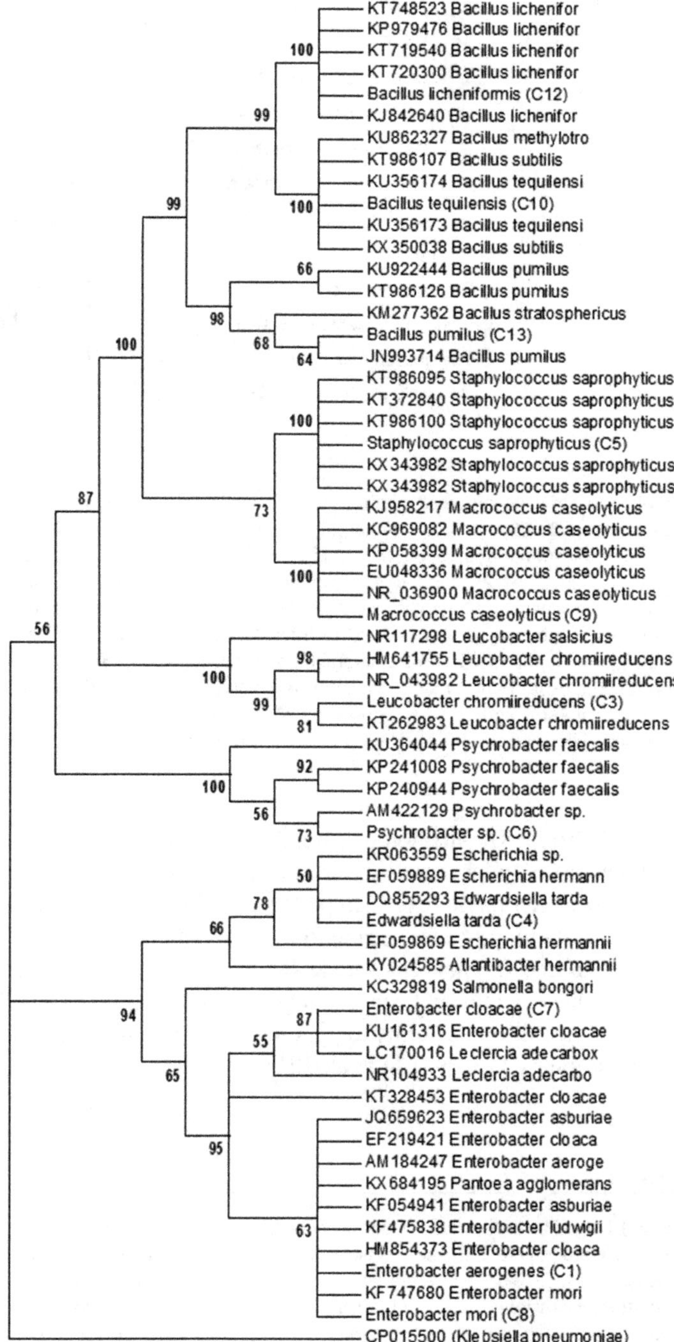

Fig. 1 Neighbour-joining dendrogram depicting the estimated phylogenetic relationships among unknown (indicated by C1, C2, etc. in parentheses) and related known species (indicated by accession numbers) based on the 16S rRNA Gene-Bank. The optimal tree with the sum of branch

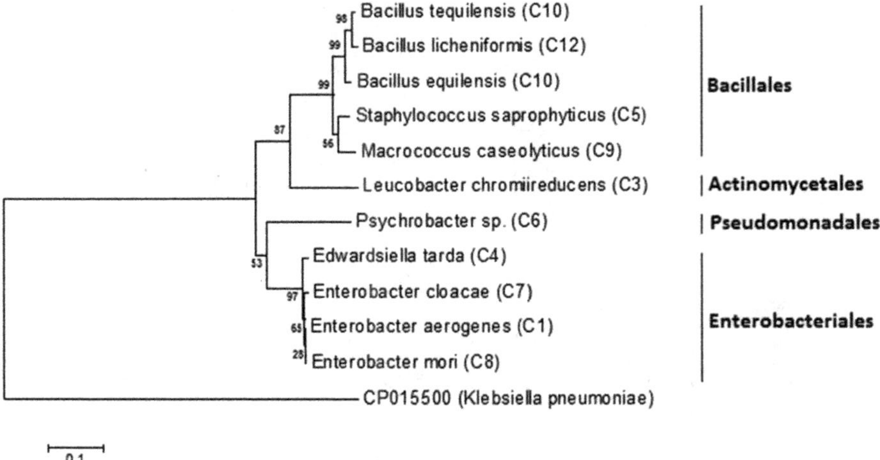

Fig. 2 Neighbour-joining dendrogram depicting the estimated phylogenetic relationships among unknown (indicated by C1, C2, etc. in parentheses) species based on the 16S rRNA Gene-Bank. The optimal tree with the sum of branch length = 1.75303059 is shown. The percentage of replicate trees in which the associated taxa clustered together in the bootstrap test (1000 replicates) are shown next to the branches (Felsenstein 1985). The tree is drawn to scale, with branch lengths in the same units as those of the evolutionary distances used to infer the phylogenetic tree. The evolutionary distances were computed using the Kimura 2-parameter method (Kimura 1980) and are in the units of the number of base substitutions per site. The analysis involved 12 nucleotide sequences. Codon positions included were 1st + 2nd + 3rd + Noncoding. All positions containing gaps and missing data were eliminated. There were a total of 893 positions in the final dataset

(IBL) (Links et al. 2012). The method of barcoding 16 s rRNA is reliable, accurate and used to track novel microbes and for further meta-genomic analysis. Until recently, the link between phylogenetic hypotheses and questions raised by fundamental ecologists pertaining to the distribution and assemblage of endosymbiontic bacteria in host species is not answered. Recently, studies have proven the co-evolution of the host and the endosymbiont gene at intra- and interspecific levels, which in turn supported the opportunity to use the endosymbiont gene as a marker to identify host evolutionary history (Liu et al. 2013).

Fig. 1 (continued) length = 1.68418054 is shown. The percentage of replicate trees in which the associated taxa clustered together in the bootstrap test (1000 replicates) are shown next to the branches (Felsenstein 1985). The tree is drawn to scale, with branch lengths in the same units as those of the evolutionary distances used to infer the phylogenetic tree. The evolutionary distances were computed using the Kimura 2-parameter method (Kimura 1980) and are in the units of the number of base substitutions per site. The analysis involved 62 nucleotide sequences. Codon positions included were 1st + 2nd + 3rd + Noncoding. All positions containing gaps and missing data were eliminated. There were a total of 881 positions in the final dataset

Table 3 Pairwise genetic distances (Kimura 2-parameter) of 11 Kuantan river fish bacteria based on 16S rRNA sequences

	C1	C8	C7	C4	C6	C10	C12	C13	C5	C9	C3
C1											
C8	0.001										
C7	0.006	0.007									
C4	0.014	0.015	0.015								
C6	0.225	0.227	0.230	0.228							
C10	0.280	0.282	0.287	0.281	0.345						
C12	0.286	0.287	0.292	0.286	0.34	0.016					
C13	0.274	0.276	0.282	0.276	0.325	0.026	0.031				
C5	0.275	0.277	0.282	0.282	0.32	0.082	0.780	0.074			
C9	0.277	0.279	0.284	0.282	0.325	0.070	0.065	0.065	0.054		
C3	0.282	0.283	0.280	0.292	0.332	0.244	0.242	0.232	0.236	0.239	

C1: *Enterobacter aerogenes*, C3: *Leucobacter chromiireducens*, C4: *Edwardsiella tarda*, C5: *Staphylococcus saprophyticus*, C6: *Psychrobacter sp.*, C7: *Enterobacter cloacae*, C8: *Enterobacter mori*, C9: *Macrococcus caseolyticus*, C10: *Bacillus tequilensis*, C12: *Bacillus licheniformis*, and C13: *Bacillus pumilus*

Table 4 Pairwise genetic distances (Kimura 2-parameter) among 4 groups (Fig. 2) of Kuantan river fish bacteria based on 16S rRNA sequences

Group	Enterobacteriales	Pseudomonadales	Bacillales	Actinomycetales
Enterobacteriales				
Pseudomonadales	0.229			
Bacillales	0.282	0.330		
Actinomycetales	0.285	0.331	0.235	

Out of 11 species, 7 were identified as *Cyclocheilichthys apogon*, 8 as *Labiobarbus festivus* and 7 as *Pristipomoides filamentosus* (Table 5). Bacterial abundance significantly varies among fish species. The overall mean bacterial abundance (CFU/g) was higher in *C. apogon* (6.68×10^3) compared to those in *L. festivus* (5.12×10^3) and *P. filamentosus* (5.20×10^3) (Fig. 3). Overall, the highest bacterial abundances were observed in fish gut (6.62×10^3), followed by fish gill (5.78×10^3) and fish skin (4.60×10^3) (Fig. 4). However, these results are different depending on fish species (Fig. 5). However, this trend was only observed in *C. apogon*. In the case of *L. festivus* and *P. filamentosus*, abundances of bacteria were lower in skin compared to gill and gut. Abundances of bacteria in gill and gut were statistically similar in both *L. festivus* and *P. filamentosus*. There is no previous study on the Kuantan River comparing the bacterial abundance in various parts of the fish body. However, Alikunhi et al. (2017) observed a similar trend (gut>gill>skin) of bacterial abundance in 13 species of commonly consumed fish collected from 3 places in the Jeddah region, Saudi Arabia, although their observed bacterial abundances in various places of fish body were higher compared to the present observation. Many factors affect the abundance of bacteria in various organs of fish body. Fish normally gets contaminated with bacteria from water, sediment

Table 5 List of bacteria identified from skin, gill and gut of kerisi, chemparus and kawan collected from the Kuantan River

	Cyclocheilichthys apogon	*Labiobarbus festivus*	*Pristipomoides filamentosus*
Skin			
Enterobacter aerogenes	×	×	×
Edwardsiella tarda			×
Leucobacter chromiireducens	×	×	
Macrococcus caseolyticus			×
Staphylococcus saprophyticus	×	×	×
Gill			
Enterobacter aerogenes	×		×
Edwardsiella tarda	×		×
Leucobacter chromiireducens		×	
Macrococcus caseolyticus	×		×
Psychrobacter sp.	×		×
Staphylococcus saprophyticus		×	×
Gut			
Edwardsiella tarda	×		
Bacillus pumilus	×	×	
Bacillus licheniformis		×	
Bacillus tequilensis		×	
Enterobacter mori			
Enterobacter cloacae			×
Macrococcus caseolyticus	×	×	
Psychrobacter sp.	×	×	×
Staphylococcus saprophyticus			×

"×" indicates presence

and food; hence, bacterial abundance in fish is largely influenced by the condition of their habitat. The observed bacterial abundances in various organs of various fishes were under the safe limit of aerobic bacterial count (5×10^5–5×10^7 cfu/g for fresh fish) (ICMSF 1986). However, regular monitoring of the bacterial load in the Kuantan river fish is suggested to keep track of any potential health risks to fish consumers.

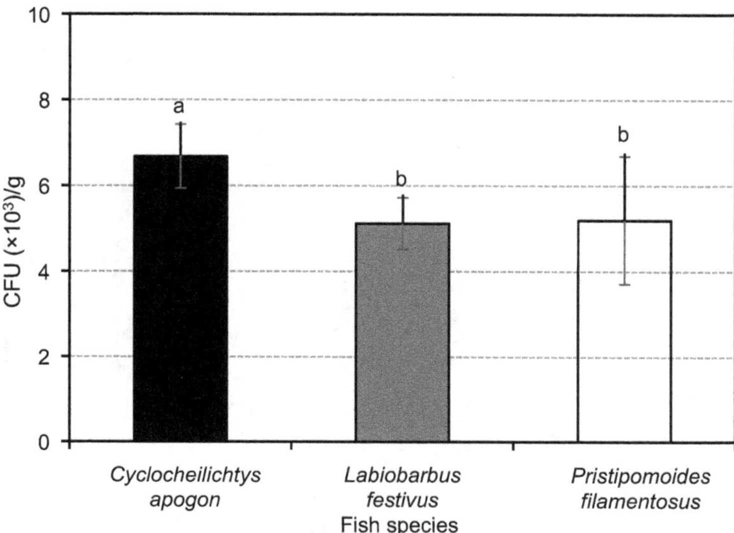

Fig. 3 Mean (bar: ±SD) number of cultivable bacteria (CFU/g) in chemparus, kawan and kerisi. Mean with no letters in common are significantly different ($P < 0.05$)

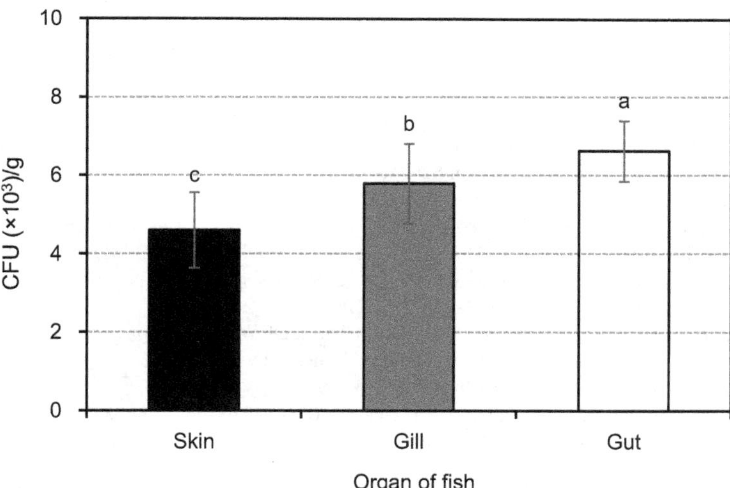

Fig. 4 Mean (bar: ±SD) number of cultivable bacteria (CFU/g) in skin, gill and gut of fishes. Mean with no letters in common are significantly different ($P < 0.05$)

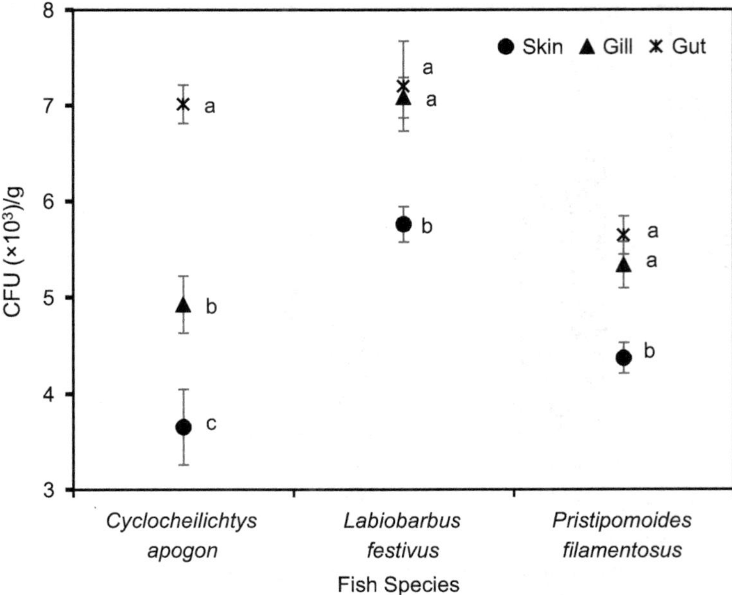

Fig. 5 Interaction between fish species and fish organ on the number (per g) of cultivable bacteria

References

Alikunhi NM, Batang ZB, AlJahdali HA, Aziz MAM, Al-Suwailem AM (2017) Culture dependent bacteria in commercial fishes: Qualitative assessment and molecular identification using 16S rRNA gene sequencing. Saudi J Biol Sci 24(6):1105–1116. https://doi.org/10.1016/j.sjbs.2016.05.017

Alperi A, Figueras MJ, Inza I, Martínez-Murcia AJ (2008) Analysis of 16S rRNA gene mutations in a subset of *Aeromonas* strains and their impact in species delineation. Int Microbiol 11:185–194

Alzahrani OM (2015) Molecular identification of Aeromonas spp. Collected from taif province of Saudi Arabia. Int J Sci Res 4:2359–2362

Austin B, Stuckey LF, Robertson PAW, Effendi I, Griffith DRW (1995) A probiotic strain of *Vibrio alginolyticus* effective in reducing diseases caused by *Aeromonas salmonicida*, *Vibrio anguillarum* and *Vibrio ordalii*. J Fish Dis 18:93–96

Begerow D, Nilsson H, Unterseher M, Maier W (2010) Current state and perspectives of fungalDNA barcoding and rapid identification procedures. Appl Microbiol Biotechnol 87:99–108

Bhattacharya M, Sharma AR, Patra BC, Sharma G, Seo EM, Nam JS, Chakraborty C, Lee SS (2016) DNA barcoding to fishes: current status and future directions. Mitochondrial DNA A DNA Mapp Seq Anal 27:2744–2752

Bienfang PK, DeFelice SV, Laws EA, Brand LE, Bidigare RR, Christensen S, Trapido-Rosenthal H, Hemscheidt TK, McGillicuddy DJ Jr, Anderson DM, Solo-Gabriele HM, Boehm AB, Backer LC (2011) Prominent human health impacts from several marine microbes: history, ecology, and public health implications. Int J Microbiol 2011:1–15

Brady C, Cleenwerck I, Venter S, Vancanneyt M, Swings J, Coutinho T (2008) Phylogeny and identification of *Pantoea* species associated with plants, humans and the natural environment based on multilocus sequence analysis (MLSA). Syst Appl Microbiol 31:447–460

Cottrell M, Waidner LA, Yu L, Kirchman DL (2005) Bacterial diversity of metagenomic and PCR libraries from the Delaware river. Environ Microbiol 7:1883–1895

de Sousa JS, Silva-Souza AT (2001) Bacterial community associated with fish and water from Congonhas River, Sertaneja, Parana, Brazil. Braz Arch Biol Technol 44:373–381

Deleetoile A, Decre D, Courant S, Passet V, Audo J, Grimont P, Arlet G, Brisse S (2009) Phylogeny and identification of Pantoea species and typing of Pantoea agglomerans strains by multilocus gene sequencing. J Clin Microbiol 47:300–331

Drancourt M, Bollet C, Carta A, Rousselier P (2001) Phylogenetic analyses of *Klebsiella* species delineate *Klebsiella* and *Raoultella* gen. Nov., with description of *Raoultella ornithinolytica* comb. nov., *Raoultella terrigena* comb. nov. and *Raoultella planticola* comb. nov. Int J Syst Evol Microbiol 51:925–932

Dunbar J, Barns SM, Ticknor LO, Kuske CR (2002) Empirical and theoretical bacterial diversity in four Arizona soils. Appl Environ Microbiol 68:3035–3045

El-Shafai SA, Gijzen HJ, Nasr FA, El-Gohary FA (2004) Microbial quality of tilapia reared in fecal contaminated ponds. Environ Res 95:231–238

Felsenstein J (1985) Confidence limits on phylogenies: an approach using the bootstrap. Evolution 39:783–791

Figueras MJ, Alperi A, Beaz-Hidalgo R, Stackebrandt E, Brambilla E, Monera A, Martínez-Murcia AJ (2011) *Aeromonas rivuli* sp. nov., isolated from the upstream region of a karst water rivulet in Germany. Int J Syst Evol Microbiol 61:242–248

Garcıa-Lopez I, Otero A, Garcıa-Lopez M-L, Santos JA (2004) Molecular and phenotypic characterization of nonmotile gram- negative bacteria associated with spoilage of freshwater fish. J Appl Microbiol 96:878–886

Han HJ, Kim DY, Kim WS, Kim CS, Jung SJ, Oh MJ, Kim DH (2011) Atypical *Aeromonas salmonicida* infection in the black rockfish, *Sebastes schlegeli* Hilgendorf, in Korea. J Fish Dis 34:47–55

Hasan MR, Hecht T, De Silva SS, Tacon AGJ (2007) Study and analysis of feeds and fertilizers for sustainable aquaculture development. FAO Fisheries Technical Paper No. 497. Rome, p 510

Hauben L, Moore ERB, Vauterinl L, Steenackers M, Mergaert J, Verdonck L, Swings J (1998) Phylogenetic position of phytopathogens within the Enterobacteriaceae. Syst Appl Microbiol 21:384–397

Hebert PDN, Gregory TR (2005) The promise of DNA barcoding for taxonomy. Syst Biol 54:852–859

Hickerson MJ, Meyer CP, Moritz C (2006) DNA barcoding will often fail to discover new animal species over broad parameter space. Syst Biol 55:729–739

Hidalgo RB, Figueras MJ (2012) Molecular detection and characterization of Furunculosis and other *Aeromonas* fish infections. In: Carvalho ED, David GS, Silva RJ (eds) Health and environment in aquaculture. InTech, Rijeka, pp 97–132

Hugenholtz P (2002) Exploring prokaryotic diversity in the genomic era. Genome Biol 3:Reviews 0003

International Commission on Microbiological Specifications for Foods (ICMSF) (1986) Microorganisms in foods 2: sampling for microbiological analysis: principles and specific applications, 2nd edn. Blackwell Scientific, Oxford, p 190

Kimura M (1980) A simple method for estimating evolutionary rate of base substitutions through comparative studies of nucleotide sequences. J Mol Evol 16:111–120

Kwon H-S, Yang E-H, Yeon S-W, Kang B-H, Kim T-Y (2004) Rapid identification of probiotic lactobacillus species by multiplex PCR using species-specific primers based on the region extending from 16S rRNA through 23S rRNA. FEMS Microbiol Lett 239:267–275

Links MG, Dumonceaux TJ, Hemmingsen SM, Hill JE (2012) The Chaperonin-60 universal target is a barcode for Bacteria that enables De novo assembly of metagenomic sequence data. PLoS ONE 7(11):e49755

Liu L, Huang X, Zhang R, Jiang L, Qiao G (2013) Phylogenetic congruence between Mollitrichosiphum (Aphididae: Greenideinae) and Buchnera indicates insect-bacteria parallel evolution. Syst Entomol 38:81–92

Ma B, Hibbing ME, Kim H-S, Reedy RM, Yedidia I, Breuer J, Breuer J, Glasner JD, Perna NT, Kelman A, Charkowski AO (2007) Host range and molecular phylogenies of the soft rot Enterobacterial genera *Pectobacterium* and *Dickeya*. Bacteriology 97:1150–1163

Maruyama A, Honda D, Yamamoto H, Kitamura K, Higashihara T (2000) Phylogenetic analysis of psychrophilic bacteria isolated from the Japan trench, including a description of the deep-sea species Psychrobacter pacificensis sp nov. Int J Syst Evol Microbiol 50:835–846

Novotny L, Dvorska L, Lorencova A, Beran B, Pavlik I (2004) Fish: a potential source of bacterial pathogens for human beings. Vet Med Czech 49:343–358

Olafsen JA (2001) Interactions between fish larvae and bacteria in marine aquaculture. Aquaculture 200:223–247

Purty RS, Chatterjee S (2016) DNA barcoding: an effective technique in molecular taxonomy. Austin J Biotechnol Bioeng 3:1–10

Rahman MM (2015a) Role of common carp (*Cyprinus carpio*) in aquaculture production systems. Front Life Sci 8:399–410

Rahman MM (2015b) Effects of co-cultured common carp on nutrients and food web dynamics in rohu aquaculture ponds. Aquac Environ Interact 6:223–232

Rahman MM, Meyer CG (2009) Effects of food type on diel behaviours of common carp *Cyprinus carpio* L. in simulated aquaculture pond conditions. J Fish Biol 74:2269–2278

Rahman MM, Verdegem MCJ (2007) Multi-species fishpond and nutrients balance. In: ven der Zijpp AJ, Verreth AJA, Tri LQ, ven Mensvoort MEF, Bosma RH, Beveridge MCM (eds) Fishponds in farming systems. Wageningen Academic Publishers, Wageningen

Rahman MM, Verdegem MCJ, Wahab MA (2008a) Effects of tilapia (*Oreochromis nilotica* L.) addition and artificial feeding on water quality, and fish growth and production in rohu-common carp bi-culture ponds. Aquac Res 39:1579–1587

Rahman MM, Verdegem M, Nagelkerke L, Wahab MA, Milstein A, Verreth J (2008b) Effects of common carp *Cyprinus carpio* (L.) and feed addition in rohu *Labeo rohita* (Hamilton) ponds on nutrient partitioning among fish, plankton and benthos. Aquac Res 39:85–95

Rahman MM, Hossain MY, Jo Q, Kim SK, Ohtomi J, Meyer CG (2009) Ontogenetic shift in dietary preference and low dietary overlap in rohu (*Labeo rohita* Hamilton) and common carp (*Cyprinus carpio* L.) in semi-intensive polyculture ponds. Ichthyol Res 56:28–36

Rastogi G, Sani RK (2011) Molecular techniques to assess microbial community structure, function, and dynamics in the environment. In: Ahmad I, Ahmad F, Pichtel J (eds) Microbes and microbial technology: agricultural and environmental applications. Springer, Heidelberg, pp 29–57

Robertson PAW, O-Dowd C, Burrells C, Williams P, Austin B (2000) Use of *Carnobacterium* sp. as a probiotic for Atlantic salmon (*Salmo salar* L.) and rainbow trout (*Oncorhynchus mykiss*, Walbaum). Aquaculture 185:235–243

Schmid M, Iversen C, Gontia I, Stephan R, Hofmann A, Hartmann A, Jha B, Eberl L, Riedel K, Lehner A (2009) Evidence for a plant-associated natural habitat for *Cronobacter* spp. Res Microbiol 160:608–614

Sebastiao FA, Furlan LR, Hashimoto DT, Pilarski F (2015) Identification of bacterial fish pathogens in Brazil by direct Colony PCR and 16S rRNA gene sequencing. Adv Microbiol 5:409–424

Smith EM, Prairie YT (2004) Bacterial metabolism and growth efficiency in lakes: the importance of phosphorus availability. Limnol Oceanogr 49:137–147

Sprber C, Mendrock U, Swiderski J, Lang E, Stackebrandt E (1998) The phylogenetic position of *Serratia, Buttiauxella* and some other genera of the family *Enterobacteriaceae*. Int J Syst Bacteriol 49:1433–1438

Stahlhut JK, Gibbs J, Sheffield CS, Smith MA, Packer L (2012) *Wolbachia* (Rickettsiales) infections and bee (Apoidea) barcoding: a response to Gerth et al. Syst Biodivers 10:395–401

Tamura K, Stecher G, Peterson D, Filipski A, Kumar S (2013) MEGA6: molecular evolutionary genetics analysis version 6.0. Mol Biol Evol 30:2725–2729

Van Horn DJ, Sinsabaugh RL, Takacs-Vesbach CD, Mitchell KR, Dahm CN (2016) Response of heterotrophic stream biofilm communities to a gradient of resources. Aquat Microb Ecol 64:149–161

Wang Y, Jiang X, Xu Z, Ying C, Yu W, Xiao Y (2016) Identification of *Raoultella terrigena* as a rare causative agent of Subungual abscess based on 16S rRNA and housekeeping gene sequencing. Can J Infect Dis Med Microbiol 2016:1. https://doi.org/10.1155/2016/3879635

Zhang Y, Qiu S (2015) Examining phylogenetic relationships of *Erwinia* and *Pantoea* species using whole genome sequence data. Antonie Van Leeuwenhoek 108:1037–1046

DNA Barcoding of Ichthyoplankton and Juvenile Fishes of a Tropical River in Malaysia

B. Akbar John, Hassan I. Sheikh, K. C. A. Jalal, B. Y. Kamaruzzaman, H. Sanower, M. Nur Hanisah, M. M. Rahman, and M. Rozihan

Abstract Taxonomic identification of early larval stages of fishes using conventional morphological keys is extremely laborious due to the overlapping characters shared between genetically closer species. Especially, species-level differentiation during their ontological development is challenging due to the paucity of information on their diagnostic features. In the present study, we aimed to use universal DNA barcoding technology to identify Ichthyoplankton and juvenile fishes of a tropical river (Kuantan River in Pahang) in Malaysia. This sampling station was chosen in order to check the distribution of juvenile and fish Ichthyoplankton samples after the recent massive flood encountered in East peninsular Malaysia during 2014. We adopted mitochondrial cytochrome oxidase C subunit 1 gene sequencing to identify fish samples. A total of 28 species from 15 families and 5 orders were identified successfully to the species level from the total of 58 DNA barcodes. Unlike the previous report, the most dominant fishes in this study belong to the Cyprinidae family followed by toxotidae, ambasidae, and eleotridae. We admit that the modified bubu light trap method adopted for larval collection in this study has its limitation to attract larvae which had negative phototactic behavior (i.e., Ariidae fishes). Phylogenetic and BLAST analysis showed accuracy of species identification with high bootstrap and percentage similarity value, respectively. Results in this study confirmed the efficiency of universal DNA barcode technology in species-level delimitation of morphologically cryptic species identification. The

B. Akbar John (✉)
Institute of Oceanography and Maritime Studies (INOCEM), Kulliyyah of Science, International Islamic University Malaysia, Kuantan, Malaysia

H. I. Sheikh
Faculty of Fisheries and Food Science, Universiti Malaysia Terengganu, Kuala Nerus, Terengganu, Malaysia

K. C. A. Jalal · B. Y. Kamaruzzaman · H. Sanower · M. Nur Hanisah · M. M. Rahman
Department of Marine Science, Kulliyyah of Science, International Islamic University Malaysia, Kuantan, Malaysia

M. Rozihan
Faculty of Agriculture, Department of Aquaculture, University Putra Malaysia, Serdang, Selangor, Malaysia

S. Trivedi et al. (eds.), *DNA Barcoding and Molecular Phylogeny*,
https://doi.org/10.1007/978-3-030-50075-7_15

data presented in this study are valuable for analyzing post-flood effect on fish distribution in Tropical River and implementing plans for future fishery resource management in Kuantan River, Pahang, Malaysia.

Keywords DNA barcoding · COX1 gene · Fish larvae · Ichthyoplankton · Kuantan River

1 Introduction

Ichthyoplankton and early juvenile stages of fishes are the most sensitive forms and their distribution is altered by the ambient environmental condition. Their assemblage is directly proportional to spawning strategies of their brooders (especially in the estuary), and driven by water currents and mass. In tropics, rivers, in general, have direct connection with the sea and their physical and topological parameters are severely altered during massive flood. For instance, recent flood incident in East peninsular Malaysia from Dec 2014 to mid-January 2015 caused massive damage to the livelihood and altered sensitive species distribution in almost all major rivers. From a regional perspective, this flood had significant impact on fish distribution and species richness that has influenced the livelihood of fishery folk in Kuantan River skirts besides planning various management measures for future sustainable fishery. On the other hand, in taxonomic point of view, accurate identification of ichthyoplankton and early juvenile stages of fishes is challenging even to an expert taxonomist. It is due to the lack of information on their early ontological development and overlapping and continuously changing morphological characters that empirical study of larval ecology is currently limited (Choat et al. 1993; Nakatani et al. 1998). Most species identification is based on the data availability on their adult subjects (Ko et al. 2013; Pegg et al. 2006) and many morpho-diagnostic characters or poorly developed larval forms especially in their egg stage (Baumgartner et al. 2004; Nakatani 2001).

The utility of standardized molecular marker in accurately identifying morphologically cryptic species using universal barcode gene (Cytochrome oxidase C subunit 1) has proven to be a promising method in species discrimination (Kochzius et al. 2010). Though the reliability of single gene sequencing to identify metazoan species is argued by many researchers (Cowan et al. 2006; Decru et al. 2016; DeSalle et al. 2005; John et al. 2016; Meier et al. 2006; Vences et al. 2005), the method is still promising in identifying early life stages and adults of many taxon (Bucklin et al. 2011; Hajibabaei et al. 2006; Hausmann et al. 2016; John et al. 2016; Spasojevic et al. 2016; Ward et al. 2005). Ko et al. (2013) have suggested that accuracy of fish larval identification using morphological, body pigmentation, and meristic count is possible with only 3–30% of samples compared to the 100% accuracy of DNA barcoding technique (Ratnasingham and Hebert 2007). The method is rapid, accurate, and helps in differentiating morphologically cryptic species and avoids inconsistency in species-level identification of sensitive forms (Frantine-Silva et al. 2015; Hubert et al. 2015; Ko et al. 2013; Overdyk et al. 2016).

The application of DNA barcoding in early larval forms (pre and post-flexion stages), eggs, and pre-juvenile stages is promising and adopted in various ecological studies in recent years (Frantine-Silva et al. 2015; Gleason and Burton 2012; Pegg et al. 2006; Valdez-Moreno et al. 2010). Many studies have even argued that molecular taxonomy is the only promising tool for identifying eggs to their corresponding species (Lewis et al. 2016). Nevertheless, the paper addresses the utility of DNA barcoding in ichthyoplankton and juvenile fish identification and their post-flood distribution in Kuantan River, Pahang, Malaysia. The data presented in this study are crucial for implementing various fishery management measures for future sustainable fishery in tropical riverine system exposed to severe flood.

2 Materials and Methods

2.1 Sample Collection

Ichthyoplankton and juvenile fish samples were collected from Kuantan River and its immediate tributaries during ebb tidal cycle (Fig. 1). Due to the bottom topology

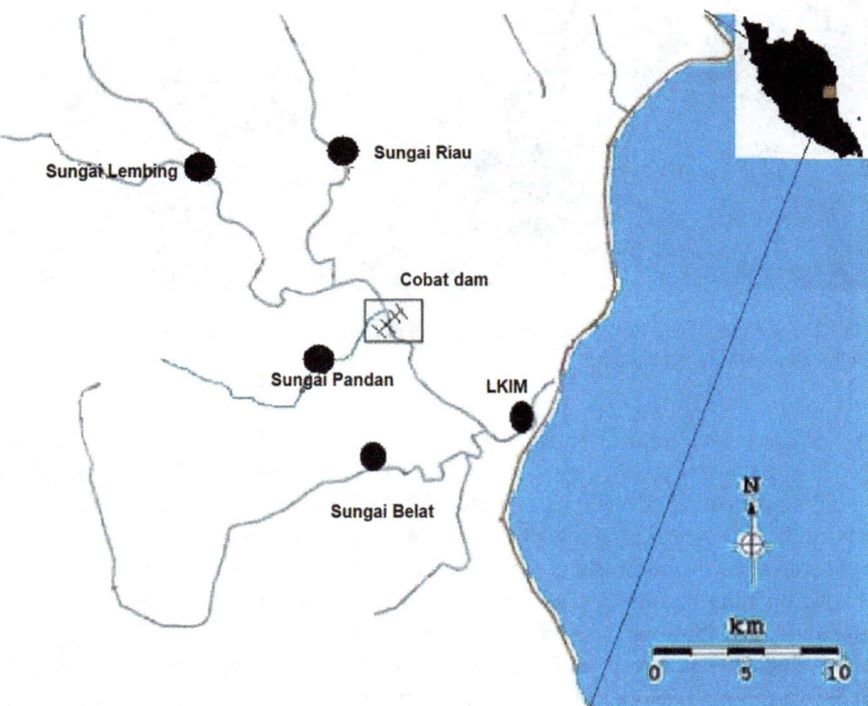

Fig. 1 Location of the sampling sites at Kuantan River

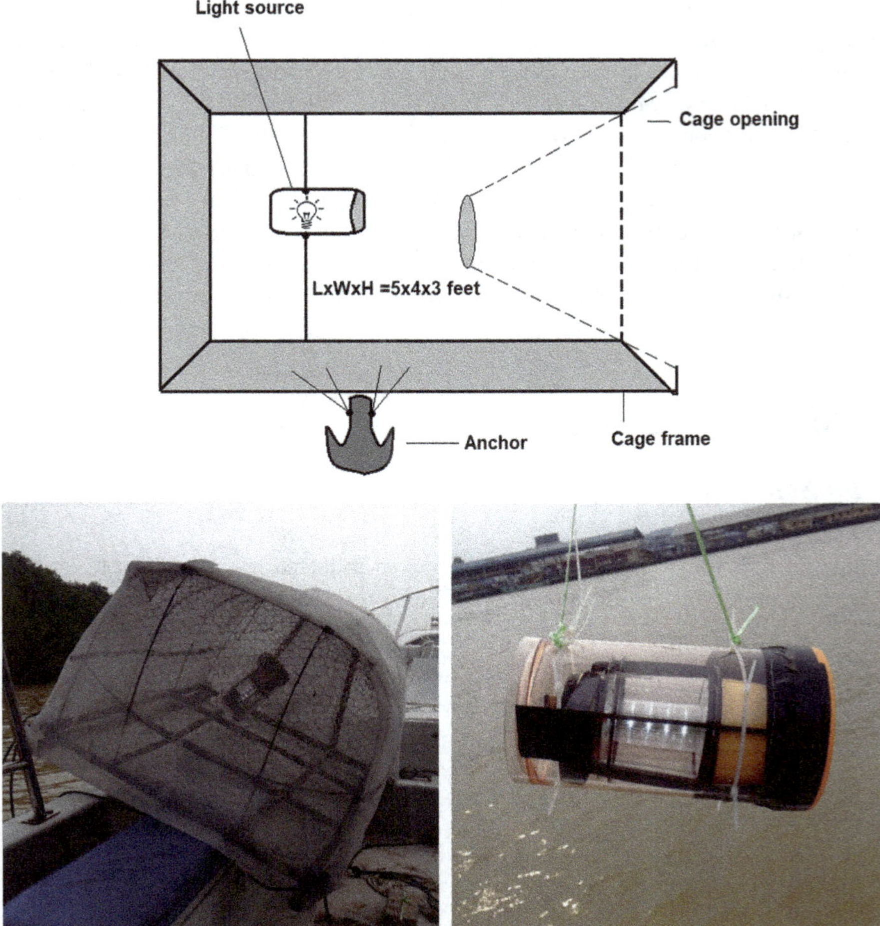

Fig. 2 Modified bubu light trap and the prepared internal light source used to attract the Ichthyoplankton during sampling

and physical condition of Kuantan riverbed, such as high turbidity, uneven water depth in Kuantan River (varied from 4 to 13 m) besides inefficiency of plankton sampling net (mesh size 500 μ), a modified bubu light trap (Fig. 2) was developed in this study and used during sampling from April to December 2015. Modified bubu light trap used in this study is 5 × 4 × 3 feet (L × W × H) size having conical-shaped opening on one side toward the interior section of the trap. The skeleton is made up of bamboo or cylindrical wood which in turn was covered completely by 2 mm thickness stainless steel wire meshes with the mesh diameter of 2.5 cm^2. An underwater light was placed in transparent plastic container and the lid was sealed with commercially available silicone gel and parafilm to ensure maximum light emission. An anchor was tied at the bottom of the cage to make sure the cage does not wash away during water current. The complete setup (modified bubu light trap) was covered by plastic window mesh sheet (mesh size <1 mm) (Fig. 2). The net was deployed under water at the depth of 4 m in sampling stations for 16 hours overnight. All samples were stored in 70% ethanol for laboratory identification.

2.2 *Sample Preparation and Identification*

Ichthyoplankton and juvenile fish samples were measured under Dino Capture 2.0 portable microscope. Samples were identified morphologically to the lowest possible taxon using standard references and taxonomically classified based on Marques et al. (2006). Size classes of larvae were tabulated and represented in Mean ± SD.

2.3 *DNA Barcoding*

Total genomic DNA was extracted using Geneaid DNA tissue isolation kitTM following manufacturer's instruction. Nanodrop 2000 UV-Vis spectrophotometer was used to quantify the total genomic DNA. Approximately 650 bp of partial cytochrome oxidase C subunit 1 gene was amplified using standard thermal cycler in polymerase chain reaction (PCR) using the primer set (Fish F1: 5′-GGTCAACAAATCATAAAGATATTGG-3′ and Fish R1: 5'--TAAACTTCAGGGTGACCAAAAAATCA-3′). The amplification reactions were performed in a final volume including 6.3 μl of molecular-grade water, 1.0 μl of 10X PCR buffer, 0.5 μl of MgCl2 (50 mM), 0.3 μl of each primer (10 mM), 0.5 μl of dNTPs (10 mM), 0.1 μl of Bioline Taq polymerase, and 1 μl of template DNA. The PCR condition includes hot start with 94 °C for 1 minute, 5 cycles of 94 °C for 30 seconds, annealing at 45 °C for 40 seconds, and extension at 72 °C for 1 minute, 35 cycles of 94 °C for 30 seconds, 51 °C for 40 seconds, and final extension at 72 °C for 10 minutes. Final Amplicon were purified by adding 0.5 μl Illustra Exo-Star 1-Step PCR Clean Up Kit (Thermo Fisher Scientific, Waltham, MA) and gel eluded in 2% China Agarose gel and photographed using Gel imager under UV light for future reference. DNA sequencing was carried at First BASE Sdn Bhd using ABI sequencer.

2.4 *Data Analysis*

DNA sequences were trimmed using sequence scanner software v2.0. The trimmed sequences with no insertion, deletion, and stop codon were subjected to Barcode of Life Database BOLD (http://www.boldsystems.org/index.php/IDS_OpenIdEngine) and NCBI BLAST (https://blast.ncbi.nlm.nih.gov/Blast.cgi?PAGE_TYPE=BlastSearch) analysis for species identification and cross-examined with morphological identification. Multiple sequence alignment (MSA) was performed in Clustal X 2.0.6v using default settings (Larkin et al. 2007). Nucleotide composition was computed using Bioedit v7.2.5 (Hall 1999). Species identification reliability was validated in evolutionary tree constructed using Neighbor-Joining (NJ) tree method using Kemura 2 parameter (K2P) model using default condition with 1000 bootstrap replicates. Genetic distance was calculated using K2P distance matrix using MEGA 6.0 V (Tamura et al. 2013).

3 Results and Discussion

During the sampling period, 372 larval/juvenile fish samples were captured in modified bubu light trap for further molecular identification. Samples with full appendages were selected for DNA isolation and for further downstream studied. Overall, 28 species of ichthyoplankton/juvenile fishes from 15 families and 5 orders were identified using DNA barcoding COI gene. Unlike previous report on adult fish distribution in Kuanan River by Jalal et al. (2012), who observed dominant fishes belong to the families such as Ariidae, Lactaridae, and Lutjanidae, the dominant families in this study belong to cyprinidae, toxotidae, ambasidae, and eleotridae with the percentage abundance of 35%, 24%, 18%, and 11%, respectively (Table 1). We have adopted the modified light trap method to collect the ichthyoplankton and a juvenile fish due to many technical constraints to use standard bongo net for sampling. It includes mainly the bottom topology of the riverbed and constant changes in river depth (3–10 m) among the immediate tributaries. Other constraints include high turbidity and organic debris on the water column which hindered the collection of the live and full form of larvae (without any appendage lost). Though this light trap method is effective to attract many species of larvae, the fishes exhibiting negative phototactic behavior such as Ariidae larvae, mouthbrooders were not trapped in the net (Mukai et al. 2008; Satoh et al. 2017). The advantage of using a light trap includes (1). Collection of un-damaged larval form for downstream application without other species DNA contamination, (2). The efficiency of light trap over plankton net with 0.5 mm mesh size, (3). Samplings in areas where turbidity and water depth are minor limiting factors.

3.1 *Database Analysis*

DNA barcode generated from 58 samples (>650 bp) were subjected to BOLD and NCBI BLAST analysis independently. A match of 95–100% was found in 72% of test sequences and 7% samples showed >90% similarity while 21% of samples were identified with more than 85% similarity value. Almost 47 out of 58 sequences exhibited higher similarity percentage with the best match to nearest neighbor indicating ambiguity of ~19% of barcode assignment to their accurate species.

Table 1 Ichthyoplankton and Juvenile fishes identified in this study using universal DNA barcode (Cytochrome oxidase C subunit 1 gene)

Family	Species Identified	IUC N Red List Status	BOLD/ BLAST similarity percentage	Kuantan River (N 03° 47′ 04.2″ E103° 19′ 05.1″)	Belat River (N 03° 46′ 15.7″ E103° 17′ 21.7″)	Panching River (N 03° 47′ 4.65″ E103° 08′ 58.5″)	Riau River (N 03° 51′ 51.6″ E103° 13′ 32.4″)	Previous study (Jalal et al. 2012)
Ambassidae	*Ambassis marianus*	Least concern	92	*				NF
Ambassidae	*Ambassis commersoni*	Least concern	85		*			NF
Ambassidae	*Parambassis siamensis*	Least concern	96			*		NF
Leiognathidae	*Secutor ruconius*	Not evaluated	100	*	*			NF
Leiognathidae	*Photopectoralis bindus*	Not evaluated	99		*			NF
Leiognathidae	*Leiognathus equulus*	Least concern	99	*				NF
Lutjanidae	*Lutjanus johnii*	Not evaluated	84	*				*
Lutjanidae	*Lutjanus russellii*	Not evaluated	100	*				*
Eleotridae	*Butis gymnopomus*	Not evaluated	100	*				NF
Eleotridae	*Prionobutis dasyrhynchus*	Not evaluated	99	*			*	NF
Toxotidae	*Toxotes chatareus*	Not evaluated	100		*			*
Toxotidae	*Toxotes jaculatrix*	Least concern	99		*			*

(continued)

Table 1 (continued)

Family	Species Identified	IUC N Red List Status	BOLD/ BLAST similarity percentage	Kuantan River (N 03° 47′ 04.2″ E103° 19′ 05.1″)	Belat River (N 03° 46′ 15.7″ E103° 17′ 21.7″)	Panching River (N 03° 47′ 4.65″ E103° 08′ 58.5″)	Riau River (N 03° 51′ 51.6″ E103° 13′ 32.4″)	Previous study (Jalal et al. 2012)
Cyprinidae	*Poropuntius smedleyi*	Not evaluated	99			*		NF
Cyprinidae	*Barbonymus schwanenfeldii*	Least concern	100				*	NF
Cyprinidae	*Barbonymus gonionotus*	Least concern	99				*	NF
Cyprinidae	*Neolissochilus stracheyi*	Least concern	98			*		NF
Cyprinidae	*Barbodes binotatus*	Least concern	99		*			NF
Cyprinidae	*Rasbora sumatrana*	Not evaluated	99			*		NF
Cyprinidae	*Rasbora dusonensis*	Not evaluated	99				*	NF
Cyprinidae	*Rasbora trilineata*	Least concern	97				*	NF
Megalopidae	*Megalops cyprinoides*	Data deficient	100	*				*
Gobiidae	*Bathygobius laddi*	Not evaluated	86	*				NF
Cichlidae	*Oreochromis niloticus*	Not evaluated	84		*			*
Percichthyidae	*Gadopsis marmoratus*	Not evaluated	85		*			NF

Monodactylidae	*Monodactylus argenteus*	Not evaluated	90		*			NF
Scatophagidae	*Scatophagus argus*	Least concern	99		*			NF
Zenarchopteridae	*Dermogenys pusilla*	Not evaluated	99		*			NF
Hemiramphidae	*Hemirhamphodon pogonognathus*	Least concern	100			*		NF
Balitoridae	*Balitora kwangsiensis*	Least concern	83				*	NF

Note: * represents the inhabiting species in each sampling station. Species identified in this study was compared with previous study where adult fishes were sampled using gill net (Jalal et al. 2012). NF represent corresponding species in the row 'Not Found' during previous during previous study

Overall, 81% of all barcodes showed <2% divergence from their possible nearest neighbors. From the barcode generated, dominant cyprinidae family represented with 8 species (*Poropuntius smedleyi, Barbonymus schwanenfeldii, B. gonionotus, Neolissochilus stracheyi, Barbodes binotatus, Rasbora sumatrana, R. dusonensis, and R. trilineata*) followed by toxotidae (*Toxotes chatareus, T. jaculatrix*), ambasidae (*Ambassis marianus, A. commersoni, and Parambassis siamensis), and* eleotridae (*Butis gymnopomus, Prionobutis dasyrhynchus).*

3.2 Phylogenetic Analysis

Phylogenetic tree constructed using NJ method and K2P distance model showed clear segregation of identified taxa to their corresponding species. Though there are some species mismatches observed in the tree [in the case of Lutjanidae (*Lutjanus russellii*) and Monodactylidae (*Monodactylus argenteus*)], a high bootstrap value in internal nods >98–100% signified reliability of constructed phylogram with a low distance between samples within the clusters formed by the species (Fig. 3). The observed nucleotide composition for all species is as follows: A (25.8%), G (18%), T (28.8), and C (27.4). The overall pairwise distance was observed to be 0.337 with an average P distance of 0.243. Genetic distance was hierarchically increased from the lowest possible taxon to the higher taxonomic level. Intraspecific genetic distance was observed between 0–2.56% compared to interspecies genetic distance (6.08–14.18%) when the proposed genetic distance threshold value of 2% for intraspecific and 6.8% for congeneric species used (Frantine-Silva et al. 2015; Pereira et al. 2013). Nucleotide substitution pattern observed among the test organisms (including all larvae/juveniles barcoded) clearly showed that transitional substitutions are very common in the gene sequence than transversional substitutions with a transition/transversion bias value of 1.54 (Table 2).

Many attempts were made to analyze ichthyoplankton assemblage changes and its composition in marine (Ko et al. 2013; Pegg et al. 2006; Valdez-Moreno et al. 2010) and inland waters of Brazil (Frantine-Silva et al. 2015) and Australia (Loh et al. 2014). However, studies on molecular identification of ichthyoplankton and early life stages from tropical river are still limited. In this study, we present first report on the efficiency of molecular identification technique to accurately discriminate species sampled from one of the major rivers in East Peninsular Malaysia. Malaysia has faced severe northeast monsoonal hit from mid-December 2014 to January 2015 with at least >60% above the normal precipitation rate. This in turn increased uncontrolled discharge of organic and inorganic substances into the river stream which are believed to be the limiting factor of sensitive organisms such as juvenile and ichthyoplankton fishes. In order to accurately identify immediate successive (i.e., postflood) juvenile and ichthyoplankton samples, we have adopted COI gene sequence similarity comparison with public databanks as proposed by Ratnasingham and Hebert (2007) and (Ward et al. 2009). Our analysis clearly showed efficiency of universal barcode gene (COI) in species identification,

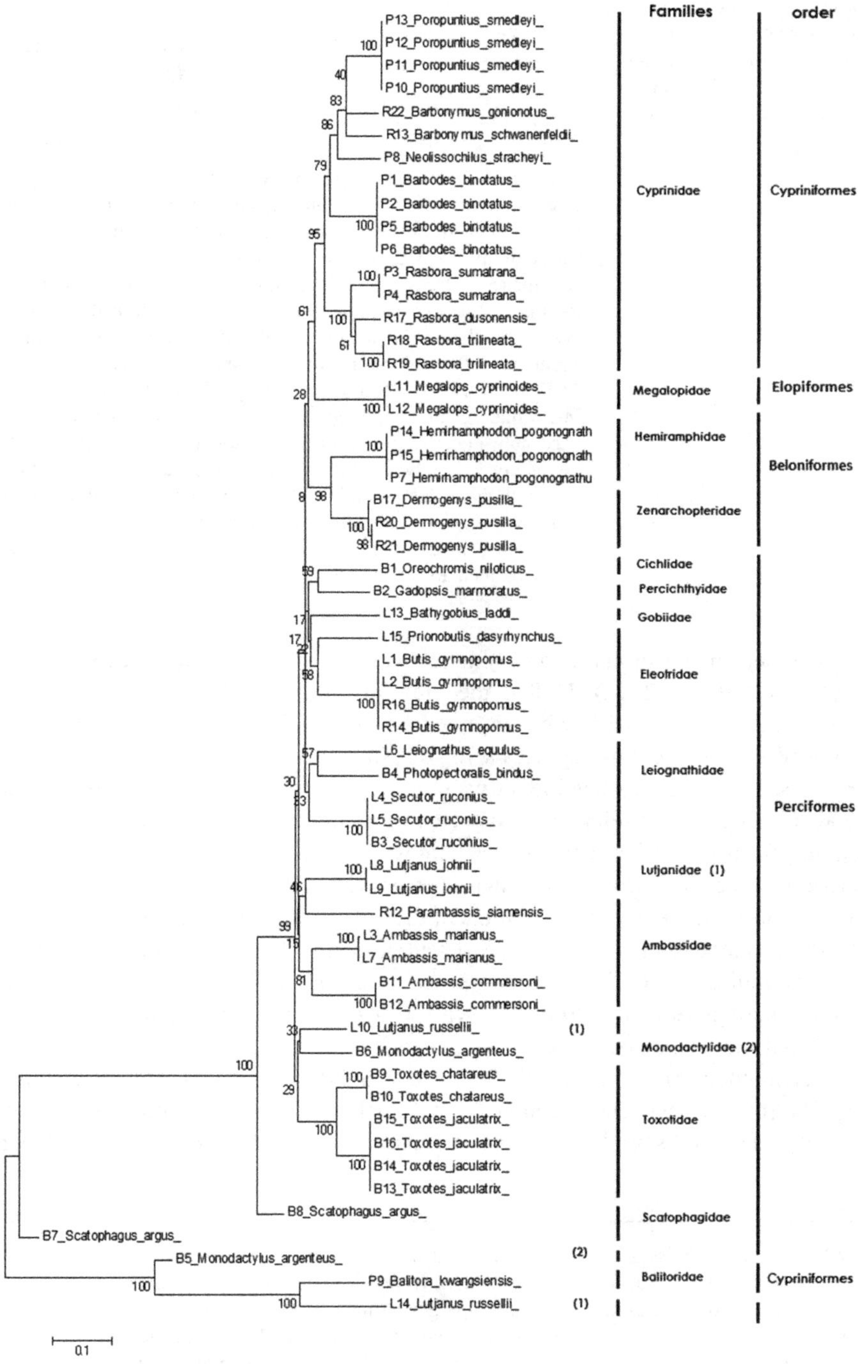

Fig. 3 Evolutionary relationships of taxa collected in this study. The evolutionary history was inferred using the Neighbor-Joining method (Saitou and Nei 1987). The optimal tree with the sum of branch length = 3.94044074 is shown. The percentage of replicate trees in which the associated taxa clustered together in the bootstrap test (1000 replicates) is shown next to the branches

Table 2 Maximum likelihood estimate of substitution matrix and pattern of nucleotide substitution observed among the test organisms

	A	T/U	C	G
A	–	*4.91*	*4.91*	**15.17**
T/U	*4.91*	–	**15.17**	*4.91*
C	*4.91*	**15.17**	–	*4.91*
G	**15.17**	*4.91*	*4.91*	–

Note: Each entry is the probability of substitution (*r*) from one base (row) to another base (column). Substitution pattern and rates were estimated under the Kimura (1980) 2-parameter model. Rates of different transitional substitutions are shown in bold and those of transversional substitutions are shown in *italics*. Relative values of instantaneous *r* should be considered when evaluating them. For simplicity, the sum of *r* values is made equal to 100. The nucleotide frequencies are A = 25.00%, T/U = 25.00%, C = 25.00%, and G = 25.00%. For estimating ML values, a tree topology was automatically computed. The maximum Log-likelihood for this computation was −10101.892. The analysis involved 57 nucleotide sequences. Codon positions included were 1st + 2nd + 3rd + Noncoding. All positions containing gaps and missing data were eliminated. There were a total of 553 positions in the final dataset. Evolutionary analyses were conducted in MEGA6

although percentage similarity and genetic distance approaches introduce some bias (Taylor and Harris 2012). Unlike the previous study by Ko et al. (2013) and Frantine-Silva et al. (2015) who observed 85 and 99% of detection efficiency of COI gene sequencing to the species level, our study depicted slightly lower ~81 and 100% species and genus-level identification, respectively. It might probably due to the availability of less reference sequences submitted especially on juvenile fishes sampled from tropical rivers. In fact, almost all samples identified in this study generally inhabit in a tropical freshwater ecosystem. Although DNA-level identification is an accurate and rapid approach, there is an urgent need for conventional taxonomic identification keys to be developed for ichthyoplankton samples for cross-examining results from DNA study. Several models have been proposed to match unknown sequences to the known species. For example, coalescent models, mutational models, distance-based models, phylogenetic, and mixed phylogenetic-coalescent models (Brown et al. 2012; David et al. 2012; Munch et al. 2008a; Munch et al. 2008b; Pons et al. 2006; Puillandre et al. 2012) that in turn require phylogenetic diversity and well-sampled intraspecific parameters (Ratnasingham and Hebert

Fig. 3 (continued) (Felsenstein 1985). The tree is drawn to scale, with branch lengths in the same units as those of the evolutionary distances used to infer the phylogenetic tree. The evolutionary distances were computed using the Kimura 2-parameter method (Kimura 1980) and are in the units of the number of base substitutions per site. The analysis involved 57 nucleotide sequences. Codon positions included were 1st + 2nd + 3rd + Noncoding. All positions containing gaps and missing data were eliminated. There were a total of 553 positions in the final dataset. Evolutionary analyses were conducted in MEGA6 (Tamura et al. 2013)

2013), similar data are still unavailable for tropical fishes, though some attempts have shown promising results (Cooper et al. 2009; Hubert et al. 2011; Westneat and Alfaro 2005). Hence, these approaches cannot be used unless the comprehensive DNA barcode library is established. With these limitations, our study using percentage similarity approach to identify generated barcodes showed initial promising results, though the limited availability of morphological characters.

4 Conclusion

In conclusion, DNA barcoding of ichthyoplankton and juvenile fishes using COI gene sequencing showed promising, rapid, and accurate identification of samples to their corresponding species level. The post-flood effect on ichthyoplankton species distribution in Kuantan River showed an apparent shift in species composition which might have a positive impact on the fishery in the near future. Though unavailability of corresponding species DNA barcode data in public databanks is one of the limiting factor for species delimitation, almost 80% of the samples were identified with higher percentage similarity (>99%). It is worth mentioning that the identification of 28 species of juvenile forms of fishes from Kuantan River compared to the previous report on conventional taxonomy (19 species) would increase our understanding on a greater diversity of species at least in Malaysian riverine system. Constant monitoring in fish composition and distribution would pave the way for various fishery management enforcement practices toward sustainable fishery in East Peninsular Malaysia, particularly in Kuantan River.

Acknowledgment The research work was sponsored under special flood Fundamental Research Grant Scheme (FRGS), Ministry of Science and Technology, Malaysia.

References

Baumgartner G, Nakatani K, Gomes LC, Bialetzki A, Sanches PV (2004) Identification of spawning sites and natural nurseries of fishes in the upper Paraná River, Brazil. Environ Biol Fish 71(2):115–125

Brown SD, Collins RA, Boyer S, LEFORT MC, MALUMBRES-OLARTE J, Vink CJ, Cruickshank RH (2012) Spider: an R package for the analysis of species identity and evolution, with particular reference to DNA barcoding. Mol Ecol Resour 12(3):562–565

Bucklin A, Steinke D, Blanco-Bercial L (2011) DNA barcoding of marine metazoa. Annu Rev Mar Sci 3:471–508

Choat HJ, Doherty PP, Kerrigan BB, Leis JJ (1993) A comparison of towed nets, purse seine and light-aggregation devices for sampling of larvae and pelagic stages of coral reef fishes. Fishery Bulletin 91:195–209

Cooper WJ, Smith LL, Westneat MW (2009) Exploring the radiation of a diverse reef fish family: phylogenetics of the damselfishes (Pomacentridae), with new classifications based on molecular analyses of all genera. Mol Phylogenet Evol 52(1):1–16

Cowan RS, Chase MW, Kress WJ, Savolainen V (2006) 300,000 species to identify: problems, progress, and prospects in DNA barcoding of land plants. Taxon 55(3):611–616

David O, Larédo C, Leblois R, Schaeffer B, Vergne N (2012) Coalescent-based DNA barcoding: multilocus analysis and robustness. J Comput Biol 19(3):271–278

Decru E, Moelants T, De Gelas K, Vreven E, Verheyen E, Snoeks J (2016) Taxonomic challenges in freshwater fishes: a mismatch between morphology and DNA barcoding in fish of the north-eastern part of the Congo basin. Mol Ecol Resour 16(1):342–352

DeSalle R, Egan MG, Siddall M (2005) The unholy trinity: taxonomy, species delimitation and DNA barcoding. Philos Trans R Soc B Biol Sci 360(1462):1905–1916

Felsenstein J (1985) Confidence limits on phylogenies: an approach using the bootstrap. Evolution 39:783–791

Frantine-Silva W, Sofia S, Orsi M, Almeida F (2015) DNA barcoding of freshwater ichthyoplankton in the Neotropics as a tool for ecological monitoring. Mol Ecol Resour 15 (5):1226–1237

Gleason LU, Burton RS (2012) High-throughput molecular identification of fish eggs using multiplex suspension bead arrays. Mol Ecol Resour 12(1):57–66

Hajibabaei M, Janzen DH, Burns JM, Hallwachs W, Hebert PD (2006) DNA barcodes distinguish species of tropical Lepidoptera. Proc Natl Acad Sci U S A 103(4):968–971

Hall TA (1999) BioEdit: a user-friendly biological sequence alignment editor and analysis program for Windows 95/98/NT. Paper presented at the Nucleic acids symposium series

Hausmann A, Miller SE, Holloway JD, deWaard JR, Pollock D, Prosser SW, Hebert PD (2016) Calibrating the taxonomy of a megadiverse insect family: 3000 DNA barcodes from geometrid type specimens (Lepidoptera, Geometridae) 1. Genome 59(9):671–684

Hubert N, Paradis E, Bruggemann H, Planes S (2011) Community assembly and diversification in indo-Pacific coral reef fishes. Ecol Evol 1(3):229–277

Hubert N, Espiau B, Meyer C, Planes S (2015) Identifying the ichthyoplankton of a coral reef using DNA barcodes. Mol Ecol Resour 15(1):57–67

Jalal K, Azfar MA, John BA, Kamaruzzaman Y, Shahbudin S (2012) Diversity and community composition of fishes in tropical estuary Pahang Malaysia. Pak J Zool 44(1):181–187

John BA, Sheikh HI, Jalal K, Zaleha K, Kamaruzzaman B (2016) Revised phylogeny of extant xiphosurans (Horseshoe Crabs) DNA barcoding in marine perspectives. Springer, Cham, pp 113–130

Kimura M (1980) A simple method for estimating evolutionary rates of base substitutions through comparative studies of nucleotide sequences. J Mol Evol 16(2):111–120

Ko H-L, Wang Y-T, Chiu T-S, Lee M-A, Leu M-Y, Chang K-Z et al (2013) Evaluating the accuracy of morphological identification of larval fishes by applying DNA barcoding. PLoS One 8(1): e53451

Kochzius M, Seidel C, Antoniou A, Botla SK, Campo D, Cariani A et al (2010) Identifying fishes through DNA barcodes and microarrays. PLoS One 5(9):e12620

Larkin MA, Blackshields G, Brown N, Chenna R, McGettigan PA, McWilliam H et al (2007) Clustal W and Clustal X version 2.0. Bioinformatics 23(21):2947–2948

Lewis LA, Richardson DE, Zakharov EV, Hanner R (2016) Integrating DNA barcoding of fish eggs into ichthyoplankton monitoring programs. Fish Bull 114(2):153–166

Loh W, Bond P, Ashton K, Roberts D, Tibbetts I (2014) DNA barcoding of freshwater fishes and the development of a quantitative qPCR assay for the species-specific detection and quantification of fish larvae from plankton samples. J Fish Biol 85(2):307–328

Marques SC, Azeiteiro UM, Marques JC, Neto JM, Pardal MÂ (2006) Zooplankton and ichthyoplankton communities in a temperate estuary: spatial and temporal patterns. J Plankton Res 28(3):297–312

Meier R, Shiyang K, Vaidya G, Ng PK (2006) DNA barcoding and taxonomy in Diptera: a tale of high intraspecific variability and low identification success. Syst Biol 55(5):715–728

Mukai Y, Tuzan AD, Lim LS, Wahid N, Raehanah S, Senoo S (2008) Development of sensory organs in larvae of African catfish *Clarias gariepinus*. J Fish Biol 73(7):1648–1661

Munch K, Boomsma W, Huelsenbeck JP, Willerslev E, Nielsen R (2008a) Statistical assignment of DNA sequences using Bayesian phylogenetics. Syst Biol 57(5):750–757

Munch K, Boomsma W, Willerslev E, Nielsen R (2008b) Fast phylogenetic DNA barcoding. Philos Trans R Soc Lond B Biol Sci 363(1512):3997–4002

Nakatani K (2001) Ovos e larvas de peixes de água doce: desenvolvimento e manual de identificação: Eletrobrás; Uem, 2001. il

Nakatani K, Baumgartner G, Latini JD (1998) Morphological description of larvae of the mapara *Hypophthalmus edentatus* (Spix)(Osteichthyes, Hypophthalmidae) in the Itaipu reservoir (Parana River, Brazil). Revista Brasileira de Zoologia 15(3):687–696

Overdyk LM, Holm E, Crawford SS, Hanner RH (2016) Increased taxonomic resolution of Laurentian Great Lakes ichthyoplankton through DNA barcoding: a case study comparison against visual identification of larval fishes from Stokes Bay, Lake Huron. J Great Lakes Res 42 (4):812–818

Pegg GG, Sinclair B, Briskey L, Aspden WJ (2006) MtDNA barcode identification of fish larvae in the southern great barrier reef–Australia. Sci Mar 70(S2):7–12

Pereira LH, Hanner R, Foresti F, Oliveira C (2013) Can DNA barcoding accurately discriminate megadiverse Neotropical freshwater fish fauna? BMC Genet 14(1):20

Pons J, Barraclough TG, Gomez-Zurita J, Cardoso A, Duran DP, Hazell S et al (2006) Sequence-based species delimitation for the DNA taxonomy of undescribed insects. Syst Biol 55 (4):595–609

Puillandre N, Lambert A, Brouillet S, Achaz G (2012) ABGD, automatic barcode gap discovery for primary species delimitation. Mol Ecol 21(8):1864–1877

Ratnasingham S, Hebert PD (2007) BOLD: the barcode of life data system (http://www.barcodinglife.org). Mol Ecol Notes 7(3):355–364

Ratnasingham S, Hebert PD (2013) A DNA-based registry for all animal species: the barcode index number (BIN) system. PLoS One 8(7):e66213

Saitou N, Nei M (1987) The neighbor-joining method: a new method for reconstructing phylogenetic trees. Mol Biol Evol 4(4):406–425

Satoh S, Tanoue H, Ruitton S, Mohri M, Komatsu T (2017) Morphological and behavioral ontogeny in larval and early juvenile discus fish Symphysodon aequifasciatus. Ichthyol Res 64(1):37–44

Spasojevic T, Kropf C, Nentwig W, Lasut L (2016) Combining morphology, DNA sequences, and morphometrics: revising closely related species in the orb-weaving spider genus Araniella (Araneae, Araneidae). Zootaxa 4111(4):448–470

Tamura K, Stecher G, Peterson D, Filipski A, Kumar S (2013) MEGA6: molecular evolutionary genetics analysis version 6.0. Mol Biol Evol 30(12):2725–2729

Taylor H, Harris W (2012) An emergent science on the brink of irrelevance: a review of the past 8 years of DNA barcoding. Mol Ecol Resour 12(3):377–388

Valdez-Moreno M, Vásquez-Yeomans L, Elías-Gutiérrez M, Ivanova NV, Hebert PD (2010) Using DNA barcodes to connect adults and early life stages of marine fishes from the Yucatan peninsula, Mexico: potential in fisheries management. Mar Freshw Res 61(6):655–671

Vences M, Thomas M, Bonett RM, Vieites DR (2005) Deciphering amphibian diversity through DNA barcoding: chances and challenges. Philos Trans R Soc Lond B Biol Sci 360 (1462):1859–1868

Ward RD, Zemlak TS, Innes BH, Last PR, Hebert PD (2005) DNA barcoding Australia's fish species. Philos Trans R Soc Lond B Biol Sci 360(1462):1847–1857

Ward RD, Hanner R, Hebert PD (2009) The campaign to DNA barcode all fishes, FISH-BOL. J Fish Biol 74(2):329–356

Westneat MW, Alfaro ME (2005) Phylogenetic relationships and evolutionary history of the reef fish family Labridae. Mol Phylogenet Evol 36(2):370–390

Molecular Identification of Reptiles from Tabuk Region of Saudi Arabia Through DNA Barcoding: A Case Study

Bishal Dhar, Mohua Chakraborty, Madhurima Chakraborty, Sorokhaibam Malvika, N. Neelima Devi, Abdulhadi A. Aloufi, Subrata Trivedi, and Sankar K. Ghosh

Abstract The deserts of Saudi Arabia provide an excellent habitat for reptiles. Although reptiles show significant vertebrate diversity, only few barcoding studies have been conducted on reptiles. In this case study, we collected different reptile species from the Tabuk region of Saudi Arabia and performed DNA barcoding in order to validate those species. We performed DNA sequencing for the COI region of 21 species belong to the order squamata. The BOLD Identification System (IDS) was used to establish species identity of the developed sequences. We searched both the private and published data in BOLD for available sequences through the "All Barcode Records" search engine. The Neighbour Joining tree of all the species under this study was constructed and the phylogenetic reconstruction was done using K2P distance model as per the standard protocol of DNA barcode. It was observed that *Chamaeleo chamaeleon* clusters with three *Diplometopon zarudnyi* sequences, of them two sequences have been generated in the lab and one sequence have been extracted from the database. *Eurylepis taeniolatus* also formd distinct branch in vicinity of three sequences of *Myrophis platyrhyncus*. This case study demonstrated the effectiveness of COI barcodes for reptile species from Saudi Araba in discriminating species recognized through prior taxonomic work contributing to the growing library of DNA barcodes of animal species of the world. Some species groups with overlapping barcodes identified in this study were good candidates for further studies of phylogeography and speciation processes. Further phylogenetic work on

B. Dhar · M. Chakraborty · M. Chakraborty · S. Malvika · N. N. Devi
Department of Biotechnology, Assam University, Silchar, Assam, India

A. A. Aloufi
Department of Biology, Faculty of Science, University of Tabuk, Tabuk, Saudi Arabia

Faculty of Science, Department of Biology, Taibah University, Medinah, Saudi Arabia

S. Trivedi
Department of Biology, Faculty of Science, University of Tabuk, Tabuk, Saudi Arabia

S. K. Ghosh (✉)
University of Kalyani, Kalyani, Nadia, West Bengal, India

S. Trivedi et al. (eds.), *DNA Barcoding and Molecular Phylogeny*,
https://doi.org/10.1007/978-3-030-50075-7_16

these species will reveal which of these highly divergent and geographically separated populations should be treated as belonging to the same species or sister species.

Keywords DNA barcoding · Reptile · Tabuk · Squamata · COI · BOLD

1 Reptiles: A Fundamental Component of Biodiversity

Reptiles are a group of vertebrate animals that comprises snakes, lizards, crocodiles, turtles, etc. These groups of animals have originated in and around 310–320 million years ago, in the late Carboniferous period (Laurin and Reisz 1995) (http://www.ucmp.berkeley.edu/carboniferous/carboniferous.php). Reptiles either have four limbs or like snakes, which had descended from four-limbed ancestors. Reptiles, contrasting to amphibians, do not have an aquatic larval stage (Sander 2012). Reptiles play an important role in the food webs of the ecosystems, filling up the critical role of both predator and prey. Reptiles have been hunted or traded, particularly as food, traditional medicines, leathers as well as decorative materials (http://www.endangeredspeciesinternational.org/reptiles3.html). Modern-day reptiles (Squamata) are the most diverse order of reptiles with more than 9600 species (Sander 2012).

Saudi Arabia Saudi Arabia occupies most of the Arabian Peninsula, with the Red Sea and the Gulf of Aqaba to the west and the Persian Gulf to the east (Figure 1). Saudi Arabia contains the world's largest continuous desert, which is known as the Rub Al-Khali or Empty Quarter. It has a land area of 2,149,690 sq. km (http://www.factmonster.com/country/saudi-arabia.html). The desert features a subtropical, hot and arid climate throughout the year, very similar to the Sahara Desert, which is actually an extension of the Sahara Desert over the Arabian Peninsula. The temperatures swing between very high heat and seasonal night time freezes. The desert of Saudi Arabia provides an excellent refuge for reptiles from the savage extremes of climate, because even a few inches of sand offer excellent insulation against heat and cold.

(http://www.saudiaramcoworld.com/issue/196805/the.toadhead.from.najad.and.other.reptiles.htm).

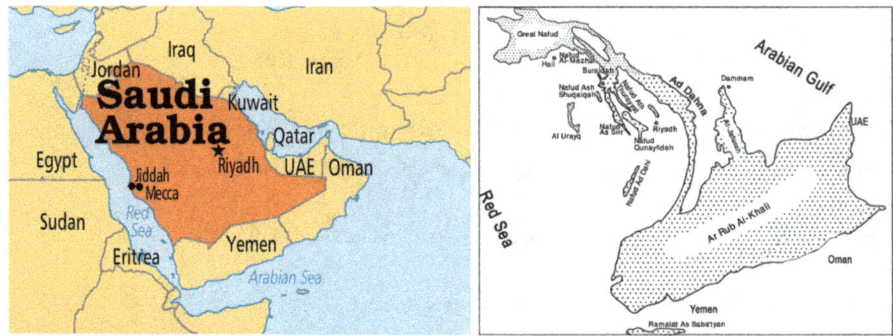

Fig. 1 Study site (Saudi Arabia) (http://www.operationworld.org/saud)

DNA Barcoding and Species Identification The ability to accurately identify and describe species is indispensable for any biological research, but the traditional morphological-based taxonomic approaches have only managed to explain 1–1.5 million species over the past 250 years (Chapple and Ritchie 2013; Mora et al. 2011), which is around 10% of the Earth's predicted eukaryotic diversity, a very meagre amount (Mora et al. 2011). It is estimated that dogging overwhelming and cumbersome approaches would not accomplish a comprehensive inventory of the world's biodiversity (Chapple and Ritchie 2013; Packer et al. 2009) and maybe for much longer given the sharp decline in the number of specialist taxonomists (Rodman and Cody 2003; Wheeler et al. 2004). The DNA barcoding approach was initiated in 2003 by Paul Hebert and his team (Hebert et al. 2003) in the University of Guelph, Ontario, as a way to overcome the existing taxonomic 'impediments' (Hebert et al. 2003). DNA barcoding has been a promising tool for the rapid and accurate identification of various species and inventorying species diversity (Hebert et al. 2003; Dawnay et al. 2007). It has been instrumental in the identification of existing species and the discovery of new species. DNA barcoding can be helpful in species diagnosis because sequence divergences are generally much lower among individuals of the same species than between species (Hebert et al. 2003). The distinction between intra- and inter-specific divergences, termed the 'barcoding gap' (Meyer and Paulay 2005), enables unknown sequences to be assigned to an existing species or flagged as a suspected new species. DNA barcoding use sequence variations in short regions (648-bp) of cytochrome c oxidase I (*COI*) to aid species identification and discovery in large assemblages of life (Hebert et al. 2003; Savolainen et al. 2005). A significant advantage of the DNA barcoding approach is that it works in situations where morphological approaches become confounding (Armstrong and Ball 2005; Chapple and Ritchie 2013), species with multiple life stages (Hebert et al. 2004) and sexual dimorphism, variable or plastic morphology (Smith et al. 2006, 2007; Burns et al. 2008). DNA barcoding is not only a powerful tool for species identification but also can play a vital role in wildlife forensics and conservation genetics (Wolinsky 2012). The occurrence of cryptic species is relatively common in nature. Cryptic species are those species that are morphologically similar but genetically distinct. DNA barcoding can be a very effective tool in the assessment of these cryptic species (Hebert et al. 2004). DNA barcoding can also be very effective for molecular phylogenic studies (Ajmal Ali et al. 2014).

2 Identification of Reptiles from Tabuk Region of Saudi Arabia through DNA Barcoding: A Case Study

2.1 BLAST Result Analysis

A total of 21 reptile sequences from the order Squamata have been collected from Tabuk Region of Saudi Arabia and sequenced. The BLAST search results of these sequences have been detailed in Table 1. A Neighbour Joining (NJ) tree has been

Table 1 Similarity match with GenBank sequences using nucleotide BLAST. The result showed the closest match with the available database sequence. The similarity of the sequences is expressed in terms of percentage of identity with *E* value

Sample code	Reptiles vouchered	Species match in BLAST	*E* value	Identity	Accession
001(F)	*Chamaeleo chamaeleon*	Diplometopon zarudnyi voucher MVZ 234273	0	99%	AY605474.1
2R(F)	*Chalcides ocellatus*	*Sceloporus virgatus* voucher AMNH herpetology 137,700	2.00E-133	82%	KU985944.1
		Sceloporus virgatus voucher AMNH herpetology 137,699	2.00E-133	82%	KU985908.1
		Hydrobates pelagicus voucher NHMO-BC33	1.00E-130	82%	GU571435.1
		Hydrobates pelagicus voucher NHMO-BC32	1.00E-130	82%	GU571434.1
3R(F)	*Scincus mitranus*	Oligosoma maccanni isolate OMA7	1.00E-125	82%	KC349736.1
		Oligosoma maccanni isolate OMA2	1.00E-125	82%	KC349722.1
		Oligosoma maccanni isolate OMA15	1.00E-125	82%	KC349720.1
5R(F)	*Eurylepis taeniolatus*	*Myrophis platyrhynchus* voucher MFL356	2.00E-133	82%	GU224964.1
		Myrophis platyrhynchus voucher MFL354	2.00E-133	82%	GU224963.1
		Myrophis platyrhynchus voucher MFL353	2.00E-133	82%	GU224956.1
7(f)	*Stellagama stellio*	Stellagama stellio voucher ZMMU R-11324	0	92%	KF691700.1
8(F)	*Stellagama stellio*	Stellagama stellio voucher ZMMU R-11324	0	91%	KF691700.1
009(f)	*Pseudotrapelus aqabensis*	Pseudotrapelus aqabensis isolate C-5-33	0	100%	KP994947.1
		Pseudotrapelus dhofarensis isolate C-4-242,743	0	91%	KP994946.1
		Pseudotrapelus jensvindumi isolate C-7-236,932	0	90%	KP994949.1
		Pseudotrapelus jensvindumi voucher CAS:225340	0	90%	KP979760.1
		Pseudotrapelus dhofarensis voucher ZISP:26351	0	90%	KP979759.1
10(F)	*Pseudotrapelus aqabensis*	Pseudotrapelus aqabensis isolate C-5-33	0	99%	KP994947.1
		Pseudotrapelus dhofarensis isolate C-4-242,743	0	91%	KP994946.1
		Pseudotrapelus jensvindumi isolate C-7-236,932	0	90%	KP994949.1
		Pseudotrapelus jensvindumi voucher CAS:225340	0	90%	KP979760.1

(continued)

Table 1 (continued)

Sample code	Reptiles vouchered	Species match in BLAST	*E* value	Identity	Accession
12(F)	*Diplometopon zarudnyi*	Diplometopon zarudnyi voucher MVZ 234273	0	99%	AY605474.1
13(F)	*Rhagerhis moilensis*	Mimophis mahfalensis voucher REPT_M12473	2.00E-167	86%	JQ909478.1
16(F)	*Cerastes gasperettii*	*Cerastes cerastes*	0	89%	EU852311.1
19(F)	*Cyrtopodion scabrum*	*Auriparus flaviceps* voucher FMNH 394359	1.00E-111	80%	DQ432755.1
		Hemidactylus pumilio voucher IBES5021	1.00E-110	80%	KU567474.1
21(F)	*Stenodactylus doriae*	*Cephalopholis cyanostigma* voucher UG0456	1.00E-111	80%	KP194176.1
22(F)	*Stenodactylus doriae*	*Cyanopica cyanus*, isolate: YIO318–10	2.00E-118	80%	AB843453.1
25(F)	*Mesalina brevirostris*	*Monasa morphoeus* voucher LGEMA-3306	1.00E-125	82%	JN801821.1
		Monasa morphoeus voucher LGEMA-9860	1.00E-120	81%	JN801823.1
26R(F)	*Acanthodactylus opheodurus*	*Conger conger* voucher CSFOM-031	1.00E-125	82%	KJ709504.1
		Conger conger voucher re 2 hg 190,506 E	1.00E-125	82%	JN231238.1
		Conger conger voucher FCFOPB064–17	1.00E-125	82%	JQ775006.1
27(F)	*Phoenicolacerta kulzeri khazaliensis*	Phoenicolacerta kulzeri	0	89%	FJ460596.1
29(f)	*Acanthodactylus opheodurus*	Monasa morphoeus voucher LGEMA-3428	2.00E-129	82%	JN801822.1
30(f)	*Hemidactylus flaviviridis*	*Hemidactylus homoeolepis* voucher CN1034	1.00E-160	85%	KU567377.1
060(f)	*Diplometopon zarudnyi*	Diplometopon zarudnyi voucher MVZ 234273	0	99%	AY605474.1
063(F)	*Stellagama stellio*	Cerastes cerastes	0	89%	EU852311.1

constructed using the developed sequences along with the downloaded BLAST hits of individual sequences. Only those BLAST hits have been considered which have the highest scores, and E_value is close to 0. Among them, only eight sequences have conspecific sequences available in the database. Remaining sequences showed a match with the closest available relative in the database like congeneric or confamilial species. In some rare cases, in the absence of true phylogenetic relative in the database, the closest hit showed random matches with species belonging to completely different taxa, like Aves and Anguilliformes. However, these cases were associated with high E-value which makes the hit false positive. As in the case of

Acanthodactylus opheodurus, in the absence of conspecific sequence, BLAST generated hit with 98% query coverage and 82% similarity with conger sequences which belongs to the phylum Aves. The E-Value of the match was however high with 1.00E-125 that showed a random match. The taxonomic details of Blast hits are given in Table 2.

3 Species Identification Using BOLD

The BOLD Identification System (IDS) was used to establish species identity of the developed sequences. This identification system for COI accepts sequences from the 5′ region of the mitochondrial Cytochrome c oxidase subunit I gene and returns a species-level identification when one is possible. We searched both the private and published data in BOLD for available sequences through the "All Barcode Records" search engine. The search returns every COI barcode record on BOLD with a minimum sequence length of 500 bp including unvalidated library and records without species-level identification. This also includes many species represented by only one or two specimens as well as all species with interim taxonomy. Further, the "Species Level Barcode Records" was used to extract a list of the nearest matches and that provided a probability of placement to a taxon.

Among the twenty-one COI barcode sequences developed in the lab, species status for only five sequences could be confirmed using the BOLD identification system. For most of the remaining sequences, conspecific sequences were not available in the BOLD database. Table 3 shows a detailed description of similarity match of the sequences using the BOLD identification system. Top five matches of the sequences using the "All Barcode Records" search were displayed for each of the sequences. In the case of 001(F), *Chamaeleo chamaeleon* fifteen COI sequences were available in the BOLD database. However, the top five similarity match did not show close identity with any of these sequences. Instead, the sequence showed 99.81% similarity with *Diplometopon zarudnyi* and IDS identified the sequence as *Diplometopon zarudnyi.* Such incongruency in the similarity may be because of the presence of hybrid sequences or mislabelled sequence. Conspecific sequences for 2R (F) *Chalcides ocellatus* were not available in the BOLD database. 3R(F) *Scincus mitranus* showed 95.4% similarity with congeneric sequence *Scinus scinus* available in a private database. Three sequences of *Eurylepis taeniolatus* were found in early release section; however, they showed an average of 87% match with the 5R (F) *Eurylepis taeniolatus.* Four sequences of *Stellagama stellio* were present in the database. They showed 88%–96% similarity with 7(F) *Stellagama stellio* and IDS did not identify species status of the sequence. However, 8(F) *Stellagama stellio* was identified up to species level as it showed 98.5% similarity with database conspecific sequence. Developed sequences of *Pseudotrapelus aqabensis* 9(F) and 10(F)) showed 99% similarity with database sequences and were identified correctly by IDS. 12(F) *Diplometopon zarudnyi* showed 99% similarity with database sequence and was identified correctly up to species level. *Rhagerhis moilensis and Mesalina*

Table 2 Taxonomic details of the BLAST hit results in NCBI

BLAST hits	Taxonomy
AY605474.1|Diplometopon_zarudnyi	Chordata; Reptilia; Squamata; Trogonophidae; Diplometopon;
DQ432755.1|Auriparus_flaviceps	Chordata; Aves; Passeriformes; Remizidae; Auriparus
EU852311.1|Cerastes_cerastes	Squamata; Viperidae;
FJ460596.1|Phoenicolacerta_kulzeri	Squamata; Lacertidae; Phoenicolacerta;
GU571434.1|Hydrobates_pelagicus	Chordata; Aves; Procellariiformes; Hydrobatidae; Hydrobates;
GU571435.1|Hydrobates_pelagicus	Chordata; Aves; Procellariiformes; Hydrobatidae; Hydrobates;
GU224956.1|Myrophis_platyrhynchus	Chordata; Actinopterygii; Anguilliformes; Ophichthidae; Myrophinae; Myrophis;
GU224963.1|Myrophis_platyrhynchus	Chordata; Actinopterygii; Anguilliformes; Ophichthidae; Myrophinae; Myrophis;
GU224964.1|Myrophis_platyrhynchus	Chordata; Actinopterygii; Anguilliformes; Ophichthidae; Myrophinae; Myrophis;
JN231238.1|Conger_conger	Chordata; Actinopterygii; Anguilliformes; Congridae; Congrinae; Conger;
JN801821.1|Monasa_morphoeus	Chordata; Aves; Galbuliformes; Bucconidae; Monasa
JN801822.1|Monasa_morphoeus	Chordata; Aves; Galbuliformes; Bucconidae; Monasa
JN801823.1|Monasa_morphoeus	Chordata; Aves; Galbuliformes; Bucconidae; Monasa
JQ909478.1|Mimophis_mahfalensis	Chordata; Reptilia; Squamata; Lamprophiidae; Psammophiinae; Mimophis
JQ775006.1|Conger_conger	Chordata; Reptilia; Squamata; Lamprophiidae; Psammophiinae; Mimophis
KC349720.1|Oligosoma_maccanni	Scincidae
KC349722.1|Oligosoma_maccanni	Scincidae
KC349736.1|Oligosoma_maccanni	Scincidae
KF691700.1|Stellagama_stellio	Chordata; Reptilia; Squamata; Agamidae; Agaminae; Stellagama
AB843453.1|Cyanopica_cyanus	Chordata; Aves; Passeriformes; Corvidae; Cyanopica;
KJ709504.1|Conger_conger	Chordata; Aves; Passeriformes; Corvidae; Cyanopica;
KP979759.1|Pseudotrapelus_dhofarensis	Chordata; Reptilia; Squamata; Agamidae; Agaminae; Pseudotrapelus;
KP979760.1|Pseudotrapelus_jensvindumi	Chordata; Reptilia; Squamata; Agamidae; Agaminae; Pseudotrapelus;
KP994946.1|Pseudotrapelus_dhofarensis	Chordata; Reptilia; Squamata; Agamidae; Agaminae; Pseudotrapelus;

(continued)

Table 2 (continued)

BLAST hits	Taxonomy
KP994947.1| Pseudotrapelus_aqabensis	Chordata; Reptilia; Squamata; Agamidae; Agaminae; Pseudotrapelus;
KP994949.1| Pseudotrapelus_jensvindumi	Chordata; Reptilia; Squamata; Agamidae; Agaminae; Pseudotrapelus;
KP194176.1| Cephalopholis_cyanostigma	Chordata; Actinopterygii; Perciformes; Serranidae; Epinephelinae; Cephalopholis;
KU567377.1| Hemidactylus_homoeolepis	Chordata; Reptilia; Squamata; Gekkonidae; Hemidactylus
KU567474.1| Hemidactylus_pumilio	Chordata; Reptilia; Squamata; Gekkonidae; Hemidactylus
KU985908.1| Sceloporus_virgatus	Chordata; Reptilia; Squamata; Phrynosomatidae; Sceloporinae; Sceloporus

brevirostris did not have any conspecific sequences available in the database. However, *Mesalina brevirostris* showed 98% similarity with *Acanthodactylus boskianus* and hence was identified as the same species. *Cyrtopodion scabrum* has a conspecific sequence available in the database but IDS did not show significant similarity with these sequence. 21(F) *Stenodactylus doriae* showed 81–89% similarity with the available database sequences while 22(F) *Stenodactylus doriae* showed 91% similarity with the sequences. 63(F) *Stellagama stellio* did not show match with any of the available database sequences.

4 Neighbour-Joining (NJ) Clustering

The Neighbour Joining tree of all the species under this study is constructed as shown in Fig. 2. The phylogenetic reconstruction was done using K2P distance model as per the standard protocol of DNA barcode. As observed in this case, 001F_*Chamaeleo chamaeleon* clusters with three *Diplometopon zarudnyi* sequences; of them, two sequences (12F) and 60(F)) have been generated in the lab and one sequence, AY605474, has been extracted from the database. Such clustering could be possible because of either the presence of mislabelled or misidentified sequence or there could be the possibility of species introgression. 2R(F) *Chalcides ocellatus* clusters separately as no conspecific sequence is available in the database. However, they align close to (KU985908, KU985944) *Sceloporus virgatus* belonging to the same order Squamata but different family Phrynosomatidae. 3RF_*Scincus mitranus* clusters separately but close to three confamilial database sequences of *Oligosoma maccanni* (KC349720, KC349736, KC349722). *Eurylepis taeniolatus* also forms distinct branch in the vicinity of three sequences of *Myrophis platyrhyncus*(GU224956, GU224963-64), which are Anguilliformes. 7R and 8R *Stellagama stellio* clusters together along with another database sequence (KF691700) of the same species. However, 63R *Stellagama*

Table 3 Species identification using BOLD-IDS (Barcode of Life Datasystem-Identification system) search engine. The developed sequences of the specimen are checked for similarity match in the Public Record Barcode Database of BOLD-IDS for comprehensive species identification

Voucher ID	Vouchered specimen	Top hit (similarity)	Status	Species Identification
001(F)	Chamaeleo chamaeleon	Diplometopon zarudnyi (99.81)	Private	No
2R(F)	Chalcides ocellatus	No match		No
3R(F)	Scincus mitranus	No match		No
5R(F)	Eurylepis taeniolatus	No match		No
7(f)	Stellagama stellio	No match		No
8(F)	Stellagama stellio	Stellagama stellio (98.51)	Early-release	Species identified
009(f)	Pseudotrapelus aqabensis	Pseudotrapelus aqabensis (98.9)	Published	Species identified
10(F)	Pseudotrapelus aqabensis	Pseudotrapelus aqabensis (99.45)	Published	Species identified
12(F)	Diplometopon zarudnyi	Diplometopon zarudnyi (99.82)	Private	Species identified
13(F)	Rhagerhis moilensis	No match		No
16(F)	Cerastes gasperettii	No match		No
19(F)	Cyrtopodion scabrum	No match		No
21(F)	Stenodactylus doriae	No match		No
22(F)	Stenodactylus doriae	No match		No
25(F)	Mesalina brevirostris	Acanthodactylus boskianus (98.38)	Early-release	No
26R(F)	Acanthodactylus opheodurus	No match		No
27(F)	Phoenicolacerta kulzeri khazaliensis	No match		No
29(f)	Acanthodactylus opheodurus	Acanthodactylus boskianus (99.46)	Early-release	Genus identified
30(f)	Hemidactylus flaviviridis	No match		No
060(f)	Diplometopon zarudnyi	Diplometopon zarudnyi (99.48)	Private	Species identified
063(F)	Stellagama stellio	No match		No

stellio clusters separately and close to 16(F) *Cerastes gasperettii*. 009F and10R *Pseudotrapelus aqabensis* clusters together with conspecific sequence KP994947 from database. Moreover, four database sequences (KP979760, KP994949, KP979759, KP994946) from three congeneric species of Pseudotrapelus clusters distinctly under the same node. As conspecific sequences are not present in the database, 13(F) *Rhagerhis moilensis* shows closest hit with *Mimophis mahfalensis,* which belong to the same family. In the NJ tree as well the two sequences form close cluster distinct from other families. 19 (F*) Cyrtopodion scabrum* forms subcluster with three sequences of Hemidactylus genus where sequences (KU567377,

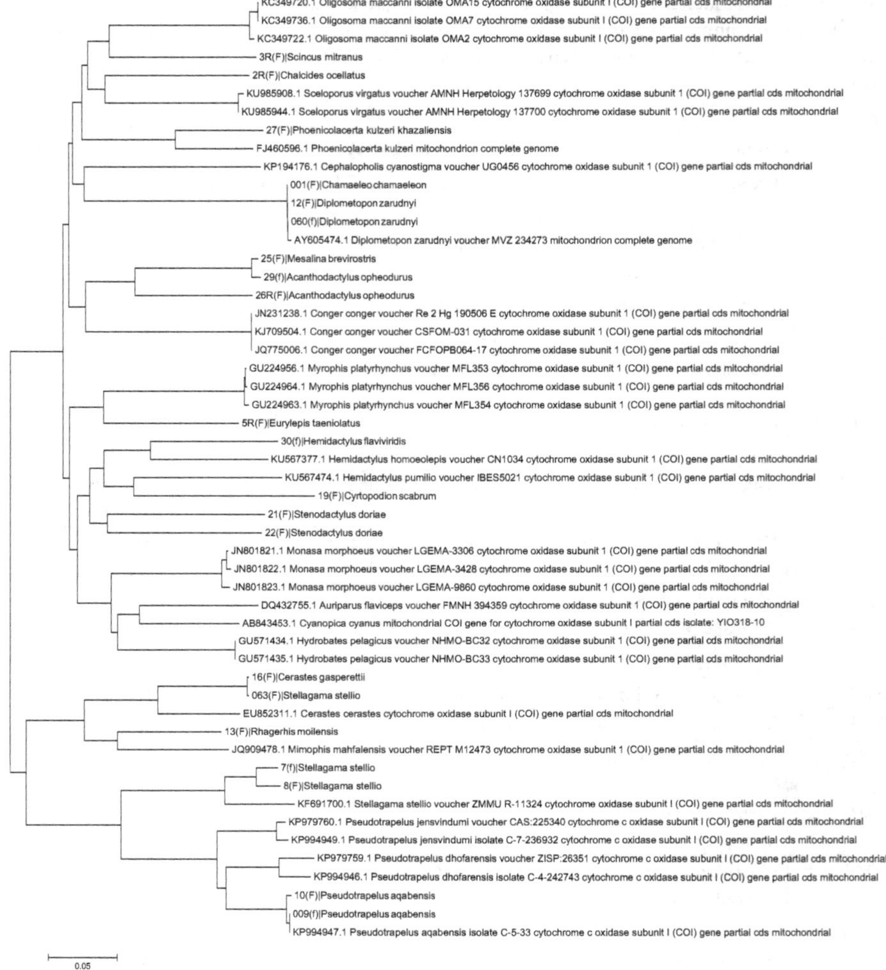

Fig. 2 Neighbour Joining tree of COI sequences of all the reptile species from Tabuk Region of Saudi Arabia along with the other database sequences as study replicates

KU567474) of two species were extracted from the database and one sequence, 30 (F) *Hemidactylys flavivirdis,* was developed in lab. Both of these genera belong to the same family Gekkonidae. 21(F) and 22(F) *Stenodactylus doriae* clusters together along with other sequences of Gekkonidae family. Species of Lacertidae family, 25 (F) *Mesalina brevirostris*, 26R (F) and 29 (F) *Acanthodactylus opheodurus* forms distinct cluster. However, 27(F) *Phoenicolacerta kulzeri khazaliensis ssp* forms separate cluster along with a conspecific database sequence FJ460596.

This case study demonstrated the effectiveness of COI barcodes for reptile species from Saudi Arabia in discriminating species recognized through prior taxonomic work contributing to the growing library of DNA barcodes of animal species of the world. The study showed that the partial COI gene enables accurate animal species identification where adequate reference sequence data exist. Some species groups with overlapping barcodes identified in this study were good candidates for further studies of phylogeography and speciation processes. Further phylogenetic work on these species will reveal which of these highly divergent and geographically separated populations should be treated as belonging to the same species or sister species.

References

Ajmal Ali M, Gyulai G, Hidvegi N, Kerti B, Al Hemaid FM, Pandey AK, Lee J (2014) The changing epitome of species identification - DNA barcoding. Saudi J Biol Sci 21(3):204–231. https://doi.org/10.1016/j.sjbs.2014.03.003

Armstrong KF, Ball SL (2005) DNA barcodes for biosecurity: invasive species identification. Philos Trans R Soc Lond Ser B Biol Sci 360(1462):1813–1823. https://doi.org/10.1098/rstb.2005.1713

Burns JM, Janzen DH, Hajibabaei M, Hallwachs W, Hebert PD (2008) DNA barcodes and cryptic species of skipper butterflies in the genus Perichares in area de Conservacion Guanacaste, Costa Rica. Proc Natl Acad Sci U S A 105(17):6350–6355. https://doi.org/10.1073/pnas.0712181105

Chapple DG, Ritchie PA (2013) A retrospective approach to testing the DNA barcoding method. PLoS One 8(11):e77882. https://doi.org/10.1371/journal.pone.0077882

Dawnay N, Ogden R, McEwing R, Carvalho GR, Thorpe RS (2007) Validation of the barcoding gene COI for use in forensic genetic species identification. Forensic Sci Int 173(1):1–6. https://doi.org/10.1016/j.forsciint.2006.09.013

Hebert PD, Cywinska A, Ball SL, deWaard JR (2003) Biological identifications through DNA barcodes. Proc Biol Sci/The Royal Soc 270(1512):313–321. https://doi.org/10.1098/rspb.2002.2218

Hebert PD, Penton EH, Burns JM, Janzen DH, Hallwachs W (2004) Ten species in one: DNA barcoding reveals cryptic species in the neotropical skipper butterfly *Astraptes fulgerator*. Proc Natl Acad Sci U S A 101(41):14812–14817. https://doi.org/10.1073/pnas.0406166101

http://www.endangeredspeciesinternational.org/reptiles3.html

http://www.saudiaramcoworld.com/issue/196805/the.toadhead.from.najad.and.other.reptiles.htm

http://www.ucmp.berkeley.edu/carboniferous/carboniferous.php The Carboniferous Period

Laurin M, Reisz RR (1995) A reevaluation of early amniote phylogeny. Zool J Linnean Soc 113 (2):165–223

Meyer CP, Paulay G (2005) DNA barcoding: error rates based on comprehensive sampling. PLoS Biol 3(12):e422. https://doi.org/10.1371/journal.pbio.0030422

Mora C, Tittensor DP, Adl S, Simpson AG, Worm B (2011) How many species are there on earth and in the ocean? PLoS Biol 9(8):e1001127. https://doi.org/10.1371/journal.pbio.1001127

Packer L, Gibbs J, Sheffield C, Hanner R (2009) DNA barcoding and the mediocrity of morphology. Mol Ecol Resour 9(Suppl s1):42–50. https://doi.org/10.1111/j.1755-0998.2009.02631.x

Rodman JE, Cody JH (2003) The taxonomic impediment overcome: NSF's partnerships for enhancing expertise in taxonomy (PEET) as a model. Syst Biol 52(3):428–435

Sander PM (2012) Reproduction in early amniotes. Science 337(6096):806–808

Savolainen V, Cowan RS, Vogler AP, Roderick GK, Lane R (2005) Towards writing the encyclopedia of life: an introduction to DNA barcoding. Philos Trans R Soc Lond Ser B Biol Sci 360 (1462):1805–1811. https://doi.org/10.1098/rstb.2005.1730

Smith MA, Woodley NE, Janzen DH, Hallwachs W, Hebert PD (2006) DNA barcodes reveal cryptic host-specificity within the presumed polyphagous members of a genus of parasitoid flies

(Diptera: Tachinidae). Proc Natl Acad Sci U S A 103(10):3657–3662. https://doi.org/10.1073/pnas.0511318103

Smith MA, Wood DM, Janzen DH, Hallwachs W, Hebert PD (2007) DNA barcodes affirm that 16 species of apparently generalist tropical parasitoid flies (Diptera, Tachinidae) are not all generalists. Proc Natl Acad Sci U S A 104(12):4967–4972. https://doi.org/10.1073/pnas.0700050104

Wheeler QD, Raven PH, Wilson EO (2004) Taxonomy: impediment or expedient? Science 303 (5656):285. https://doi.org/10.1126/science.303.5656.285

Wolinsky H (2012) Wildlife forensics. Genomics has become a powerful tool to inform conservation measures. EMBO Rep 13(4):308–312. https://doi.org/10.1038/embor.2012.35

Closing Shots: DNA Barcoding and Molecular Phylogeny

Subrata Trivedi, Hasibur Rehman, Shalini Saggu, Chellasamy Panneerselvam, and Sankar K. Ghosh

Abstract This chapter is a brief discription of all the chapters included in this book. It includes different topics related to DNA barcoding and molecular phylogeny along with some unique case studies. They include topics related to implications of DNA barcoding, invasive align species identification through DNA barcoding, microbial DNA barcoding, applications of DNA barcoding in clinical microbiology and plant-animal interactions, DNA barcoding in relation to fish and fisheries management, DNA barcoding of red alga, ruminant mammals, elasmobranchs, aves and reptiles.

Keywords DNA barcoding · Molecular phylogeny · Invasive alien species · Ruminant · Reptile · Aves · Fisheries management

DNA barcoding has become a major focus of research in recent times and has gained global attention. In this book, we have covered broad aspects of DNA barcoding and molecular phylogeny along with some case studies.

The first chapter of this book deals with the implications of DNA barcoding. DNA barcoding has become a promising tool for validating the different species and also has several implications like detection of mislabeling in the food industry, safety assessment, controlling agricultural pests, detection of disease vectors, monitoring water quality, etc.

S. Trivedi (✉) · C. Panneerselvam
Department of Biology, Faculty of Science, University of Tabuk, Tabuk, Saudi Arabia

H. Rehman
Department of Urology, The University of Alabama at Birmingham (UAB), Birmingham, Alabama, USA

S. Saggu
Department of Cell, Developmental and Integrative Biology (CDIB), The University of Alabama at Birmingham (UAB), Birmingham, Alabama, USA

S. K. Ghosh
University of Kalyani, Kalyani, Nadia, West Bengal, India

S. Trivedi et al. (eds.), *DNA Barcoding and Molecular Phylogeny*,
https://doi.org/10.1007/978-3-030-50075-7_17

The second chapter deals with the prospects and limitations of DNA barcoding in birds. Museums contain many extinct birds both as whole specimens and as skin and feathers. DNA from such specimen is used to identify avian species along with presently available species.

Invasive alien species (IAS) is a topic of great interest not only for ecological consequences but also for the massive economical damage caused by the introduction of these species. The third chapter of this book is devoted to this very important topic on invasive species identification through DNA barcoding. This chapter not only highlights how DNA barcoding can promptly identify the invasive species but also on the management process of invasive species.

It is well known that microbes are available in every corner of our planet and even in extreme conditions. Chapter "Microbial DNA Barcoding: Prospects for Discovery and Identification" elucidates the prospects of discovery and identification of microbes through DNA barcoding.

The next chapter (Chapter "DNA Barcoding on Bacteria and its Application in Infection Management") deals with a topic in medical science—how DNA barcoding of bacteria can help in infection management. This chapter deals with the application of DNA barcoding in clinical microbiology.

Various ecological processes are involved in plant–animal interactions. Any disruption in this interaction can be detrimental not only for the species involved but also other species in the community. Traditionally, these interactions have been studied by field observations. Chapter "DNA Barcoding: Implications in Plant–Animal Interactions" of this book demonstrates how DNA barcoding has revolutionized the field of community ecology.

Red alga has an immense ecological and economical value. Chapter "A Molecular Assessment of Red Algae with Reference to the Utility of DNA Barcoding" is devoted to the molecular identification of red alga through DNA barcoding.

Identification of rays is a challenging task due to overlapping morphological characters. Chapter "DNA Barcoding of Rays from the South China Sea" is related to DNA barcoding of rays from South China Sea. This chapter additionally addresses the management plan of elasmobranch fishery in Malaysia.

The next chapter is on the complete molecular phylogeny of elasmobranchs with emphasis on the phylogeny for framing conservation strategies.

Fish is an important protein course and fishery industry provides huge employment worldwide. Chapter "A Review on DNA Barcoding on Fish Taxonomy in India" is a review on the DNA barcoding and fish taxonomy in India.

Chapter "Applications of DNA Barcoding in Fisheries" of this book is on the applications of DNA barcoding in fisheries.

The next chapter is regarding the taxonomically wide biogeographic study of birds in the context of DNA barcoding.

Ruminant mammals differ from other mammals by having a four-chambered stomach. Chapter "Molecular Characterisation of Ruminant Mammals Using DNA Barcodes" is on the molecular characterization of ruminant mammals. This chapter depicts how the *COI* gene can be used for the molecular identification of ruminants.

Chapter "DNA Barcoding and Molecular Phylogeny of Indigenous Bacteria in Fishes from a Tropical Tidal River in Malaysia" is an important case study in which DNA barcoding and molecular phylogeny are done on the bacterial species isolated from different tissues from three commercial tropical tidal fishes of Malaysia.

The next chapter is on DNA barcoding and molecular phylogeny of ichthyoplankton and juvenile fishes of Kuantan River in Malaysia. This study may be valuable for analyzing post-flood effect on fish distribution in tropical river and implementing plans for future fishery resource management.

The deserts of Arabian Peninsula are an ideal habitat for several species of reptiles. As compared to other vertebrate groups, reptiles are less studied in the context of DNA barcoding. The chapter "Molecular Identification of Reptiles from Tabuk Region of Saudi Arabia through DNA Barcoding: A Case Study" provides important information on DNA barcoding of reptiles.

Although DNA barcoding is a major topic for research worldwide, only a few books are available at present. Our previous book *DNA Barcoding in Marine Perspective* published by Springer International focused only on marine habitat. This book covers a broader arena and will be useful for academics and researchers worldwide.

CPSIA information can be obtained
at www.ICGtesting.com
Printed in the USA
LVHW051314250121
677420LV00001B/77